D1259570

SCIENCE IN DISPUTE

SCIENCE IN DISPUTE

Volume 3

NEIL SCHLAGER, EDITOR

Produced by
Schlager Information Group

Detroit • New York • San Diego • San Francisco • Cleveland • New Haven, Conn. • Waterville, Maine • London • Munich

Science in Dispute, Volume 3
Neil Schlager, Editor

Project Editor
Brigham Narins

Editorial
Mark Springer

Permissions
Kim Davis

Imaging and Multimedia
Leitha Etheridge-Sims, Mary K. Grimes, Lezlie Light, Dan Newell, David G. Oblender, Christine O'Bryan, Robyn V. Young

Product Design
Michael Logusz

Manufacturing
Rhonda Williams

For more information, contact
The Gale Group, Inc.
27500 Drake Road
Farmington Hills, MI 48331-3535
Or you can visit our Internet site at
http://www.gale.com

ISBN: 0-7876-5767-0
ISSN: 1538-6635

Printed in the United States of America
10 9 8 7 6 5 4 3 2 1

CONTENTS

Life Science

Mathematics and Computer Science

Medicine

Physical Science

ABOUT THE SERIES

Overview

Welcome to *Science in Dispute*. Our aim is to facilitate scientific exploration and understanding by presenting pro-con essays on major theories, ethical questions, and commercial applications in all scientific disciplines. By using this adversarial approach to present scientific controversies, we hope to provide students and researchers with a useful perspective on the nature of scientific inquiry and resolution.

The majority of entries in each volume of *Science in Dispute* cover topics that are currently being debated within the scientific community. However, each volume in the series also contains a handful of "historic" disputes. These include older disputes that remain controversial as well as disputes that have long been decided but that offer valuable case studies of how scientific controversies are resolved. Each historic debate is clearly marked in the text as well as in the Contents section at the beginning of the book.

Each volume of *Science in Dispute* includes approximately thirty entries, which are divided into seven thematic chapters:

- Astronomy and Space Exploration
- Earth Science
- Engineering
- Life Science
- Mathematics and Computer Science
- Medicine
- Physical Science

The advisory board, whose members are listed elsewhere in this volume, was responsible for defining the list of disputes covered in the volume. In addition, the board reviewed all entries for scientific accuracy.

Entry Format

Each entry is focused on a different scientific dispute, typically organized around a "Yes" or "No" question. All entries follow the same format:

- **Introduction:** Provides a neutral overview of the dispute.
- **Yes essay:** Argues for the pro side of the dispute.
- **No essay:** Argues for the con side of the dispute.
- **Further Reading:** Includes books, articles, and Internet sites that contain further information about the topic.
- **Key Terms:** Defines important concepts discussed in the text.

Throughout each volume users will find sidebars whose purpose is to feature interesting events or issues related to a particular dispute. In addition, illustrations and photographs are scattered throughout the volume. Finally, each volume includes a general subject index.

About the Editor

Neil Schlager is the president of Schlager Information Group Inc., an editorial services company. Among his publications are *When Technology Fails* (Gale, 1994); *How Products Are Made* (Gale, 1994); the *St. James Press Gay and Lesbian Almanac* (St. James Press, 1998); *Best Literature By and About Blacks* (Gale, 2000); *Contemporary Novelists*, 7th ed. (St. James Press, 2000); *Science and Its Times* (7 vols., Gale, 2000-2001); and *The Science of Everyday Things* (4 vols., Gale, 2002). His publications have won numerous awards, including three RUSA awards from the American Library Association, two Reference Books Bulletin/Booklist Editors' Choice awards, two New York Public Library Outstanding Reference awards, and a *CHOICE* award for best academic book.

Comments and Suggestions

Your comments on this series and suggestions for future volumes are welcome. Please write: The Editor, *Science in Dispute,* Gale Group, 27500 Drake Road, Farmington Hills, MI 48331.

ADVISORY BOARD

LIST OF CONTRIBUTORS

Linda Wasmer Andrews
Freelance Writer

William Arthur Atkins
Freelance Writer

Adi R. Ferrara
Freelance Science Writer

Maura C. Flannery
Professor of Biology, St. John's University

Katrina Ford
Freelance Writer

Donald Franceschetti
Distinguished Service Professor of Physics and Chemistry, University of Memphis

Natalie Goldstein
Freelance Science Writer

Jeffrey C. Hall
Assistant Research Scientist, Associate Director, Education and Special Programs, Lowell Observatory

Amanda J. Harman
Freelance Writer

Judson Knight
Freelance Writer

Philip Koth
Freelance Writer

Brenda Wilmoth Lerner
Science writer

K. Lee Lerner
Professor of Physics, Fellow, Science Research & Policy Institute

Eric v.d. Luft
Curator of Historical Collections, SUNY Upstate Medical University

Charles R. MacKay
National Institutes of Health

Lois N. Magner
Professor Emerita, Purdue University

Leslie Mertz
Biologist and Freelance Writer

M. C. Nagel
Freelance Science Writer

Lee Ann Paradise
Science Writer

Cheryl Pellerin
Independent Science Writer

David Petechuk
Freelance Writer

Laura Ruth
Freelance Writer

Jens Thomas
Freelance Writer

Marie L. Thompson
Freelance Writer/Copyeditor

David Tulloch
Freelance Writer

Rashmi Venkateswaran
Senior Instructor, Undergraduate Laboratory Coordinator, Department of Chemistry, University of Ottawa

Elaine H. Wacholtz
Medical and Science Writer

Stephanie Watson
Freelance Writer

ASTRONOMY AND SPACE EXPLORATION

Is it appropriate to allow the use of nuclear technology in space?

Viewpoint: Yes, nuclear technology offers the only feasible method for many future space missions.

Viewpoint: No, nuclear technology in space is a dangerous proposition that should be avoided.

Few topics generate as much impassioned debate as the use of nuclear energy—the extraction of energy from radioactive nuclei by fission (splitting of the atomic nuclei) or fusion (slamming two lighter nuclei together to produce a heavier one and an attendant release of energy). Nuclear energy is efficient and clean and produces markedly greater energy return per unit amount of fuel than do other methods of energy generation.

Nuclear energy was first harnessed as a means of developing and delivering weapons of mass destruction: the A-bombs that were dropped on Hiroshima and Nagasaki on August 6 and 9, 1945, to end World War II. Although even this action has generated fierce debate (whether it was better to end the war catastrophically but quickly or to allow it to continue as a conventional conflict), mass killing remains the introduction of nuclear technology to human history. Decades of nuclear saber rattling by various world powers have served as a constant reminder of this fact. Since Hiroshima, nuclear technology has been put to more benign means, and today nuclear power plants deliver efficient energy in many parts of the world. Widely publicized power plant accidents, however, such as those at Three Mile Island, near Harrisburg, Pennsylvania, and the far worse accident at Chernobyl, Ukraine, are an ongoing source of concern about the use and abuse of nuclear power. It is therefore not surprising that debate has surrounded the use of nuclear propellant and power systems in spacecraft.

The so-called space age began in 1957, when the Soviet Union launched the tiny 10-inch satellite *Sputnik* ("Explorer") into Earth orbit. Since then, the question of providing power to spacecraft that need it has been a constant challenge for engineers.

The bulk motion of a spacecraft is not the problem. Once launched into orbit or onto an interplanetary trajectory by a rocket, a spacecraft needs little fuel for propulsion. With no atmospheric friction to slow it, the spacecraft coasts solely under the influence of the gravity of the Sun and that of any planets it happens to be near. Planetary gravitational fields frequently are used to accelerate spacecraft on their journeys—the ultimate example of clean energy use. Onboard propellant is used only for small course corrections. Some form of control is essential, however, as is power for the spacecraft's control systems and any scientific instrumentation it carries.

Critical to the choice of power generation, and to almost every other design decision for a spacecraft, is weight. Even in the age of the space shuttle, most satellites and spacecraft are launched from Earth with various types of rockets. The Delta rocket is commonly used, and the Titan can be used for larger payloads. The rocket has to launch out of Earth's gravitational well not

only the weight of its payload but also its own weight. Although the idea of "staging" rockets—discarding portions of the rocket in which fuel is spent to reduce the remaining weight—works well, most rockets are limited in their payloads. Every ounce not spent on propulsion and power equipment can be used for instrumentation and communication, a vital design goal for interplanetary spacecraft intended to perform ongoing scientific observations of other planets.

To send spacecraft to explore the outer parts of the solar system, using the Sun as a power source becomes more difficult. Many spacecraft in Earth's orbit, such as the 1970s *Skylab*, the *Hubble* space telescope, and the extraordinarily long-lived *International Ultraviolet Explorer*, have used solar panels to harness the Sun's energy as a power source. In the cold, dark outer reaches of the solar system, the Sun's feeble energy is a less satisfactory option, and radioisotopic thermal generators (RTGs) are preferred.

Although recent debate has centered on the RTGs aboard the *Cassini* spacecraft due to reach Saturn in 2004, the general issue of use of nuclear energy sources in spacecraft has been a political hot button since these fuels were first used.

Concerns about the use of nuclear technology in spacecraft center on the possibility of dispersal of radioactive material into the atmosphere after a failed launch or after reentry and breakup of a satellite bearing the material. Although the subject matter of many disputes in science leads them to be of interest to the scientific community only, the issue of spacecraft power sources has boiled over to the general public and the political arena. The dispute stems from understandable long-term unease about the widespread destruction that would result from a full nuclear exchange and the history of nuclear accidents since 1945. The concerns have served to cloud with rhetoric the rational statements of both sides. The proponents of nuclear power in space occasionally dismiss their counterparts as naive fools while sidestepping the opponents' questions and supplying no reasoned arguments of their own. There also have been loud and at times almost violent denouncements from protesters of nuclear power. The following articles examine the evidence of successful and failed missions in which nuclear power was used, the payload and efficiency reasons that lead many scientists to support the use of nuclear energy aboard spacecraft, and the likely consequences of failed missions involving the amounts of radioactive material commonly involved. Debates such as this one, in which firmly held personal beliefs are involved, are the most difficult ones to evaluate methodically. The essays provide a foundation for readers to do so. —JEFFREY C. HALL

Viewpoint:

Yes, nuclear technology offers the only feasible method for many future space missions.

In the early years of exploration of outer space, small fuel cells, batteries, and solar modules provided energy for space missions. Even though only 100 to 500 watts of power were needed for these first-generation spacecraft, increasing energy requirements in space travel soon became obvious as increasingly sophisticated missions were flown. Increased power requirements caused engineers to investigate a variety of options, such as chemical, solar, beamed space energy, and nuclear sources. Today, future manned Moon and Mars missions will likely require one to two million watts (commonly called megawatts) of power.

Nuclear Space Technology The assertion of this essay is that the use of nuclear technology in space is appropriate; that is, nuclear propellant and power systems (PPSs) meet the essential requirements needed for space missions: minimum weight and size and maximum efficiency, durability, and reliability all attained with appropriate levels of safety. In this article, *power system* refers to the on-board energy requirements of a spacecraft (components such as heaters and communications), and *propulsion system* refers to the energy storage, transfer, and conversion requirements to propel or maneuver a spacecraft (components such as the space shuttle's main engines). Nuclear technology can be applied to both onboard power systems and to the propulsion of spacecraft (for example, nuclear rockets).

With the projected growth in energy required by increasingly complex and extended missions, nuclear PPS technology is the only feasible performance option when it is compared with chemical PPS technology (the only other realistic possibility and the one most often used in the past).

NASA Nuclear Agenda For its fiscal 2003 budget, the National Aeronautics and Space Administration (NASA) is providing additional funds for nuclear systems because agency officials realize that for the near future only nuclear systems will adequately propel and power spacecraft for the exploration of the solar system and beyond. Administrator Sean O'Keefe has declared that NASA has been restricted in the

past in its ability to explore long distances in relatively short times because of its reliance on current chemical propulsion technology. O'Keefe contends that nuclear propulsion and power capability is needed to overcome this distance and time problem.

The strength of nuclear propulsion is that it is more efficient than traditional chemically propulsion. Stanley Borowski, a nuclear and aerospace engineer at NASA's Glenn Research Center in Cleveland, Ohio, states that rockets launched with nuclear propulsion have twice the propellant mileage of chemical rockets currently in use.

After NASA's announcement regarding expanded use of nuclear PPSs, experts from agencies such as the Department of Energy (DOE), Los Alamos National Laboratory, and Sandia National Laboratories (SNL) expressed satisfaction that nuclear systems would onceagain be used on a regular basis. Scientist Roger Lenard of SNL asserts that space travel into unexplored parts of the solar system requires development of a better PPS.

A NERVA model on display in Chicago in 1967.
(© Bettmann/Corbis. Reproduced by permission.)

Past Nuclear Research and Development

Research and development of nuclear PPSs have progressed markedly in the two major space powers: the United States and Russia. During the last three decades of the twentieth century, both countries used nuclear power sources to meet the electrical and thermal energy requirements of selected spacecraft. These power requirements include operation of onboard scientific experiments, spacecraft maintenance and monitoring, temperature control, and communication. Nuclear fuel has proved an ideal energy source in space because of its high power,

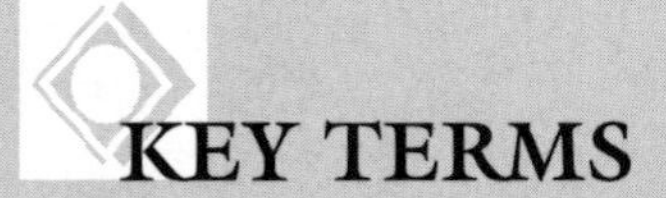

KEY TERMS

ENERGY: The capacity for doing work.

FISSION: The action of dividing or splitting something into two or more parts.

ISOTOPE: Each of two or more forms of the same element that contain equal numbers of protons but different numbers of neutrons in their nuclei.

ISP: See *specific impulse.*

PLUTONIUM: A synthetic element that emits alpha radiation. Plutonium is used in nuclear weapons and as an energy source. Chemical symbol, Pu.

POWER: The rate of doing work.

RADIOACTIVE: Emitting or relating to the emission of ionizing radiation or particles.

RADIOISOTOPIC THERMAL GENERATOR (RTG): A small nuclear reactor onboard a space probe. The RTG generates energy for the probe's instruments.

RADIOISOTOPIC: Related to a radioactive isotope.

SOLAR CELL: The device used to convert the sun's energy to electricity.

SPECIFIC IMPULSE (ISP): The pounds of thrust produced per pound of propellant consumed per second; measured in seconds.

acceptable weight and volume, and excellent reliability and safety in systems such as RTGs.

RTGs were first developed in the 1960s for the U.S. Space Nuclear Auxiliary Power (SNAP) program. The United States launched RTGs for test satellites in the 1960s, the Apollo lunar missions of the early 1970s, and missions to the planets past Mars (such as the *Galileo*, *Voyager*, and *Cassini* missions). RTGs have safely and effectively been used in robotic missions beyond the orbit of Pluto, where use of other systems would have proved extremely cumbersome and exorbitantly expensive.

RTGs have proved to be highly reliable and almost maintenance-free power supplies capable of producing as much as several kilowatts of electrical power. These generators have operated efficiently for decades under adverse conditions, such as deep space exploration. Even at lower-than-normal power levels, communication with *Pioneer 10* (launched in 1972) has been maintained for 30 years. The *Voyager 1* and *Voyager 2* spacecraft (both launched in 1977) have operated for 25 years and are expected to function for at least 15 years more. The reliability of RTGs stems from the ability to convert heat to electricity without moving parts. That is, RTGs convert thermal energy generated from radioisotopic decay into electricity. The radioisotope consists of plutonium-238 oxide fuel used with static electrical converter systems. As of 2002, the DOE has provided 44 RTGs for use on 26 space missions to provide some or all of the onboard electric power. RTG reliability is extremely important in space applications in which there is a large investment and equipment repair or replacement is not feasible. Nuclear power systems have been largely ignored in the United States as effective power generation options because of political pressure and antinuclear sentiment.

Both the United States and Russia have developed prototypes of rockets that have their exhaust gases heated not by a chemical reaction ("burning") but with a reaction in which a nuclear fuel such as uranium is heated and turned from a liquid into a high-temperature gas. These nuclear rockets have been shown to be much more efficient than conventional chemical rockets.

Research and development of U.S. nuclear-propelled rockets was a major endeavor from the mid 1950s to the early 1970s. In 1956 a project named Rover began at the Los Alamos Scientific Laboratory. The aim of Rover was to develop and test nuclear reactors that would form the basis of a future nuclear rocket. Between 1959 and 1972, Los Alamos scientists built and tested 13 research reactors. These scientists found that one of the main advantages of a nuclear rocket is its high exhaust velocity, which is more than two times greater than that of chemical rockets. The higher velocity translates into lower initial fuel mass and shorter travel times in space. A spin-off of Rover was NERVA (nuclear engine for rocket vehicle application). Initiated by NASA in 1963, NERVA was designed to take the graphite-based reactor built in the Rover program and develop a working rocket engine for space flight. The system designed from NERVA was a graphite-core nuclear reactor located between a liquid hydrogen propellant tank and a rocket nozzle. The nuclear reactor would heat hydrogen to a high temperature, and the gas would be expelled out the rocket nozzle. The Rover/NERVA nuclear rocket program, in which six reactor–nuclear engines were tested, successfully demonstrated that a nuclear reactor could be used to heat liquid hydrogen for spacecraft propulsion.

Chemical-Based Rocket Propulsion To put the performance of nuclear propulsion in perspective, it is instructive to examine the basic design and performance of chemical rocket systems, which currently dominate all types of rocketry. Chemical rocket technology typically entails use of a *fuel* and an *oxidizer.* For example, the space shuttle's main engines operate through a chemical reaction between liquid oxygen (oxidizer) and liquid hydrogen (fuel). The combustion of the oxygen-hydrogen mixture releases heat in the form of steam and excess hydrogen. These hot gases are then expelled through a thermodynamic nozzle and provide thrust to lift the rocket.

The chemical propulsion system has limited effectiveness because of its specific impulse (ISP). ISP is the pounds of thrust produced per pound of propellant consumed per second, measured in seconds. High ISP, like a high "miles per gallon" for a car, is desirable to minimize propellant consumption, maximize payload, and increase spacecraft velocity, which translate to shorter travel time. The problem with chemical rockets is that most of the vehicle's weight is propellant; the result is a low ISP. To increase ISP, chemical rockets are built in stages—each stage is ejected once its propellant is consumed. Even with this improvement in reducing weight (that is, use of rocket staging), a maximum ISP for chemical engines usually is 400 to 500 seconds, which is relatively small compared with the ISP of nuclear-based propulsion systems.

ISP effectiveness is limited in chemical rockets because the same materials are used for heat source and propellant. The most efficient chemical rockets burn hydrogen and oxygen (an element of relatively high molecular weight) as the heat source to form superheated steam as the exhaust gas. The molecular formula of steam is H_2O (water). Because the oxygen atom (O) is 16 times heavier than the hydrogen (H) atom, a low-molecular-weight propellant, the water molecule has 18 times the weight of the hydrogen atom and nine times the weight of the H_2 molecule, the usual form of hydrogen. Such chemical systems must rely on heavier materials, which directly contribute to a lower ISP, and limit the amount of energy released. Finally, chemical rockets can generate high power levels only for short periods. It is impractical to dramatically increase "burn time" because the resulting rocket would become too massive and too expensive to launch.

Nuclear Propulsion: Overcoming Chemical Specific Impulse Limitations Nuclear propulsion systems overcome the ISP limitations of chemical rockets because the sources of energy and propellant are independent of each other. In current designs, the energy source comes from a nuclear reactor in which neutrons split isotopes, such as uranium-235. The heat produced in this process is used to heat a low-molecular-weight propellant, such as hydrogen. The propellant is then accelerated through a thermodynamic nozzle, as it is in chemical rockets. At the same temperature attainable in chemical rockets, the propellant molecules of nuclear rockets (because of their lower molecular weights) move many times faster than those used in chemical rockets. Nuclear propulsion systems, therefore, allow higher ISPs, thousands of seconds as opposed to the 400 to 500 seconds in chemical systems.

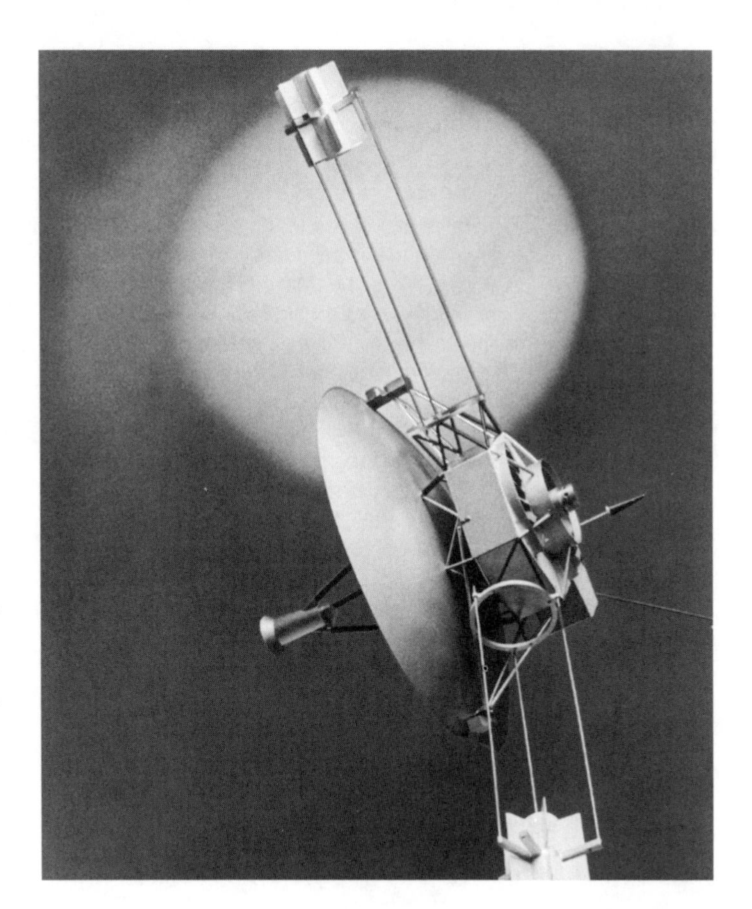

A model of *Pioneer 10*, a nuclear-powered probe sent to Jupiter.
(© Bettmann/Corbis. Reproduced by permission.)

The heat released in this nuclear process can be used for propelling a spaceship over long periods to high speeds and can be converted directly to electricity, either through static processes (nonmoving devices, such as a thermoelectric device) or dynamic processes (moving devices, such as a turbine). Such processes can provide approximately one megawatt of continuous power each day with only one gram of uranium.

Promising Nuclear Propulsion Nuclear propulsion holds much promise for many future applications in space exploration. Nuclear propulsion will enable entire classes of missions that are currently unattainable with chemical systems. Nuclear propulsion systems are classified as being either *thermal* or *electric*.

The graphite-core nuclear reactors built in the NERVA program are nuclear thermal reactors. The nuclear reactor heats a propellant to a high temperature, and the propellant is exhausted out of an engine's nozzle, attaining an ISP of approximately 1,000 seconds. Although the NERVA program was terminated in 1973, nuclear consultant Michael Stancati of Science Applications International Corporation in Schaumburg, Illinois, recently declared that

ANTIMATTER FOR FUTURISTIC SPACECRAFT

Antimatter is one of the most recognized energy sources in science fiction literature. Every particle in the universe is said by physicists to possesses a mirror image of itself, an antiparticle, except that its charge is reversed. When matter meets antimatter, the two substances obliterate themselves and are converted into pure energy. The famous fictional spaceship the *Enterprise* from the television show *Star Trek* uses engines with an antimatter-powered warp drive to journey through interstellar space. Although warp drive is quite impossible with current and near-future technologies, power from antimatter is possible.

NASA is considering antimatter as a rocket propellant to travel throughout the solar system. One gram of antimatter would carry as much potential energy as that now carried on board approximately 1,000 space shuttle external tanks. Although antimatter-powered human space flight could work from a physics point of view, it currently (and in the near future) is unfeasible from the engineering and economic standpoints.

Gamma rays, which are given off by antimatter, are deadly to humans. The necessary shielding for human protection from gamma rays would be so heavy that it would offset the enormous energy gains of using antimatter. Also, at $100 billion per milligram, less than one grain of sand, antimatter is the most expensive material in the universe.

Pennsylvania State University physicist Gerald Smith says antimatter can be used in alternative forms. Instead of using antimatter as the sole source of energy, Smith's group is investigating the use of tiny amounts of antimatter as "triggers" to begin a nuclear fusion reaction with pellets of hydrogen. Smith comments that using antimatter in this way is like using "a lot of little hydrogen bombs." For example, the Pennsylvania State research team is designing a high-speed ion compressed antimatter nuclear (ICAN-II) engine that would use antiprotons to implode pellets with nuclear fusion particles at their cores. Located between the ICAN-II engine and the crew compartment would be massive shock absorbers that would cushion the ship as a series of small blasts propelled it through space.

Harold P. Gerrish, Jr., of the Marshall Space Flight Center propulsion laboratory says that such future spacecraft will use only a few billionths of a gram of antimatter to propel, for example, a 400-ton spacecraft to Mars and back in as little as three months. The same trip with traditional chemical engines would take eight to ten months.

—William Arthur Atkins

the nuclear thermal reactor is a very credible option for long-duration space missions.

Nuclear electric propulsion (NEP) is another possible system. In this method, the propellant is heated to a very high temperature and becomes plasma (an ionized gas). It is then accelerated by electrostatic or electromagnetic fields to increase the exhaust velocity. Because ISP is a function of exhaust velocity, a higher ISP results. Such a nuclear electric plasma rocket (similar to an ion rocket) could attain an ISP ranging from 800 to 30,000 seconds. Specialists claim NEP can reduce the travel time to Mars from nine months to three to five months.

NASA's Future Nuclear Program NASA's future nuclear plans consist of two parts: one developed and one not. Initially, existing RTG technology will be further developed. Without RTGs, NASA and DOE leaders believe that the ability to explore neighboring planets and deep space will not occur in accordance with current project requirements.

Further into the future, several advanced technologies are envisioned for nuclear space systems. One new technology, called Stirling radioisotope generator (SRG), will be used as the dynamic part of nuclear reactors. SRG has the potential to be a high-efficiency power source for missions of long duration, eventually replacing the low-efficiency RTGs. Edward Weiler, NASA's head of space sciences, has said that NASA's nuclear initiative will conduct research in nuclear power and propulsion in areas such as nuclear fission reactors joined with ion drive engines (similar to NEP).

Safety Concerns Nuclear PPSs are not without risk. Safety concerns have always reduced their usability and in many cases has stopped them from being deployed. Although the risk from nuclear PPSs is greater than that from chemical systems, former space shuttle astronaut and senior NASA scientist Roger Crouch says that people's fears about nuclear materials are not grounded on a realistic risk assessment.

More than 30 years have been invested in the engineering, safety analysis, and testing of RTGs. This proven technology has been used in 26 U.S. space projects, including the *Apollo* lunar landings, the *Ulysses* mission to the Sun's poles, the *Viking* landings on Mars, the *Pioneer* missions to Jupiter and Saturn, and the *Cassini* mission to Saturn and Titan. The RTGs have never caused a spacecraft failure. Three accidents with spacecraft that contained RTGs have occurred, but in each case the RTGs performed as designed, and the malfunctions involved unrelated systems.

Summary Because of its many advantages, nuclear energy will continue to provide propulsion and power on space missions well through the twentieth-first century, whether as RTGs, other advanced generators, or nuclear reactors. For spacecraft missions that require up to a few kilowatts of power, an RTG could be the most cost-effective solution. However, for spacecraft missions with large power requirements (in the range of megawatts, or thousands of kilowatts, of power), a nuclear reactor is a better alternative.

Wesley Huntress, president of The Planetary Society, the world's largest space advocacy group, said recently that the development of nuclear propulsion and power technology will make exploration of the solar system more accessible with much shorter flight times and more powerful investigations of the planets. Huntress compared the hundreds of nuclear submarines that ply the world's oceans with the nuclear-powered spaceships that will one day explore outer space.

Even though nuclear space technology was essentially scrapped in the 1980s owing to public fears about its safety, the space community was, and still is, clear that the capabilities of this technology will dramatically reduce flight time to the planets and provide almost unlimited power for operation in space and on the planets. With a return to nuclear technology, the two limiting problems with exploring space—travel time and available power—will be solved.

Compared with the best chemical rockets, nuclear PPSs are more reliable, more efficient, and more flexible for long-distance, complex missions that require high amounts of power for long amounts of time. The American Nuclear Society supports and advocates the development and use of radioisotopic and reactor-based nuclear systems for current and future use to explore and develop space with both manned and robotic spacecraft. —WILLIAM ARTHUR ATKINS

Viewpoint:
No, nuclear technology in space is a dangerous proposition that should be avoided.

"Remember the old Hollywood movies when a mad scientist would risk the world to carry out his particular project? Well, those mad scientists have moved to NASA." This quote would be amusing if not for its source. Dr. Horst Albin Poehler worked as a scientist for NASA contractors for 22 years, 15 of those as senior scientist. Poehler was talking about the *Cassini* mission NASA launched in 1997. The nuclear-powered probe, headed for Saturn, carries 72 pounds (32.7 kg) of plutonium-238 to power its instruments.

Plutonium-238 is a reliable source of energy, and the amount required to produce this energy is not prohibitively large. Because plutonium-238 emits alpha radiation, no heavy shielding of electronics is required on board a spacecraft. (Alpha radiation, actually decay, is the process by which a particle consisting of two neutrons and two protons is ejected from the nucleus of an atom. The particles lose energy rapidly and thus are readily absorbed by air.) For all these reasons, and because it is easy to obtain, plutonium-238 is a favorite of engineers looking for inexpensive, sustainable power sources.

Plutonium-238 is also 280 times more radioactive than is plutonium-239, the plutonium isotope used in nuclear weapons. Its oxide, the compound produced when plutonium-238 and oxygen combine, is extremely dangerous to humans. A fraction of a gram of plutonium-238 is considered a carcinogenic dose if inhaled. Plutonium-238 also can be transported in the blood to the bone marrow and other parts of the body, causing further destruction. The RTGs aboard a spacecraft carry plutonium dioxide.

The Record to Date If we are to accept the use of nuclear technology in space, we must recognize that there are inherent risks in the use of such technology. Atomic energy, although attractive on many levels, can cause untold suffering in successive generations when things go wrong. We accept this risk with nuclear power plants situated too close to populated areas. We believe scientists when they tell us that sending 72 pounds of plutonium-238 to space is a risk-free proposition. This amount is the equivalent of 17,000 pounds (7.7 metric tons) of plutonium-239 (the plutonium used to build nuclear bombs), according to nuclear physicist Richard E. Webb, author of *The Accident Hazards of Nuclear Power Plants.*

Dr. Roger Bonnet holds up a model of the *Rosetta* orbiter, produced by the European Space Agency.
(© AFP/Corbis. Reproduced by permission.)

Should We Really Be So Trusting? The United States launched 26 missions to space with nuclear material on board. Three missions resulted in accidents. The most famous, though not the worst, mishap was the *Apollo 13* accident. The lunar excursion module that returned to Earth and was deposited by NASA in the Pacific Ocean carried an RTG containing more than 8 pounds (3.6 kg) of plutonium-238. Water sampling in the area revealed no radioactivity.

The worst accident involving a U.S. RTG in space occurred in 1964. A satellite powered by an RTG designated SNAP 9-A failed to achieve orbit. The burn-up on reentry caused the release and dispersal of the plutonium core, in a fine dust, the way NASA had planned. At the time the belief was that dispersing the contamination was preferable to a concentrated rain of plutonium. The SNAP 9-A accident, according to Dr. John Gofman of the University of California at Berkeley, contributed to a high global rate of lung cancer. Gofman is a physicist who worked on the Manhattan Project. He earned a medical degree and became a world authority on the dangers of low-level radiation. NASA admits that radioactivity from SNAP 9-A was found in the soil and water samples taken by the Atomic Energy Commission. After the SNAP accident, NASA changed its RTG design to achieve "full fuel containment." This change came about, however, only after SNAP 9-A contamination had been detected on almost every continent and people had died because of the contamination.

Russia, and the Soviet Union, has had more than 40 launches involving nuclear power. Six accidents have occurred, most recently in 1996, when a Mars probe fell to earth and burned over Chile and Bolivia. The amount of radiation released is labeled "unknown," and no attempt has been made to retrieve the radioactive core of the probe or its remains.

In 1978 the Soviet satellite *Cosmos 954* disintegrated over the Northwest Territories in Canada. Scattered across a vast area were thousands of radioactive particles, pieces of the satellite's nuclear power core that survived reentry. Dr. Michio Kaku, professor of theoretical physics at the City University of New York declared, "If Cosmos 954 had sprayed debris over populated land, it would have created a catastrophe of nightmarish proportions." Later estimates were that close to 75% of the radioactive material was vaporized on reentry and dispersed over the planet.

This safety record on two continents shows that we keep paying the price for trusting nuclear technology. It is a lesson that we failed to learn after Three Mile Island and Chernobyl.

Claims for Safety In the early 1980s NASA estimated the chance of a catastrophic space shuttle accident was one in 100,000. After the

Challenger accident on the twenty-fifth shuttle mission, January 28, 1986, NASA lowered its estimate to one in 76. Nobel laureate physicist Richard Feynman, a member of the Rogers Commission that investigated the *Challenger* accident, declared in an appendix to the Rogers report that NASA management "exaggerates the reliability of its product, to the point of fantasy." NASA ignored incidents on previous shuttle missions that clearly pointed to a design flaw or design problem. Feynman points out that NASA took the fact that no accidents had occurred as "evidence of safety." In fact, NASA had known since 1977 there was a design flaw in the O-ring. Independent risk assessment put the probability of a solid rocket booster (SRB) failure at one in 100 *at best* (it was the O-ring in the SRB that failed). But before the *Challenger* accident, NASA stuck by its numbers, claiming the shuttle was safe simply because it had not crashed yet.

One would have hoped that the seven lives on the *Challenger* had not been lost in vain. But the NASA public relations machine overrode common sense when it came to *Cassini*. First came the claims that the *Titan 4* rocket, which lifted *Cassini* into space, was safe. This rocket in fact had a spotted record over the years, memorably exploding in 1993, destroying a one-billion-dollar satellite system it was launching.

NASA reassured the public that even if the rocket were to explode in the atmosphere, the RTG would remain intact. NASA chose to ignore a test conducted by GE, the manufacturer of the RTG, that resulted in destruction of the generator. NASA claimed the test conditions were not realistic. Michio Kaku, however, showed quite clearly that under the extreme conditions of a launch-pad explosion, the RTG was very likely to rupture or be destroyed.

NASA then estimated that any plutonium released during a launch accident would remain in the launch area. As Kaku put it, "NASA engineers have discovered a new law of physics: the winds stop blowing during a rocket launch."

If we cannot trust NASA to honestly inform us of risks, should we trust them to launch plutonium dioxide over our heads?

The Human Error Factor Even if the hardware is safe and the plutonium is well shielded, one factor in technology that cannot be resolved is that the likelihood of human error has no statistical value. As Kaku points out, "[O]ne can design a car such that the chances of an accident approach a million to one, with air bags . . . etc. However, this does not foresee the fact that someone might drive this car over a cliff."

Three Mile Island. Chernobyl. The *Hubble* space telescope. The *Exxon Valdez*. All these events were the results of human error, sometimes a chain of errors compounding one another. We can design the safest RTG, launch it with the safest rocket in the world, and still end up with plutonium raining down on us because of a simple human error. The Mars climate orbiter was lost because of a failure to convert pounds of thrust (an English unit of measure) to newtons (a metric unit). Two nuclear power plants in California were installed backward. Human error is inevitable.

The Alternative Proponents of nuclear energy in space say that deep space missions require nuclear power because solar energy in deep space is insufficient. The fact is that solar energy cells and concentrators have improved vastly over the years. In 2003 the European Space Agency will be launching *Rosetta*, a deep space mission that will study the nucleus of comet 46P/Wirtanen. *Rosetta* will reach its target in 2012, with solar-powered instruments. The probe is designed to conserve energy and achieve its longevity in part by putting the electrical instruments on board into hibernation mode for a lengthy period. *Rosetta* will use solar energy and fuel; no nuclear material will be on board.

Dr. Ross McCluney, an optical physicist, pointed out in 1997 that NASA has several alternatives to nuclear power in space. Newer solar cells are highly efficient, so fewer are needed. New electronic components require less power to do their work. Instead of launching a massive probe, as in *Cassini*, the mission will be split into smaller probes, which will be less expensive and faster to operate than previous missions.

McCluney argues that hybrid power systems can be used for propulsion of probes going as far as nine astronomical units from the sun. Beyond that, McCluney reasons, we can wait for technology that does not require nuclear power. There is no reason to rush into very deep space if it entails risk to people on Earth.

Weapons in Space The U.S. Space Command considers space the future battlefield. The George W. Bush administration has denounced the 1972 Antiballistic Missile Treaty, which among other things forbids antimissile tests. Although the 1967 Outer Space Treaty forbids weapons of mass destruction in space, the United States is researching combat technology in space and is testing space antiballistic systems. The effect of a U.S. military presence in space will most likely be an arms race with many other countries. In addition to the inherent dangers of sending plutonium into space in vehicles and containers of doubtful safety, the specter of a space nuclear arms race has entered the picture.

The twenty-sixth *Challenger* mission would have carried the *Ulysses* space probe, with 24.2 pounds (11 kg) of plutonium. Such a mission would have been far more dangerous than the one in which seven people died. Richard Feynman wrote in his appendix to the Rogers report: "When playing Russian roulette the fact that the first shot got off safely is little comfort for the next." Nuclear technology in space is a game of Russian roulette that must be stopped before the bullet hits us. —ADI R. FERRARA

Further Reading

Clark, Greg. "Will Nuclear Power Put Humans on Mars?" Space.com, Inc. May 21, 2000 [cited July 29, 2002]. <http://www.space.com/scienceastronomy/solarsystem/nuclearmars _000521.html>.

Clark, J. S. *A Historical Collection of Papers on Nuclear Thermal Propulsion.* Washington, DC: AIAA, 1992.

El-Genk, M. S. *A Critical Review of Space Nuclear Power and Propulsion 1984–1993.* New York: American Institute of Physics Press, 1994.

Feynman, Richard P. "Personal Observations on Reliability of Shuttle." In *Report of the Presidential Commission on the Space Shuttle Challenger Accident.* Vol. 2, Appendix F. June 6, 1986 [cited July 29, 2002]. <http://history.nasa.gov/rogersrep/v2appf.htm>.

Grossman, Karl. *The Wrong Stuff: The Space Program's Nuclear Threat to our Planet.* Monroe, ME: Common Courage Press, 1997.

"Interstellar Transport." *Sol Station.* Sol Company. 1998–2000 [cited July 29, 2002]. <http://members.nova.org/~sol/station/interste.htm>.

Kaku, Michio. "Dr. Michio Kaku's Speech on *Cassini* at the Cape Canaveral Air Force Station Main Gates, July 26 1997" [cited July 29, 2002]. <http://www.lovearth.net/mkaku.htm>.

———. "A Scientific Critique of the Accident Risks from the *Cassini* Space Mission." August 9, 1997, modified October 5, 1997 [cited July 29, 2002]. <http://www.animatedsoftware.com/cassini/mk9708so.htm>.

Kulcinski, G.L. "Nuclear Power in Space: Lecture 25." March 25, 1999 [cited July 29, 2002]. <http://silver.neep.wisc.edu/~neep602/lecture25.html>.

Lamarsh, J.R. *Introduction to Nuclear Engineering.* 2nd ed. Reading, MA: Addison Wesley, 1983.

Leonard, David. "NASA to Go Nuclear; Spaceflight Initiative Approved." Space.com, February 5, 2002 [cited July 29, 2002]. <http://www.space.com/news/nasa_nuclear_020205.html>.

McCluney, Ross. "Statement on the Solar Power Alternatives to Nuclear for Interplanetary Space Probes." August 14, 1997 [cited July 29, 2002]. <http://www.animatedsoftware.com/cassini/rm9708s.htm>.

McNutt, Ralph L., Jr. "Plutonium's Promise Will Find Pluto Left Out in the Cold." *SpaceDaily.* February 20, 2002 [cited July 29, 2002]. <http://www.spacedaily.com/news/outerplanets-02b.html>.

"Space Nuclear Power System Accidents: Past Space Nuclear Power System Accidents." *NASA Space Link.* September 19, 1989 [cited July 29, 2002]. <http://spacelink.nasa.gov/NASA.Projects/Human.Exploration.and.Development.of.Space/Human.Space.Flight/Shuttle/Shuttle.Missions/Flight.031.STS-34/Galileos.Power.Supply/Space.Nuclear.Power.System.Accidents>.

Newman, David. "Antimatter: Fission/Fusion Drive: Antimatter Catalyzed Micro Fission/Fusion (ACMF)" [cited July 29, 2002]. <http://ffden-2.phys.uaf.edu/213.web.stuff/Scott%20Kircher/fissionfusion.html>.

"Nuclear Rockets." Los Alamos National Laboratory Public Affairs Office [cited July 29, 2002]. <http://www.lanl.gov/orgs/pa/science21/NuclearRocket.html>.

"President's Budget Cancels Current Outer Planets Plan in Favor of Developing Nuclear Propulsion and Power for Mars Landers and Future Outer Planets Probes." The Planetary Society. February 4, 2002 [cited July 29, 2002]. <http://www.planetary.org/html/society/press/budget_03.htm>.

"Reusable Launch Vehicles." *World Space Guide.* FAS Space Policy Project [cited July 29, 2002]. <http://www.fas.org/spp/guide/russia/launch/other.htm>.

Sandoval, Steve. "Memories of Project Rover Come Alive at Reunion." *Reflections* 2, no. 10 (November 1997): 6–7.

"Space Nuclear Power Technology." *NASA Space Link.* [cited July 29, 2002]. <http://spacelink.nasa.gov/NASA.Projects/Human.Exploration.and.Development.of.Space/Human.Space.Flight/Shuttle/Shuttle.Missions/Flight.031.STS-34/Galileos.Power.Supply/Space.Nuclear.Power.Technology>.

"Space, Future Technology, Nuclear Propulsion." Discovery Channel [cited July 29, 2002]. <http://www.spaceref.com/directory/future_technology/nuclear_propulsion/>.

Sutton, G. *Rocket Propulsion Elements.* 6th ed. New York: John Wiley and Sons, 1992.

"Vision for 2020." United States Space Command. February 1997 [cited July 29, 2002]. <http://www.spacecom.mil/visbook.pdf>.

"When Isaac Met Albert." Marshall Space Flight Center. November 12, 1997 [cited July 29, 2002]. <http://science.msfc.nasa.gov/newhome/headlines/msad12nov97_1.htm>.

Willis, Christopher R. *Jobs for the 21st Century.* Denton, TX: Android World, 1992.

Historic Dispute: In his classic debate with Albert Einstein, was Niels Bohr correct in his approach to interpreting the world in light of the newly discovered field of quantum mechanics?

Viewpoint: Yes, Bohr's interpretation of the world in light of quantum mechanics was correct, and new applications of his interpretation are being determined with the passage of time.

Viewpoint: No, while the physics community came to accept the arguments of Bohr, some of the questions raised by Einstein remained unsatisfactorily resolved.

One of the great triumphs of human thought was Isaac Newton's formulation of his laws of motion and his law of universal gravitation. These laws laid the foundation for the development of classical mechanics, the branch of physics that describes the motion of slowly moving objects under the influence of external forces. An "object" might be a ball rolling down a hill, or the Moon orbiting Earth. The power of Newton's laws was that they were predictive: given an object of known mass and velocity, and given an environment such as an inclined surface or a gravitational field, it was possible to predict the subsequent motion of that object. Continuing refinement of the methods and techniques used to solve classical mechanical problems led to highly precise predictions of motions.

The careful reader, however, will notice a crucial caveat in the definition of classical mechanics above. The objects in question must be *slowly moving,* and in this context, "slow" means relative to the speed of light. It is now well known that classical mechanics does not predict the motion of rapidly moving objects or particles correctly, and that classical mechanics in general does not provide a full description of the physical universe as we currently observe it. Does this mean classical mechanics is wrong? Certainly not; it means that classical mechanics is valid only within certain limits, and those limits happen to encompass everything we experience in our daily lives. We do not move at half the speed of light, nor do we observe phenomena at the atomic level without specialized equipment.

For Newton and the scientists of his day, the principles of classical mechanics fully described the world as they perceived it. In the nineteenth century, however, problems began to emerge. James Clerk Maxwell and Michael Faraday achieved the next great advance in physics after Newton's laws, the mathematical description of electricity and magnetism as complementary elements of a unified concept, sensibly called *electromagnetism.* Accelerated charged particles were discovered to emit electromagnetic radiation (light), and this led to severe problems with the "classical" model of atoms, in which electrons orbited the nucleus of the atom like little moons orbiting a planet. In classical mechanics, orbiting particles are constantly accelerated, so it seemed like the radiating electrons should quickly lose all their energy and crash into the nucleus. Obviously that didn't happen in real atoms, so physicists quickly realized that the laws of classical mechanics simply didn't work when applied to atomic particles.

These and other problems led to the defining development of twentieth-century physics: quantum mechanics. The great problem prior to quantum mechanics was that matter and light had been treated as fundamentally different phenomena: matter was interpreted as classical particles, while light was treated as waves. In quantum mechanics, this distinction is blurred: matter and light each exhibit characteristics of particles and waves. This concept does not exist within the limits of classical mechanical theory. Development of this concept by the great physicists of the day—Max Planck, Albert Einstein, Louis de Broglie, Niels Bohr, Erwin Schrödinger, and Werner Heisenberg, among others—led to the mature theory of quantum mechanics that underpins modern physics.

Quantum mechanics emerged at the start of the twentieth century, and it was a concept that instigated a radical change in our understanding of the universe. As with any profound concept, its implications rippled from hard science into broader philosophical issues.

In particular, quantum theory introduced a *probabilistic* view of the universe. Previous formulations of mechanics and the structure of atoms employed a *deterministic* viewpoint, in which the characteristics of the particles and systems in question were described with definite numerical values. Two physicists, Erwin Schrödinger and Werner Heisenberg, proposed a radically different model. Instead of specifying the precise location of, for example, an electron, Schrödinger described it in terms of its probable location. Put in terms of a simple statement, Schrödinger's argument was that we cannot say "The electron is in this location at this point in time," but we instead say "It is most likely that the electron is in this location at this point in time." Heisenberg continued this idea with his famous uncertainty principle, which states that complementary characteristics of a particle, for example position and momentum, cannot be simultaneously determined with absolute precision. If we know the location of an electron with high certainty, its momentum is uncertain, and vice versa. The classic mental image of this is the process of measuring the position of an electron. How do we do it? We can only observe things by shining light on them. But photons (quanta of light) carry momentum sufficient to move an electron. The classical analogy might be like trying to determine the location of a billiard ball by striking it with the cue ball—we hear a click, so we know roughly where the balls struck, but now the target ball is moving off somewhere else. By analogy, when we shine light on the electron to determine its location, we introduce an uncertainty into its location by the very act of performing the experiment.

This was the crux of the great debate between Albert Einstein and Niels Bohr, and it is one of the finest examples in human intellectual discourse of the ideas and motivations that guide the evaluation of radical new theories. Einstein resisted the probabilistic description of the atom, probably because of deeply held religious convictions that dictated to him a sense of order and absoluteness about the universe. Einstein's famous "God does not throw dice" statement cuts to the center of his reservations about a "fuzzy" universe in which a fundamental uncertainty governed the basic nature of particles. Bohr, on the other hand, was more comfortable with the ideas of Schrödinger and Heisenberg, and he and Einstein debated these ideas at length. The details of these ideas are discussed in the articles that follow.

The essential lesson of the Bohr-Einstein debate is that science, however rigorous and dispassionate, remains a thoroughly human process. Some scientists are deeply religious, while others are atheists—just as might be found in any other profession or culture. Some scientists are conservative, while some are liberal. Although scientists are trained to judge evidence objectively, it is all too easy for these and other personal beliefs, or the career-long development of a given idea, to affect a neutral, objective evaluation of the data. This is not to say that Einstein was religious and Bohr was not; that would be a gross oversimplification. The split between Einstein and Bohr centered on profound philosophical differences, and because they were two of the most brilliant minds of the twentieth century, they were each able to construct detailed arguments supporting their points of view. The debate was also respectful and civil; given Einstein and Bohr's respective achievements, there was no need for acrimony. Their debates about the nature of the atom and, by extension, of the underlying structure of the universe, remain one of the essential examples of rigorous testing and cross-examination of a new and controversial idea. —JEFFREY C. HALL

KEY TERMS

ATOM: Originally believed to be the smallest indivisible particle of which matter was composed, it is now known to consist of protons, neutrons, and electrons, in addition to at least 40 other subatomic particles.

ELECTRON: A negatively charged subatomic particle. Its mass is approximately one thousandth that of the proton. In a neutral atom, the number of electrons equals the number of protons. The electron is found in an area around the nucleus determined by the orbit it occupies.

HIDDEN VARIABLES: Proposed as a means of putting reality back into physics. Rather than a wave-particle merely being a statistical probability pattern until observation, it has been suggested that there are hidden variables not explained by quantum theory so that a particle has actual properties independent of observation.

LOCALITY: The local area around an object is the area that light could travel away from it from a given time. You can imagine it as an expanding sphere that grows over time. Objects outside that sphere are non-local, as, according to Einstein's theory of Relativity, nothing can move faster than light. Therefore, if locality holds true no object outside the local area can influence the behavior of an object.

PHOTOELECTRIC EFFECT: The observation that light of great intensity but the wrong wave length had insufficient energy to remove an electron from the outermost energy level of an atom. If light of the correct energy was aimed at the atom, the electron received sufficient energy to be removed. Increasing the intensity increased the number of ejected electrons. Increasing the energy of the light caused the ejected electrons to have higher energy and hence velocity.

QUANTUM: A discrete quantity or amount, from the Greek word *quanta.* Coined by Max Plank in 1900 in his explanation of black body radiation, the idea that energy was quantized, rather than continuous, was at first resisted, but later developed into a new theory of physics at the atomic scale.

QUANTUM MECHANICS: The most current model that describes the atom. Quantum mechanics uses wave functions to describe the region of greatest probability of finding the electrons in an atom.

UNCERTAINTY PRINCIPLE: The realization that if a system is defined by two complementary variables, the product of the uncertainties of the two variables has a limit of a very small number (usually Planck's constant). In other words, if two variables defining a system are complementary, the determination of one with certainty results in the inability to determine the other with certainty.

WAVE FUNCTION: A mathematical solution to the wave equation developed by Schrödinger. The wave function mathematically contains the limitations originally set out by Bohr to describe the energy states of electrons in an atom.

Viewpoint:

Yes, Bohr's interpretation of the world in light of quantum mechanics was correct, and new applications of his interpretation are being determined with the passage of time.

The Effects of Quantization Max Planck (1858–1947) proposed that rather than having continuous values (like any height if you ascend a ramp), energy could only come in discrete packets (like specific heights such as when you climb stairs). He called these packets *quanta.* One person who immediately saw the validity and consequently the applications of Planck's quanta was Albert Einstein (1879–1955). He applied the concept of the quantum to the solution of yet another problem in physics at the time, known as the *photoelectric effect.* He saw that if, as Planck stated, energy had to consist of discrete packets, then the energy of the light was more important than intensity. As energy is directly related to the frequency, a beam of higher frequency (blue) and low intensity would contain the necessary energy to remove the electrons. On the other hand, a beam of lower frequency (red), no matter how intense, could not remove the electrons. Further, he reasoned, if the energy of the light was more than required to actually remove the electron, the remainder of the energy would be transferred to the electron that was removed. This was experimentally shown and the quantum gained acceptance.

Niels Bohr (1885–1962) was a quiet, introspective Danish scientist. He was fascinated by the discoveries of Planck and Einstein, both of whom he held in great regard. He was intrigued and dissatisfied by the model of the atom proposed by Rutherford, under whom he worked. Upon the advice of a colleague, he studied the most recent results in the field of *spectroscopy*, the study of the frequencies of light emitted or absorbed by substances (mostly atoms or simple molecules). He noticed that unlike for white light (where all frequencies or colors are observed blending together), atoms and molecules had *discrete spectra*. In other words, they emitted or absorbed only very specific frequencies of light. Bohr brilliantly put this information together with Planck's quantum to arrive at the first working model of the atom. He theorized that the electron that orbited the nucleus could not simply assume any energy value. He realized that the solution lay in having orbits of specific energy values in which the electrons could remain. If energy was added to the atom/molecule in the form of radiation or a collision, the electron could be "bumped" into a higher energy level. It would eventually lose that energy (in the form of radiation, seen in the spectrum of the atom/molecule), but the electron could not continue to spiral into the nucleus. The electron would return to its original (and lowest) energy level. This explanation satisfied all the physical requirements. It provided a stable model for the atom, and it explained the emission/absorption spectra of atoms and simple molecules. It was a landmark discovery, heralded because of Bohr's ability to determine an equation for the energy levels that gave almost exact values for the energies of the hydrogen atom. Unfortunately, Bohr could not make his model work for any atoms other than hydrogen and one form of helium.

Up to this point, Einstein was a great supporter of the quantum theory, so named because it used the fact that energy in atoms was limited or quantized. Further revelations, however, caused Einstein to change his mind in an almost complete turnabout.

Probabilistic Nature of Quantum Mechanics

Bohr had been able to make deterministic calculations of the atom (that is definite numerical values), in spite of the fact that his success was limited. Shortly after Bohr presented his model, great work was done in another field that had a direct impact on the direction in which quantum mechanics would develop.

Prince Louis de Broglie (1892–1987) provided yet another piece of the puzzle when he stated that if light could possess the properties of a particle, matter should possess the properties of a wave! This dual nature of matter and light came to be known as *wave-particle duality*.

Niels Bohr

(© Corbis. Reproduced by permission.)

Erwin Schrödinger (1887–1961) was now able to formulate his wave-equations that, to date, are the most accurate representation of the atom. His model of the atom describes the location of the electrons in atoms using a probability function rather than a defined orbit. He essentially postulated that while it was possible to know where the electron was most likely to be, it was impossible to know exactly where it would be. Werner Heisenberg (1901–1976) delivered the final blow to the deterministic scientists. He declared that it was impossible to determine two complementary qualities of an atomic system exactly. If one was known exactly, the other could not be known except in the most general terms. This limitation came to be known as the *Heisenberg Uncertainty Principle*. It was the uncertainty principle that made Einstein turn completely away from the quantum mechanical theory of the atom. Until the day he died, he refused to accept quantum mechanics as the complete solution of the atom.

What, then, was the nature of the disagreement between Bohr and Einstein? Essentially, until Schrödinger and Heisenberg, science was deterministic in nature. Physical quantities could be determined with great accuracy, and the behavior of a system could be predicted if the nature of the system was known. Hence, the laws of classical physics allowed for the determination of the acceleration, momentum, velocity, and position of a ball to be known if it was thrown in the air. Einstein pushed back the borders of classical physics with the Special and then General Theories of Relativity. He could not, at that time, perform actual physical experiments that would demonstrate the validity of his ideas, but he constructed what are known as *thought experiments.* In a thought experiment, an "imaginary" experiment was performed that HAD to conform to the known laws of physics. In other words, just because it was not performed in a laboratory did not mean that the physical laws could be ignored. Einstein's theories of relativity were indeed hailed as extraordinary because his thought experiments were subsequently proven with the advent of further developments in technology. Einstein's use of Planck's quantum to explain blackbody radiation was validated by the observed energies of the blackbody. Hence, Einstein accepted the quantization of energy with no hesitation.

However, Schrödinger moved away from the concept of absolute values with the introduction of the wave equation. The wave equation essentially described electrons using a probability function. Unlike Bohr, who assigned electrons to orbits having quantized energy values, Schrödinger deemed it necessary to calculate the *probability* of finding a certain electron in a particular region. The wave equation described the shape of the region where it was most probable that the electron could be found.

It is most likely that Einstein understood that the basic nature of this new field would no longer be deterministic. He accepted and extended the work of Satyendranath (S. N.) Bose (1894–1975) on statistical mechanics (the modeling of the average behavior of a large number of particles) to gases. The Bose-Einstein statistics was the first indication that exact behavior of each and every particle in a large assembly of particles was not necessarily a possibility. Rather, the behavior of the particles was averaged according to statistical probabilities. In spite of this awareness, Einstein was not satisfied with Schrödinger's wave equations. Bohr, on the other hand, immediately saw the necessity of the application of probabilities to describe wave functions.

It was while working with Bohr that Heisenberg formulated the uncertainty principle. While it is a simple mathematical formulation that limits the extent to which two complementary physical quantities can be determined, its implications were immediately understood by both Bohr and Einstein. The uncertainty principle was, in effect, a statement that the observer and the observed could not remain distinct at the atomic level.

What does this mean? If a ball is thrown into the air, the act of observing its motion (using vision or any measuring instrument) does not disrupt the motion of the ball, nor does it change its behavior in any way. The same is true of a car in linear motion, of an asteroid in circular motion, or a planet in elliptical motion. In all of classical physics, observation of a system has no measurable effect on the system and does not affect its subsequent behavior. However, as we approach the infinitesimal atomic systems, this observation is no longer true. It is not possible to "see" atomic systems visually. It is necessary to use sophisticated instruments, such as an electron microscope or an x-ray diffractometer. Such devices use energy, or light, or high-speed small particles (in some ways these are all synonymous) to study the system. The energy/light/high-speed small particle reflects off the system and delivers a signal that is read by the device.

However, the energy/light/high-speed small particle, in sending back information, also *disrupts* the system! (The closest analogy might be trying to determine the trajectory of a baseball by throwing another baseball at it.) If an effort was being made to determine the momentum and the position of an electron in an atom, then Heisenberg maintained that if the momentum of the electron was determined with some precision, the position was only vaguely known. Similarly if the position was determined with some precision, then the momentum was only vaguely known. If the position was determined precisely, it was not at all possible to determine the momentum and vice versa.

Bohr realized that this limitation was inherent in the nature of quantum mechanics. Einstein maintained that if such a situation existed, quantum mechanics could not be the final solution to the atom and was most likely an intermediate solution.

Bohr-Einstein Debates Einstein's abrupt turnabout was a blow to the scientific community headed by Bohr, who firmly believed that Heisenberg's revolutionary statement of the nature of the atom was correct. When they met at the Solvay conference, a place where the great minds of the time all assembled to discuss the astonishing developments of the past few years, Einstein and Bohr had the first open debate about the validity of the quantum theory.

In order to refute the contention that only one of two complementary variables could be determined exactly, Einstein designed a thought experiment. The experiment consisted of a single electron being passed through a plate containing a slit and striking a photographic plate. Einstein maintained that once the electron had exposed a spot on the photographic plate, there was no doubt in its position or its path. He thus maintained that quantum mechanics dealt with the average behavior of particles and hence the uncertainty arose due to the averages. Bohr revised Einstein's experiment to include a second plate between the plate with the slit and the photographic plate. The second plate contained two slits that lay on either side of the original slit. Bohr showed that if a strong beam of electrons was passed through such a system, an interference pattern was produced on the photographic plate. Then he suggested reducing the intensity of the beam until it was so weak that only one electron at a time could go through the series of slits. He agreed that the first electron, as Einstein had suggested, would strike a particular spot, but if other such "single" electrons were allowed to pass through, they would eventually produce the same interference pattern! Thus, indirectly, the first electron DID have an effect on the behavior of the subsequent electrons. Bohr used this experiment to indicate that the system could not be completely independent of the method of observation.

Three years later, at the next Solvay conference, Einstein proposed yet another thought experiment called the photon in the box. He designed a system consisting of a box containing a clock and some photons (light particles). The box had a small hole. The clock, at a certain time, would allow the release of a single photon. The mass of the box before and after the escape of the photon would allow for calculation of the mass of the photon (exactly). Once its mass was known, then the energy of the photon could be determined exactly using Einstein's own equation, $E = mc^2$. Bohr took one night to come up with a response to Einstein. He indicated to Einstein that in order to determine the mass of the box, it would be attached to a spring scale. The initial measurement was accurate without question. However, once the photon escaped, the box would move due to the mass change. According to Einstein's own theory of relativity, such a movement would cause a change in the rate of the clock in the box. The longer you took to read the mass (spring scales take time to settle), the less certainly you knew the time. The more quickly you measured the time, the less certainly you knew the mass. Since the mass is directly related to the energy, the energy of the photon and time are complementary values and this again reduced to an equation of the form of Heisenberg's uncertainty principle. After this exchange with Bohr, Einstein did not publicly debate this issue with him. While he accepted that Bohr's logic was valid, he could never bring himself to accept the possibility that the Universe could be anything other than deterministic.

Deeper Meaning Behind the Bohr-Einstein Debate The crux of the issue for Einstein was that he believed that the universe is ordered and has a logical nature. He believed that based on physical observations, one should be able to predict the nature of a system. Most importantly, he felt that the scientist was the observer and the system was the observed.

Werner Heisenberg
(© Hulton-Deutsch Collection/CORBIS. Reproduced by permission.)

If Heisenberg was right, however, it meant that there was no order in the universe, and there was no predictability. Probability is about chance, not certainty. It meant that a system could ultimately not be known because in the final analysis, trying to analyze the system caused the system to change. In other words, the observer was part of the system, along with the observed. Bohr realized that it was possible to determine the values for two complementary parts of system; it was just not possible to determine them simultaneously. The observer determined what the observed would be by choosing the system and once chosen, until the course of the measure was completed, it could not be changed.

Deep down, in spite of his professed disbelief in any religious system, Einstein realized that Heisenberg's uncertainty principle affected his moral and religious core. Christianity, Judaism, and Islam, all religions that follow the Old Testament, essentially believe in that which is self-evident. Order exists in the universe, the order comes from a source, and the source is God. If science showed that the fundamental nature of the universe was not ordered, but random (based on probability), what did that say of

God? Quantum mechanics touched on the very sensitive core of a number of scientists who professed atheism, yet who were not capable of truly accepting it in their work. Even if they didn't believe in God, they believed in the concept of order and a source (by any other name). Einstein's true dilemma was his inability to accept that the science he loved pointed to a world that was not ordered. Could such a world have been created by God? Creation, by its very nature, supports order. As long as Einstein's science supported this order, he accepted it. The moment the science veered from the path of order, Einstein could not accept it. His faith was deeper than he himself thought.

This is not to say that Bohr was without faith. However, he did not hesitate to carry his scientific convictions to the logical conclusion. If that meant that the concept of an orderly creation of God was disproved, then, so be it. However, the eastern philosophies are much broader in their way of thinking and it was to these philosophies that the great quantum scientists, such as Bohr and Schrödinger, turned. The Hindu philosophy believes in the created as part of the creator, not such a different philosophy from that derived in quantum physics.

An interesting anecdote is offered by Léon Rosenfeld (1904–1974), a contemporary and a close friend and supporter of Bohr. Rosenfeld met with Hideki Yukawa (1907–1981) in Kyoto, Japan, in the early 1960s. Yukawa's work with the meson and the complementary concepts of elementary particle and field of nuclear force derived largely from Bohr's insight. When Rosenfeld asked Yukawa whether the Japanese had difficulty convincing their scientists of the validity of the concept of complementarity, Yukawa's response was "No, Bohr's argumentation has always appeared quite evident to us. You see, we in Japan have not been corrupted by Aristotle."

Science today has more than validated the concept of complementarity, and has found numerous other examples of complementary quantities that exhibit the limitations of the uncertainty principle. In spite of his quiet voice, and soft demeanor, Bohr's science and strength of conviction have firmly established themselves in a world where quantum mechanics is a part of daily life. There is little doubt that Bohr's interpretation of the world of quantum mechanics is here to stay until, perhaps, the next great scientific revolution. —RASHMI VENKATESWARAN

Viewpoint:
No, while the physics community came to accept the arguments of Bohr, some of the questions raised by Einstein remained unsatisfactorily resolved.

Between 1927 and 1936 a series of debates took place between Niels Bohr (1885–1962) and Albert Einstein (1879–1955) over the details of quantum physics and the nature of reality. While the physics community came to accept the arguments of Bohr, some of the questions raised by Einstein remained unsatisfactorily resolved. Bohr's view became the standard used by physicists, and all experimental results since have confirmed quantum theory. Yet Bohr's interpretation leads to many uncomfortable consequences in the philosophy of physics.

At the very least Albert Einstein's criticisms of quantum theory were important in that they forced Bohr and others to face uncomfortable paradoxes inherent in the Copenhagen interpretation of quantum physics, and refine and clarify their arguments. Bohr himself wrote that "Einstein's concern and criticism provided a most valuable incentive for us all to reexamine the various aspects of the situations as regards the description of atomic phenomena." Yet the popular opinion among the scientific community at the time was that Einstein's objections were naïve, and that he was struggling to understand the new physics. Over time there has grown a greater appreciation for Einstein's opposition to Bohr's interpretation, and their debates are now considered some of the most important in the history of science.

Duality and Uncertainty Two of the ideas essential to the Copenhagen interpretation are the ideas of wave-particle duality and the Heisenberg uncertainty principle, developed by Werner Heisenberg (1901–1976) in 1927. Particles, atoms, and even larger objects possess the strange property of exhibiting both wave and particle-like properties, depending on how they are observed. This can create some strange situations. For example, if Thomas Young's (1773–1829) famous two-slit interference experiment (which first showed the wave-like nature of light) is modified so that only one photon of light is passed through one at a time, bizarre quantum effects can be generated. If the experiment is allowed to continue as normal then the expected wave interference pattern is built up one photon at a time. The quantum wave function that represents the photon has the same chance of passing through either slit, and so in effect interferes with itself to produce the pattern. However, if a detector is placed at one of the slits then the photon "particle" either passes through the slit or it does not, and suddenly the interference pattern vanishes. Setting up the experiment to detect the photon as a particle produces a result that suggests it is a particle, whereas setting up the experiment to look for wave properties (the interference pattern) gives results that exhibit wave properties.

The uncertainty principle states that it is impossible to measure certain pairs of properties with any great accuracy. For example, you cannot know both the position and momentum of a particle within certain well-defined limits. The more accurately you determine the position of the particle, the more uncertain the value of the momentum will become, and vise versa. There are other uncertainty pairs of properties aside from position and momentum, such as energy and time.

Einstein's criticisms of Bohr's quantum theory exploited the strangeness of wave-particle duality and attempted to circumvent the uncertainty principle, and thereby show quantum theory to be incorrect. However, while Einstein may have failed in these attempts, in doing so other fascinating questions were raised.

Bohr developed a theory based on the experimental evidence that contradicted classical Newtonian physics. He believed he had discovered a consistent interpretation of quantum mechanics. Bohr suggested it was meaningless to ask what an electron is. Physics is not about what is, but what we can say to each other concerning the world. Since we can only really understand classic physics concepts, such as position and momentum, the role of quantum theory was to provide a mechanism with which these could be communicated. However, that in no way implies that objects at the quantum level actually have these values. In Young's experiment we can either leave the wave-particles alone, and observe an interference pattern, or take a peek at the wave-particles as they go through the experiment, as in doing so lose the pattern. The two situations are not contradictory, but complementary.

The Solvay Conferences Taking Bohr's viewpoint, the world has no independent existence, but is tied to our perceptions of it. Einstein feared that physics as defined by Bohr was becoming wholly statistical and probabilistic in nature, and strove to keep some level of reality in physics by attempting to find an underlying determinism that was not explained by quantum theory. In 1927, during informal meetings at the fifth Solvay Conference in Brussels, Einstein attempted to offer logical arguments against the wave interpretation of quantum theory. He proposed a number of imaginary experiments to Bohr, attempting to outflank the uncertainty principle, however, Bohr managed to resolve all of Einstein's objections.

In 1930, at the sixth Solvay conference, Einstein attempted to attack the uncertainty principle again. He asked Bohr to imagine a box with a hole in its side which would be opened and closed by a clockwork shutter inside the box. The box contains a radiation source which emits a single photon through the shutter which would be opened for a short time. Einstein argued you could weigh the box before and after the photon was emitted and the change in mass would reveal the energy of the photon, using the equation $E=mc^2$. You could then read the time the photon was emitted from the clockwork inside the box, thereby determining both energy and time to an accuracy that would defy the uncertainty principle.

Albert Einstein
(Archive Photos. Reproduced by permission.)

Many commentators have noted that Bohr's reply is difficult to follow, due to the condensed form in which he gave it, but nevertheless ingeniously showed that the uncertainty principle is still supported, and is a function of the measurement of the change in mass. Einstein accepted Bohr's reply as correct.

Is Quantum Theory Complete? Einstein then modified the photon-box thought experiment in an article with R. C. Tolman and Boris Podolski (1896–1966). After the photon was gone, they suggested, you could either measure the mass change, thereby determining its energy, or read the time of its release from the internal mechanism (but not both) according to the principles of quantum theory. You could then predict either the time it would arrive at a certain location, or its energy at that location (but not both).

However, either of these can be done without disturbing the photon. The point, which was not made clear in the paper, was that if you can determine either property without disturbing the photon then they are not mutually interfering properties, and the process must actually have an exact time of occurrence and the photon must have an exact energy, even if we cannot observe them together. This would mean that quantum theory could not provide a complete description for the world.

Bohr's response was that the setup is not one, but two, experiment systems. The weighing device and the clock work shutter are two separate systems and so there are two different quantum phenomena to be measured. Therefore quantum uncertainty is saved.

Einstein's most serious attack on quantum physics was published in 1935 in a paper with Podolski and Nathan Rosen (1909–1995). The EPR paper, as it came to be known, suffered from an argument that was not as strong as Einstein first intended, and it was unclear in places. Bohr's reply was also confused. Nonetheless, most scientists at the time thought that Bohr had got the better of Einstein, and Bohr's interpretation became the orthodox interpretation. However, lying in wait in the EPR paper was the germ of an argument that would take 50 years to be resolved.

Imagine a single stationary particle that explodes into two halves. The conservation of momentum implies that the momentum of one particle can be used to deduce the momentum of the other. Or, symmetrically, you could measure the position of one particle, which would reveal the other particle's position. Like Einstein's previous thought experiment, the key is that you can therefore deduce something about a particle without observing it. However, this time the experiment definitely remains one quantum system.

Quantum theory suggests that until either particle is observed it is a probabilistic wave function. This wave function will have a wide range of possibilities, only one of which will be revealed when the particle is observed and the wave collapses. It does not, however, have an actual position or momentum until observation. This causes problems if the two particles in the EPR experiment are separated by a large distance. When both are observed simultaneously they must both give the same result as per the conservation of momentum. If the particles have no actual values until observation, how does one particle "know" what value the other wave function has collapsed to? They would appear to need faster-than-light communication, which Einstein referred to as "spooky forces" at a distance. Einstein's solution was to argue that there were hidden variables underneath quantum mechanics which gave the particles an actual position and momentum at all times, independent of observation. Bohr rejected Einstein's argument by saying that the particles must be viewed against the total context of the experiment, and that a physical property cannot be ascribed to a particle unless the situation is a meaningful one. This reasoning still allows this spooky conspiracy between the two particles, and it appears that even Bohr was unsatisfied with his answer. Certainly many commentators since have suggested that Bohr's published answer to the EPR paper does not make much sense.

Aftermath After 1936 the public debate over quantum physics between Bohr and Einstein ceased. The Copenhagen interpretation remained dominant, and is happily used by physicists who ignore the bizarre philosophical implications simply because it works so well. However, the EPR argument was too powerful to disappear completely, and later developments brought it back into the spotlight.

In 1965 John Bell (1928–1990) published a highly mathematical paper that suggested a method for determining experimentally between Bohr and Einstein's interpretation of reality. Bell's inequality theorem examined the logic that governs the process of measurement in two-particle systems. The paper established a theoretical limit to the level of correlation for simultaneous two-particle measurement results. However, if quantum physics was correct then experimental results should sometimes exceed Bell's limit.

It was not until the early 1980s that sufficiently accurate experiments were carried out to test Bell's inequalities, and therefore determine between Bohr and Einstein's views of quantum physics. The most important experiments were carried out by Alain Aspect and his colleagues. Their experiments, confirmed by many others since, produced observations which cannot be predicted by a theory in which the particles in the EPR experiment do not influence each other. This appears to have settled the argument between Bohr and Einstein in Bohr's favor, and has some profound consequences for the philosophy of physics.

If Einstein had been basically correct, then quantum effects would be the result of the behavior of real particles lying "underneath" the quantum level. There would also be no faster-than-light signaling between the particles, which is usually referred to as locality. However, the work of Bell, Aspect, and others strongly suggests that Einstein was wrong, and that as Bohr suggested either there is no underlying reality behind quantum physics, or there are non-local effects that violate Einstein's theory of relativity.

Einstein was right to question the Copenhagen interpretation, and it should still be questioned. That is, in many ways, the nature of scientific inquiry. The philosophical problems that come with the Bohr's theory of quantum physics are considered too great for many thinkers, and in many quarters there is a profound "unhappiness" with the Copenhagen interpretation. There have been many other attempts to reinterpret quantum physics in a manner that removes these problems, such as the many-worlds interpretation, the many-minds interpretation, and non-local hidden variables, so Einstein was by no means alone in his philosophical objections. It is hoped that these alternative interpretations will one day be tested experimentally, just as Einstein's has been, but for now they remain different only in the philosophical, not physical, consequences they predict.

While many contemporaries saw the objections of Einstein as the reactions of a conservative unable to cope with the rapidly developing theory, many historians of science have come to regard the debate as one of the most important in the philosophy of science. It is important to realize that Einstein was not trying to discredit quantum physics per se, rather he disagreed with the philosophical direction Bohr's interpretation was taking.

The questions raised by the EPR paper have only recently been tested experimentally. Hidden variables and locality have now been shown to be incompatible, but there is still some room for debate, and the possibility of non-local hidden variables has not been ruled out. Indeed, in attempting to resolve Einstein's objections to the Copenhagen interpretation, many more questions, with some extreme philosophical consequences, have been raised. Bohr's interpretation of physics may have many challenges ahead in the years to come. —DAVID TULLOCH

Further Reading

Beller, Mara, Robert S. Cohen, and Jurgen Renn, eds. *Einstein in Context.* Cambridge: Cambridge University Press, 1993.

Bernstein, Jeremy. *A Comprehensible World: On Modern Science and Its Origins.* New York: Random House, 1967.

Bohr, Niels. *Atomic Physics and Human Knowledge.* New York: Science Editions, Inc., 1961.

———. *Essays 1958–1962 on Atomic Physics and Human Knowledge.* Great Britain: Richard Clay and Company, Ltd., 1963.

Cohen, R. S., and John J. Stachel, eds. "Selected Papers of Léon Rosenfeld." *Boston Studies in the Philosophy of Science* 21 (1979).

Davies, P. C. W., and J. R. Brown, eds. *The Ghost in the Atom.* Cambridge: Cambridge University Press, 1986.

Fine, Arthur. *The Shaky Game: Einstein, Realism and the Quantum Theory.* Chicago: The University of Chicago Press, 1986.

Gamow, George. *Biography of Physics.* New York: Harper & Brothers, Publishers, 1961.

Maudlin, Tim. *Quantum Non-Locality and Relativity.* Oxford: Blackwell, 1994.

Murdoch, Dugald. *Niels Bohr's Philosophy of Physics.* Cambridge: Cambridge University Press, 1987.

Petruccioli, Sandro. *Atoms, Metaphors and Paradoxes.* Cambridge: Cambridge University Press, 1993.

"The Sciences." *New York Academy of Sciences.* 19, no. 3 (March 1979).

Snow, C. P. *The Physicists.* Great Britain: Morrison & Gibb Limited, 1981.

Is the emphasis on the construction of very large ground-based telescopes (6–10 meters) at the expense of smaller telescopes (1–4 meters) a poor allocation of resources?

Viewpoint: Yes, the emphasis on the construction of large telescopes is a poor allocation of resources, as smaller telescopes offer accessibility and affordability that larger telescopes cannot match.

Viewpoint: No, the emphasis on the construction of large telescopes is not a poor allocation of resources; on the contrary, such large telescopes are indispensable to the research questions at the heart of modern astronomy.

At the time of its invention 400 years ago, the telescope was a crude and arcane instrument known to only a few brilliant scientists such as Galileo. Today, it is a commonplace item available in department stores for less than $100. Small department store telescopes, however, often carry misleading advertising in the form of huge print on the box proclaiming "Magnifies 880X!" This means that with the proper eyepiece in place, the telescope will magnify an image 880 times—but it is hardly the most relevant property of the instrument.

Telescopes are most usefully characterized by the diameter of their primary light-gathering optical element, whether a lens or mirror. This main optical element is the telescope's "light bucket," with which it collects light from distant objects. The size of this element is therefore the real indicator of a telescope's power, and when astronomers talk about the size of a telescope, they invariably are referring to this quantity.

A very small telescope (with a 50 mm [2 in] objective lens, for example) collects only a small amount of light. Such an instrument cannot be productively used at high magnification, because it does not collect enough light to produce a bright, magnified image. To observe distant, faint objects in detail, there is no substitute for a larger light-gathering area. Telescopes have therefore grown over the years from the small refractor used by Galileo to huge reflecting telescopes with light collecting areas 10 m (33 ft) in diameter or more.

The size of the telescope used for modern research in astronomy obviously depends on the science in question. Some projects can be carried out with 1-m (39-in) class telescopes—that is, telescopes with a primary mirror on the order of 1 m (39 in) in diameter. If the observations being carried out are primarily of bright objects, such a telescope is perfectly adequate. Because the area of a circular mirror increases as the square of the diameter, however, a 4-m (157-in) telescope collects 16 times as much light as a 1-m (39-in) telescope, enabling it to take observations of fainter objects, or to observe comparably bright objects 16 times as rapidly. In addition, larger telescopes can provide greater resolution—the ability to see fine detail—than smaller ones. A very simple example is that of imaging two very close stars, which in a small telescope might look like a single point of light, but which a large telescope would resolve as two separate points.

Because of these two reasons—ability to observe fainter objects, and the ability to see greater detail—increasingly large telescopes have figured prominently in cutting-edge astronomical discovery. When one is suddenly able to observe objects that could not be observed before, exciting science invariably results.

The catch, of course, is the price. The cost of a telescope does not increase proportionally to the diameter of its collecting area. A 50-mm telescope (i.e., with a lens roughly 2 inches in diameter) might be purchased for $70, but a 200-mm (8-in) telescope costs not $280, but closer to $1,000. Multiply the mirror diameter by two again (400 mm, or 16 in), and you'll spend at least $15,000, while a high-quality, computer controlled 16-in (41-cm) telescope can be purchased and installed for $75,000.

Continuing into the realm of large research-grade telescopes, the costs rise sharply. A modern 4-m (157-in) telescope has a primary mirror 80 times as large as the little 50-mm (2-in) telescope's, but such an instrument will cost $20 to $30 million to build. Finally, the cost of the gargantuan 8–10-m (315–394-in) class instruments can easily reach nine figures: $100,000,000 or more. As will be seen in the articles that follow, these telescopes are now beginning to exploit multiple, segmented mirrors to produce a large whole, interferometric techniques to achieve extremely high resolution, and adaptive optics to counteract the degradation of image quality that results as light traverses Earth's atmosphere. This technology is extremely expensive, and one of these large telescopes therefore costs about as much as three brand-new, fully functional Boeing 737s.

Such staggering expenses are typically much more than a single educational institution can afford. A university with an astronomy department wishing to build a large telescope frequently must join a consortium of institutions with similar interests, who can collectively come up with the entire sum. Adding more cooks to stir the proverbial pot adds the usual difficulties of administration and competing agendas, often putting projects behind schedule and over budget. In some cases, private foundations have spearheaded the construction of large new telescopes. And there are large telescopes available for use, on a competitive basis, at nationally funded observatories such as Kitt Peak near Tucson, Arizona. In any of these cases, there is a finite amount of money, and frequently the high cost of constructing and then running a large telescope have precluded the construction or maintenance of smaller facilities. The question naturally arises: if you can build one 10-m (394-in) telescope, or 100 1-m (39-in) telescopes, for about the same amount of money, which is the proper way to go?

Certainly a 10-m (394-in) telescope can carry out projects that are simply out of the reach of a 1-m (39-in) instrument. But this is not to say there is no longer any useful science to be done with smaller instruments. Large telescopes offer unparalleled light-gathering capability, but there are plenty of poorly understood aspects of our own Sun, from which we receive no shortage of light. Examples of fundamental problems in astronomy involving bright, easily observed objects abound. In the light of perennial budget limitations, what should we be doing? In the following articles, we will examine the tradeoffs between pushing the limits of ever-larger light buckets that peer to the edge of the observable Universe, and maintaining the small observatories that gather fundamental data about brighter or more nearby objects, at a small fraction of the cost. —JEFFREY C. HALL

Viewpoint:

Yes, the emphasis on the construction of large telescopes is a poor allocation of resources, as smaller telescopes offer accessibility and affordability that larger telescopes cannot match.

Many large, national observatories are retiring smaller ground-based telescopes in order to increase funding for larger ground-based telescopes. However, small telescopes can effectively work both alone and in conjunction with larger telescopes if they are adapted to specialized or dedicated projects. For this reason, the emphasis on the construction of large, ground-based telescopes (from 6–10 m) at the expense of smaller telescopes (from 1–4 m) is a poor allocation of resources. (Telescopes are typically compared to one another by the diameter of their primary mirror, which in this article will be measured in meters; one meter is approximately equal to 3.28 feet.)

Scientific Impact The scientific impacts of the various sizes of telescopes are often compared on the basis of their contributions within the astronomical community. Chris R. Benn and Sebastian F. Sánchez of the Isaac Newton Group of Telescopes in Santa Cruz de La Palma, Spain, found that ground-based small telescopes, from 1–4 m, had a strong prevalence in their contributions to the 1,000 most-often-

KEY TERMS

ELECTROMAGNETIC: A disturbance that propagates outward from any electric charge that oscillates or is accelerated; far from the electrical charge, it consists of vibrating electric and magnetic fields that move at the speed of light and are at right angles to each other and to the direction of motion.

GAMMA RAY BURSTS: Intense blasts of soft gamma rays originating at extreme distances from Earth, which range in duration from a tenth of a second to tens of seconds, and which occur several times a year from sources widely distributed over the sky.

GROUND-BASED: Pertaining to instruments or devices (such as telescopes) that are fixed to Earth's surface.

INTERFEROMETRY: The design and use of optical interferometers (instruments in which light from a source is split into two or more beams, and subsequently reunited after traveling over different paths, in order to display interference).

OPTICAL: Relating to objects (such as stars) that emit waves in the visible portion of the electromagnetic spectrum; also, instruments (such as telescopes) that gather and manipulate light.

QUASARS: Massive and extremely remote celestial objects that emit exceptionally large amounts of energy; they typically have a star-like appearance in a telescope.

RADIATION: The emission and propagation of waves transmitting energy through space or through some medium.

SPECTROSCOPY: The branch of science concerned with the investigation and measurement of spectra produced when matter interacts with or emits electromagnetic radiation.

SUPERNOVAE: Stars that suddenly increase greatly in brightness because of catastrophic explosions that eject the vast majority of their masses.

TELESCOPE: An instrument for viewing celestial objects by the refraction (bending) of light via lenses, or by the reflection of light via mirrors.

cited research papers and to the 452 astronomy papers published in *Nature* from 1991 to 1998.

One conclusion of the Benn-Sánchez study with regards to both large and small, ground-based telescopes was that their usefulness was proportional to their light-gathering area and approximately proportional to their cost. In other words, the smaller cost of building a small telescope is as well spent as the larger cost of building a large telescope, thus supporting the stance that telescopes in the 1–4 m category should continue to be built. Benn and Sánchez further concluded that small- and mid-size telescopes should not be deactivated to help pay for building larger new ones. They stated that "cutting-edge" advancements by the smaller class of telescopes support the continuing impact that small telescopes make in the scientific community.

Scientific Discovery Gopal-Krishna and S. Barve, members of the Astronomy Society of India, stated in 1998 that after analyzing 51 astronomy papers published in *Nature* from 1993 to 1995 that were based on data from ground-based optical telescopes, 45% of the researchers solely used data from telescopes with mirror diameters less than 2.5 meters. The authors suggested that many of the leading scientific researches being performed within astronomy are still done by smaller-sized telescopes, further highlighting the essential role that these telescopes perform in astronomical research.

Current Trends A 10-year review conducted at the end of the twentieth century of astronomy priorities by the National Research Council (NRC) barely mentioned the critical function of small telescopes in modern scientific research, a viewpoint held by many astronomers. The report, instead, focused on large projects such as a proposed 30-meter segmented-mirror telescope that would have a construction cost of at least $500 million for the United States and its international partners. Even though many astronomers hold a firm belief in the value of small telescopes, the unfortunate trend, as shown in the NRC study, is toward emphasizing the large telescope.

The National Optical Astronomy Observatories (NOAO) of Tucson, Arizona, recently transferred its 1-meter telescopes to private consortia of universities. NOAO, one of the largest astronomy-based organizations in the United States, provides U.S. research astronomers with access to ground-based telescopes that image in the optical-and-infrared (OIR) portions of the spectrum. The effect of such transfers is to remove smaller telescopes from access to the general astronomy research community.

Although small telescopes have not garnered the attention that larger telescopes have, and have moreover been removed from operations in favor of larger telescopes, there is continuing evidence that the leaders of some organizations realize that small telescopes are critical to robust astronomical research. As an example, consider the actions of the NOAO. It has been unable to maintain its expertise in all aspects of OIR astronomy due to budgetary constraints. This, in turn, has caused NOAO to limit its areas of concentration so that it can maintain its scientific leadership (as described previously,

The Gemini North 8-meter telescope on Mauna Kea, Hawaii. *(© AFP/Corbis. Reproduced by permission.)*

NOAO has shed its one-meter telescopes). As a result, NOAO's twin 8-meter Gemini telescopes were given its highest priority with respect to science, operations, and instrumentation. However, NOAO's second priority was to support its smaller telescopes (those from 2–4 meters) that show the best possible capabilities. NOAO stated that such smaller telescopes are essential in order to (1) support the Gemini program, (2) provide other national capabilities (such as projects within wavelength bands other than optical and infrared), and (3) support the scientific programs of its researchers and students. The NOAO will continue to technologically upgrade and maintain its remaining small telescopes with dedicated instruments (such as charge-coupled device [CCD] imagers) in order to fully support the activities of its astronomers and staff. Such activities go contrary to the widely held belief that only large telescopes should be built and maintained for the astronomical community.

NOAO will be forced to focus its attention on fewer tasks, but these critically selected tasks will involve telescopes of both small and large sizes in order to accomplish successful science projects in the most efficient way possible. Even though many other organizations are eliminating smaller telescopes, the NOAO is proving that small telescopes are important, even essential, to future astronomical research and development.

Specialty Purposes. As an example of a specialty use for small telescopes, a new 3.5-meter WIYN telescope (owned and operated by the WIYN Consortium, consisting of the University of Wisconsin, Indiana University, Yale University, and the NOAO) at Kitt Peak National Observatory will complement the large Gemini North telescope in the areas of intermediate field-of-view, wide-field multi-object spectroscopy, high-resolution imaging, and near ultraviolet.

John Huchra of the Harvard-Smithsonian Center for Astrophysics in Cambridge, Massachusetts, says that small telescopes can be very cost effective when used for projects that are not suitable for larger telescopes. Huchra states that in some cases small telescopes are assuming a more important role as they complement the work of larger telescopes. Indeed, Janet Mattei, director of the American Association of Variable Star Observers, in Cambridge, Massachusetts, stated that small telescopes are the "backbone" of astronomy even though they rarely make headline news.

Availability. Astronomers realize that the availability of large telescopes may only be a few nights each month and that time is scheduled months or even up to a year in advance. So the smaller telescopes allow astronomers to fill their need for observation time, and to monitor rare celestial events or track the long-term behavior of astronomical objects.

Automation. Fully automated small telescopes are frequently used to perform jobs that need to be done on a regular basis, and done more accurately than possible with human effort. These preprogrammed systems are designed, for instance, to automatically open the viewing dome when observational/recording conditions are appropriate. Alex Filippenko, an astronomer at the University of California, Berkeley, says that smaller, robotic telescopes, unlike large telescopes, are ideal for performing repetitive tasks that require many observational nights. Filippenko uses a 0.75-meter Katzman Automatic Imaging Telescope to find nearby supernovae. The telescope records images of several thousand galaxies, one by one, and then repeats its observations every three to five clear nights.

Another robotic small telescope system is the Robotic Optical Transient Search Experiment at Los Alamos National Laboratory, New Mexico. It uses small telescopes that can swivel quickly to any portion of the sky. Astronomers built the system to pursue the optical flashes associated with gamma ray bursts. Such quick response to random events is normally very difficult with large telescopes, because nearly all their observational time is already reserved.

Unattended automation of small telescopes, such as the two previously mentioned, has proven to be very effective in increasing the efficiency of observations and in reducing operational cost. However, the major motivating factor for automation of small telescopes is the new types of science that are permitted, which are otherwise unavailable under either conventional scheduling policies or large telescopes. Unattended automation has proven to be an effective operational style of smaller U.S. research telescopes.

Long-term Observations. Greg Henry of Tennessee State University, Nashville, uses the Fairborn Observatory, located near the Mexican border in Arizona. Henry's observations illustrate a big advantage that small telescopes have over large telescopes. Small telescopes can be dedicated to performing long-term observations that watch for changes over time. Henry's four small telescopes are used to study long-term changes in the brightness of stars that are similar to the Sun. The results of studying those Sun-like stars can be used to understand how the Sun influences Earth's climate. Such observations would be impossible with large telescopes because observation time is constantly in demand by many researchers, and must be reserved months ahead. Henry states that he has gotten 50 to 100 years worth of data, which is much more than he could have gotten manually or with the infrequent use of large telescopes. Moreover, Henry asserts that the cost of obtaining his data was a small fraction of what he would have incurred had he used a large telescope.

As another example highlighting the advantages of small telescopes over large ones, astronomers are using worldwide networks to conduct continuous studies of specific stars. One such group of small telescopes is the Whole Earth Telescope project, directed by astronomer Steve Kawaler at Iowa State University, Ames. The network consists of more than 60 international astronomers located in 16 countries around the world (such as China, South Africa, Lithuania, and Brazil). The telescopes range from a 0.6-meter telescope in Hawaii to an almost 3-meter telescope in the Canary Islands. Studying dwarf stars around the clock, in order to predict what the Sun might become in a few billion years, is impossible for a single telescope because of Earth's rotation, and would be impractical for several large telescopes because such dedicated use for long periods of time is not practical for large, expensive telescopes. Instead, the strategy is to use numerous small telescopes from around the world to continuously observe the dwarf stars, and then send all the collected data to the projects' Iowa headquarters.

Education. The 0.6-meter telescope on the campus of Wesleyan University, Middletown, Connecticut, contributes to astronomy education. Wesleyan is representative of smaller educational institutions with only a few astronomers and an emphasis on undergraduate education. Such institutions are important for producing astronomy applicants for graduate study. The value of this educational telescope is that it: (1) stimulates student interest in astronomy, (2) teaches undergraduate science using actual observations and measurements in laboratory settings, (3) trains advanced students in the use of telescopes and other related instruments within astronomical observatories, and (4) supports research programs that cannot be carried out in larger facilities. These reasons demonstrate the need for small telescopes in the training of prospective members of the astronomical community. Large telescopes are not available for such training purposes because they are inevitably booked months ahead for research astronomers and scientists.

Research. The SARA (Southeastern Association for Research in Astronomy) facilities promote all areas of observational astronomy. A SARA observatory located on Kitt Peak refurbished a 0.9-meter telescope formerly operated by NOAO. With declining numbers of publicly available small research telescopes, this act greatly helped to improve access to smaller telescopes. The SARA telescope is equipped with a four-port instrument selector that allows use of several instruments during a given night of observation. It is fully computer-controlled with remote robotic scheduling of the telescope through the Internet. The goal of the SARA project is to make the data gathered available to classroom students, as well as to on-site observers.

SARA intends to take an increasing role in helping to represent the interests of astronomers who need smaller telescopes to conduct their research. In fact, SARA has recently become the host institution for the North American Small Telescope Cooperative (NASTeC). The purpose of NASTeC is to emphasis the wide distribution and availability of small-to-medium research telescopes and to coordinate collaboration with astronomy-based projects. The diminishing availability of small telescopes at large, national facilities require that new operating models be developed in order to properly educate the next generation of astronomers. Consortia of small universities such as SARA are one model for meeting these goals.

Future Support Since many projects involving small telescopes are being terminated in preference to larger telescopes, many members of the astronomical community are actively voicing their support for small telescopes. The advantages of small telescopes began to be told in earnest in October 1996 when astronomers met at the Lowell Observatory, Flagstaff, Arizona. Since then, the American Astronomical Society has featured at least one session devoted to small telescope issues at its meetings.

The momentum, however, clearly belongs to the growing number of large (such as 8–10-meter) telescopes. NOAO director Sidney Wolff concedes that complicated projects involving multi-institutional collaborations continue to demand larger telescopes. However, Wolff also says that small telescopes are making exciting contributions to frontline research, including (1) measuring changes to the expansion rate of the universe, (2) finding optical counterparts to gamma-ray bursts, (3) making precise observations of Sun-like stars in order to understand solar influences on climate change, (4) making direct detection of extrasolar planets, (5) conducting the first deep all-sky survey in the near infra-red, and (6) discovering near-Earth asteroids.

Concerned astronomers have succeeded in saving many of the small telescopes that appeared to be facing retirement during the 1990s. The struggle to keep them open continues as large telescopes dominate the field. Harvard astronomer John Huchra says that, in practice, small telescopes continue to carry much of the workload in astronomy. When used in conjunction with large telescopes and for unique, dedicated purposes ideal for its small size, the small telescope will continue to be a valuable asset to the scientific community. Huchra continues by saying that when small telescopes are used for projects uniquely geared toward their strengths, the small telescope can be incredibly cost effective.

PROFOUND QUESTIONS ASKED BY ASTRONOMERS

There are many profound questions that astronomers hope to someday answer with the use of some of the world's most powerful telescopes. Scientists at the National Research Council recently stated that resolution of such questions could allow science to take giant leaps forward in knowledge about the physics of the universe.

Astronomers currently can account for only about 4% of ordinary matter in the universe. *What is dark matter?* is a question that, once answered, will tell what types of "dark matter" make up the missing universe. The leading dark matter candidates are neutrinos and two fairly unknown particles called neutralinos and axions. These particles are predicted by some physics theories but have never been detected.

What is dark energy? is an associated question to dark matter that also must be answered to explain the structure of the universe. After adding up all the potential sources for ordinary matter and dark matter, the total mass density of the universe comes up about two-thirds short. The missing energy could mean that the vacuum of space is not a true emptiness (or void). Instead, space might be filled with vacuum energy, which is a low-grade energy field created when ordinary particles and their antimatter sidekicks periodically burst into and out of existence.

When Albert Einstein asked the question *What is gravity?* he extended Isaac Newton's concept of gravity by taking into account extremely large gravitational fields and objects moving at velocities close to the speed of light. However, Einstein did not consider quantum mechanics, an area of extremely small gravitational fields that has never been experimentally observed. For instance, near the center of black holes gravitational forces are very powerful at extremely small distances. These energies violate the laws of quantum mechanics and cannot totally explain the concept of gravity.

The four dimensions of space and time are easily observed. The current string-theory model of the universe combines gravity with 11 dimensions. Astronomers cannot see these 11 dimensions because their instruments are too crude to answer the question: *Are there additional dimensions?*

Additional questions such as *How did the universe begin?* will be answered, along with many more, only through the use of larger and larger telescopes. These super-telescopes will help astronomers and scientists explain the physics of the universe.

—William Arthur Atkins

Astronomers have learned that for monitoring broad areas of the sky, imaging the same object night after night, creating worldwide networks of telescopes, automating projects, and other such dedicated uses, small telescopes play an essential role that cannot be filled by their larger counterparts. As long as a mix of both large and small telescope sizes is maintained, then the largest variety of important objectives in astronomical research can continue to be met. However, if the construction of new, large, ground-based telescopes means that smaller telescopes are retired or otherwise abandoned, then our nation's research funds are indeed being poorly allocated. —WILLIAM ARTHUR ATKINS

Viewpoint:

No, the emphasis on the construction of large telescopes is not a poor allocation of resources; on the contrary, such large telescopes are indispensable to the research questions at the heart of modern astronomy.

The invention of the telescope in 1608 is credited to the Dutch spectacle-maker Hans Lipperskey. Italian mathematician, physicist, and astronomer Galileo Galilei (1564–1642) is credited with making the first celestial observations with a telescope in 1609, when he used a "spyglass" of about 0.05 m (2 in) in diameter.

Classifications Since Galileo, telescopes that collect electromagnetic waves (such as radio, microwave, infrared, light [visible], ultraviolet, x rays, and gamma radiation) have generally been classified as either: (1) refractors (where radiation is bent, or refracted, as it passes through an objective lens), (2) reflectors (where radiation is

The foundation for the Subaru Telescope under construction in 1994.
(© Roger Ressmeyer/Corbis. Reproduced by permission.)

reflected by a concave mirror and brought to a focus in front of the mirror), or (3) as a catadioptric system (where radiation is focused by a combination of lenses and mirrors). The telescopes considered in this article are optical; that is, they are used for intercepting visible light (and often also for infrared and ultraviolet "light"). Over the years optical telescopes have gotten bigger and better through the use of increasingly sophisticated structural hardware and software (for example, computer-control of the telescope).

Size Matters Many of the most recent and important discoveries in astronomy are due to a quest that began soon after the invention of the telescope: to increase the amount of light that is captured and brought to a focus so that astronomers can study increasingly fainter objects. The size of the primary mirror of a telescope determines the amount of radiation that is received from a distant, faint object. The size, quality, and ability to sharply define the details (or the resolution) of an image are all very important characteristics of professional telescopes.

Optical telescopes have circular (or in a few cases, very nearly circular) primary mirrors (the primary mirror is a telescope's main component for intercepting and focusing light). The area of the primary mirror is related to its diameter by the formula $A=(\pi/4)d^2$, where "A" is the mirror's area, "d" represents the mirror's diameter, and "π" is a constant (approximately equal to 3.142). This formula demonstrates that a telescope's light-collecting area increases fourfold if its diameter is doubled. For example, using the above equation for the area of a circle, a 4-meter-diameter telescope has an area of about 12.5 square meters, while an 8-meter-diameter telescope (double the diameter of the 4-meter) has an area of approximately 50 square meters. So an 8-m (315-in) telescope has four times the radiation collecting capacity over a 4-m (157-in) telescope. The largest optical telescopes in the world are the twin 10-m (394-in) Keck telescopes on Mauna Kea in Hawaii. In less than 400 years of telescope technology, the Keck telescope has increased the collection of light by 40,000 times over that of Galileo's telescope.

Some of the most important unanswered questions in astronomy have to do with cosmology (the science and origin of the universe). For example, astronomers want to know when, how, and why the galaxies were formed. In order to solve problems like these astronomers need to be able to analyze the radiation coming from the furthest and the faintest objects in the sky. To observe such objects, very large telescopes are indispensable.

Twentieth-Century Telescope Technology During the twentieth century, advanced telescopes continued to grow in size. Of the two principal types of optical telescopes—refracting and reflecting—the reflecting type long ago won out in the race to build ever-larger telescopes. The largest refractor in existence, with an objective lens of 1.02 m (40 in) in diameter, is located at the Yerkes Observatory in Williams Bay, Wisconsin. A 0.91-m (35-in) refractor is located at the Lick Observatory in California and a 0.84-m (33-in) refractor is located at Meudon, France. These telescopes represent the practical limit to the size of refracting telescopes, whose technological apex was reached near the end of the nineteenth century. In the 300 years of their reign, however, they provided enormously important information about stars, galaxies, and other celestial bodies.

Mount Palomar—1948. A reflecting telescope of monumental importance in the construction of large telescopes was the 5.1-m (17-ft) Hale Telescope on Mount Palomar, outside San Diego, California. The Hale telescope enabled a string of fundamental discoveries about the cosmos, including critical data on the evolution of stars and the existence of quasars (or "quasi-stellar radio sources"), thought to be the farthest known objects in the universe yet imaged, billions of light-years away (a light-year is the distance that light travels in vacuum in one year).

Largest Reflectors at Present. The largest conventional (that is, possessing a single, monolithic primary mirror) reflecting optical telescope is the 8.3-m (27-ft) Subaru telescope at the Mauna Kea Observatory. Other very large reflectors are the 8.1-m (26-ft) Gemini telescope, also at Mauna Kea, the 6-m (20-ft) telescope in the Special Astrophysical Observatory near Zelenchukskaya, in the Caucasus Mountains; and the 5.1-m (17-ft) telescope in the Palomar Observatory in California.

Largest Composition Reflectors. The largest of the reflecting telescopes with segmented or multi-mirror reflectors are the twin W.M. Keck telescopes at the Mauna Kea Observatory in Hawaii. Each has a segmented primary mirror, composed of 36 separate hexagonal pieces. Each segment is about 1.8 m (6 ft) across, creating a 10-m (33-ft) diameter primary mirror. The Keck telescopes began operations in the early 1990s to probe the chemical composition of the early universe, and added to the knowledge that the universe is not only expanding, but also that the expansion is accelerating.

State of the Art When the 5.1-m (17-ft) Hale Telescope first began to look at the universe in 1948 it was assumed that this size telescope was the largest that could be built without Earth's unsteady atmosphere negating the value of any larger aperture or any better optics. Science correspondent and former program leader of the Artis Planetarium in Amsterdam Govert Schilling wrote that ground-based telescopes languished for several decades at this construction limit. Being outside the Earth's blurry atmosphere, more interest was directed towards space-based telescopes, such as the 2.4-m (8-ft) Hubble Space Telescope (HST) project that started development in the late 1960s.

Beginning in the 1970s Frederic Chaffee, Director of the Multiple Mirror Telescope Observatory, wrote that the computer revolution spawned one of the most intense periods in the design and construction of telescopes. Computers and other new technologies have revitalized the use of large telescopes with current ones being developed in the 6–10 m (20–33 ft) class. The need for the construction of larger ground-based telescopes has to do with the never-ending questions that astronomers pose concerning the deepest mysteries of the cosmos; questions such as "When did galaxies first come into existence?"; "What is the elusive 'dark matter' whose mass is thought to dominate the universe?"; "How many stars have planets?"; and "Do alien worlds possess life?" The list of open questions goes on, tantalizing both astronomers and the general public alike. And many of these questions can only be answered with the use of advanced new large telescopes.

New Large Telescope Technology Older telescopes, such as the Hale Telescope, contain obsolete technology. These telescopes contain a mirror that has been described as "a huge hockey puck of glass." For instance, the Hale possesses a bulky 26-in-(66-cm)-thick mirror that weighs 20 tons. An enormous support structure is necessary to hold it up, while at the same time forced to adjust to the constant movements necessary to keep up with the Earth's rotation.

By the 1980s new technologies and designs were revitalizing the ground-based telescope field. University of Arizona astronomer Roger Angel cast huge mirrors that were mostly hollow, with a honeycomb-like structure inside to guarantee stiffness. University of California astronomer Jerry Nelson combined 36 smaller sheets that would act as one unit under a computer's control. Another essential technology common to all these new designs is the essential role of sophisticated computers both controlling the telescope and its component parts, as well as aiding in the digital processing of the observed light.

New Large Telescopes and Their Impact What follows is an overview of just some of the new giant telescopes that are currently in operation—or soon will be—along with some of the uses that each is (or will be) employed to research.

The 10-meter Keck Telescope as seen from its twin, Keck II.
(© Roger Ressmeyer/Corbis. Reproduced by permission.)

GTC. The Gran Telescopio Canarias (GTC) at the Roque de los Muchachos Observataory on La Palma in the Canary Islands (Spain) is a high-performance, segmented 10.4-m (34-ft) telescope that will become the largest optical telescope on European territory when it is completed in early 2003. The GTC will consist of a coordinated set of subsystems—including interconnected computers, electronic equipment, cameras, spectroscopes, sensors, and actuators—that will allow the GCS to find extra-solar planets (planets circling stars other than the Sun) and proto-stellar objects (newly forming stars). The GTC will be equipped with an adaptive-optics system that can reduce—and at times eliminate—the Earth's atmospheric blurring.

LBT. The Large Binocular Telescope (LBT) is being built on Arizona's Mount Graham at a cost of $84 million. When finished in 2004 it will consist of twin 8.4-m (27-ft) telescopes that will ride on a single mounting and be linked by ***interferometry*** (to yield the equivalent light-gathering power of a single 11.8-m [39-ft] instrument). Because of its binocular arrangement—a revolutionary optical design—the telescope will have a resolving power (ultimate image sharpness) corresponding to a 22.8-m (75-ft) telescope. The LBT will thus have a collecting area larger than any existing or planned single telescope. It will provide unmatched sensitivity for the study of faint objects.

SALT. The Southern African Large Telescope (SALT) will be completed around December 2004 as the Southern Hemisphere sister to the Hobby-Eberly Telescope at the McDonald Observatory in Texas. Its innovative design consists of an 11-m (36-ft) wide segmented (hexagonal) mirror array that will have a clear aperture of 9.1 m (30 ft). At a cost of $30 million it will be a specialized instrument designed only for spectroscopy (the investigation and measurement of the radiation spectrum). SALT will be able to record distant stars, galaxies, and quasars a billion times too faint to be seen with the unaided eye.

Mauna Kea. As discussed previously, on Mauna Kea in Hawaii the Subaru Telescope contains a mirror more than 8.3 m (27 ft) across; the Gemini North Telescope is 8.1 m (26 ft) across; and the twin Keck telescopes possess light-gathering surfaces 10 m (33 ft) in diameter. Geoff Marcy, from the University of California, has so far discovered 35 planets orbiting Sun-like stars, in part, using the Keck telescope. Marcy says that lesser telescopes would be unable to detect the gravitational wobble that he looks for when scouring the sky for planets. So far he has found a planet the size of Saturn, and with the use of even larger telescopes hopes to soon discover smaller planets the size of Neptune.

The Keck telescopes, as segmented mirror structures where adjacent segments stay aligned to millionth-of-an-inch optical tolerances, have made an enormous number of discoveries. California Institute of Technology professor George Djorgovski has concentrated on gamma-ray bursts—mysterious flashes of high-energy radiation—that appear to be billions of light-years away. Djorgovski says that the Keck telescopes have allowed him to solve the mystery of these bursts. University of California astronomer Andrea Ghez has used the Kecks for her exploration of the Milky Way's core. An electronic camera sensitive to infrared radiation that is attached to the telescopes has allowed Ghez to penetrate the dust that surrounds the center. Through this combination of instruments Ghez has been able to measure the motion of stars that lie 100 times as close to the galactic core as the nearest star (Proxima Centauri) is to the Sun. Ghez has declared that these stars are traveling 100 times as fast as Earth orbits the Sun (at nearly 1,600 mi [2,574 km] per second).

Adaptive Optics and Interferometry The technology called adaptive optics (AO) is a process that "de-twinkles" starlight. Stars and galaxies twinkle because the Earth's turbulent atmosphere acts to deform incoming starlight. AO, through the use of computers, measures the amount of twinkle and cancels it out by deforming the surface of a flexible mirror. Astronomers have used this advanced technology at both the Keck and Gemini telescopes to take pictures as clear as the Hubble Space Telescope. Senior Editor for *Sky & Telescope* Roger W. Sinnott asserts that the Arizona's Large Binocular Telescope should reveal the universe *ten times more sharply* than the space-based Hubble due to the use of AO!

The resolution of telescopes can be improved with a process called interferometry. Light is combined from widely separated telescopes to result in precision never before seen in telescopes. By sampling incoming light at two or more different vantage points, it is possible to simulate the resolution of a single gigantic telescope, whose diameter equals the distance between the outer edges of the farthest-spaced mirrors. This is the technology that will be employed with the Large Binocular Telescope.

Super-giant Telescopes New technology is allowing bigger, lighter mirrors to be made more quickly than ever before. In addition, the limitations on the sharpness of telescope images are also being overcome. Technologies to cut through the distortion of the Earth's atmosphere is promising to allow the study of the universe to ever fainter, more distant limits and with unprecedented clarity. John Huchra, professor at the Harvard-Smithsonian Center for Astrophysics, has said, "What's been happening in the telescope game is incredible." For the foreseeable future, these advances in viewing new celestial phenomena can only be accomplished with the new class of large telescopes.

But telescope designers are not stopping at 6–10-m (20–33-ft) telescopes. Designs are being contemplated for the next generation of ground-based telescopes, now being called super-giant telescopes, with a projected range of size from 30 m (100 ft) to a staggering 100 m (330 ft). These football-field-sized telescopes—in reality hundreds of individual mirrors aligned to make a single giant telescope—could have 100 times the light-gathering capability of the Keck telescopes.

All these new and envisioned technologies for large and ultra-large telescopes are revolutionizing astronomy. Large telescopes are taking advantage of the latest sensors, fastest computers, adaptive optics, and other technologies in order to achieve resolutions and image sharpness at levels far exceeding that of even the Hubble space telescope. Though smaller telescopes definitely have their place in astronomy, the really big, new findings in astronomy are being achieved with the latest generation of large, ground-based telescopes. Allocating the lion's share of resources in astronomy to the operation and construction of this new class of large telescopes—even at the reduction or retirement of their smaller cousins—is not only advisable, it is indeed essential to the progress of the science of astronomy, and to the continuing effort to understand our place in the universe. —PHILIP KOTH

Further Reading

Benn. C. R., and S.F. Sánchez. "Scientific Impact of Large Telescopes." *The Astro-*

nomical Society of the Pacific 113, no. 781 (March 2001): 385–96.

"Big Scopes Aren't Always Best." *Sky and Telescope* 101, no. 2 (February 2001): 24.

Chaffee, Frederic. "Telescopes for the Millennium." *Technology Review* 97, no. 2 (February/March 1994): 52.

De Grasse Tyson, Neil. "Size Does Matter." *Natural History* 108, no. 7 (September 1, 1999): 80.

Henry, Gregory W., J. C. Barentine, and F. Fekel. "Small Telescopes, Big Results." *Science* 289, no. 5480 (August 4, 2000): 725.

Irion, Robert. "Telescopes: Astronomers Overcome 'Aperture Envy.'" *Science* 289, no. 5476 (July 7, 2000): 32–34.

Moore, Patrick. *Eyes on the Universe: The Story of the Telescope.* New York: Springer-Verlag, 1997.

"NOAO in the Gemini Era." *A Strategy for Ground-Based Optical and Infrared Astronomy.* National Academy Press [cited July 23, 2002]. <http://www.nap.edu/readingroom/books/gboi/chap4.html>.

Ringwald, F. A., Rebecca L. Lovell, Sarah Abbey Kays, Yolanda V. Torres, and Shawn A. Matthews. *The Research Productivity of Small Telescopes and Ground-Based versus Space Telescopes.* California State University and Florida Institute of Technology, American Astronomical Society 197 (January 2001).

"The Role of Small Telescopes in Modern Astronomy, October 14–15, 1996, Lowell Observatory." The First Annual Lowell Observatory Fall Workshop [cited July 23, 2002]. <http://www.noao.edu/aura/stma/>.

Rutten, Harrie G. J., and Martin A. M. Van Venrooij. *Telescope Optics: Evaluation and Design.* Richmond, VA: Willmann-Bell, 1988.

Schilling, Govert. "Astronomy: Telescope Builders Think Big—Really Big." *Science* (June 18, 1999).

———. "Giant Eyes of the Future." *Sky and Telescope* 100, no. 2 (August 1, 2000): 52.

Sinnott, Roger W., and David Tytell. "A Galaxy of Telescopes: Coming of Age in the 20th Century." *Sky and Telescope* 100, no. 2 (August 1, 2000): 42.

Sung-Chu, Ling-Uei. "Telescope Project Unites Astronomers." *Iowa State Daily Online Edition.* Iowa State University, Ames, Iowa [cited July 23, 2002]. <http://www.iowastatedaily.com/vnews/display.v/ART/2002/03/05/3c8463bf7e153?in_archive=1>.

"Telescopes." Royal Observatory Greenwich, National Maritime Museum [cited July 23, 2002]. <http://www.rog.nmm.ac.uk/leaflets/telescopes/telescopes.html>.

"Telescopes for the Millennium." *Technology Review* 94, no. 2 (February/March 1994): 52.

Wilson, Raymond N. *Reflecting Telescope Optics: Basic Design Theory and Its Historical Development.* New York: Springer-Verlag, 1996.

EARTH SCIENCE

Was the timing of the rise in Earth's atmospheric oxygen triggered by geological as opposed to biological processes?

Viewpoint: Yes, the timing of the rise in Earth's atmospheric oxygen was triggered not by biological processes but by geological processes such as volcanic eruption, which transported elements (among them oxygen) from Earth's interior to its atmosphere.

Viewpoint: No, the theories based on geological principles accounting for the timing of the rise in Earth's atmospheric oxygen have insufficient data to supplant biological processes as the cause.

As most people know, oxygen is essential to most forms of life, with the exclusion of anaerobic or non-oxygen-dependent bacteria. But when, and from where, did this life-giving oxygen arise during the course of Earth's history? The first question, regarding the point at which oxygen appeared on the planet, is answered with relative ease by recourse to accepted scientific findings. According to the best knowledge available at the beginning of the twenty-first century, oxygen first appeared between 2.2 and 2.4 billion years ago. This would place the appearance of oxygen somewhere between 2.1 and 2.3 billion years after Earth's formation from a spinning cloud of gases that included the Sun and all the future planets and satellites. Yet though the "when" question is less fraught with controversy than the "how" question, there are still complications to this answer.

First of all, there is the fact that any knowledge of events prior to about 550 million years ago is widely open to scientific questioning. The term *scientific* is included here in recognition of the situation, which is unique to America among all industrialized nations, whereby a substantial body of the population rejects most scientific information regarding Earth's origins in favor of explanations based in the biblical book of Genesis. Despite what creationists might assert, the fact that there is dispute between scientists regarding the exact order of events in no way calls into question the broad scientific model for the formation of Earth, its atmosphere, its seas, and its life forms through extremely lengthy processes. In any case, most knowledge concerning these far distant events comes from readings of radiometric dating systems, which involve ratios between stable and radioactive samples for isotopes or elements such as potassium, argon, and uranium, which are known to have extremely long half-lives.

In addition, the answer to the "when" question is problematic due to the fact that scientists know only that Earth experienced a *dramatic increase* in oxygen levels 2.2 to 2.4 billion years ago. Logically speaking, this leaves open the possibility that oxygen may have existed in smaller quantities prior to that time—a position maintained by Hiroshi Ohmoto at Pennsylvania State University, whose work is noted in one of the essays that follow. Furthermore, there is evidence to suggest that oxygen levels may have fluctuated, and that the development of oxygen may not have been a linear, cumulative process; by contrast, like the evolution of life forms themselves, it may have been a process that involved many false starts and apparent setbacks. Certainly oxygen may have existed for some time without being in the form of usable atmospheric oxygen, since it is an extremely reactive chemical element, meaning that it readily bonds with most other elements and subsequently must undergo another chemical reaction to be released as usable atmospheric oxygen.

Given the fact that the "when" question is, as we stated earlier, the simpler of the two, this only serves to underscore the difficulty in answering the "how" question. Broadly speaking, answers to the question of how oxygen arose on Earth fall into two categories. On one side is the biological or biogenic position, based on the idea that the appearance of oxygen on the planet resulted from the growth of primitive life-forms. The bodies of the latter served as chemical reactors, releasing to the atmosphere oxygen previously trapped in various chemical compounds. On the other side is the geological position, which maintains that the growth of oxygen in the atmosphere resulted from processes such as volcanic eruption, which transported elements (among them oxygen) from Earth's interior to its atmosphere.

As we shall see from the two competing essays that follow, there is much to recommend each position, yet each faces challenges inherent to the complexity of the questions involved, and the reasoning and selection of data required to answer them. For example, there is the fact that changes in the atmosphere seem to have corresponded with the appearance of relatively complex multicellular life-forms on Earth. At first this seems to corroborate the biological position, but the reality is more complicated.

The first cells to form were known as prokaryotic cells, or cells without a nucleus, which were little more than sacs of self-replicating DNA—much like bacteria today. These early forms of bacteria, which dominated Earth from about 3.7 billion to 2.3 billion years ago, were apparently anaerobic. By about 2.5 billion years ago, bacteria had begun to undergo a form of photosynthesis, as plants do today, and this would necessarily place great quantities of oxygen into the atmosphere over long stretches of time. These events produced two interesting consequences: first of all, the formation of "new" cells—spontaneously formed cells that did not come from already living matter—ceased altogether, because these were killed off by reactions with oxygen. Second, thereafter aerobic respiration would become the dominant means for releasing energy among living organisms.

But does this mean that biological rather than geological forms were responsible for the great increase of oxygen in the atmosphere? Not necessarily. After all, attempts to establish a biological connection may be a case of finding causation where none exists. (This is rather like a person issuing an order to a house cat that happens to subsequently display the demanded behavior—not, however, because of the command, but rather because it chose to do so of its own whim.) Furthermore, there is abundant evidence linking volcanic and other forms of tectonic activity—i.e., activity resulting from movement of plates in Earth's crust and upper mantle—with the spreading of carbon and other life-giving elements from Earth's interior to its surface. It is no accident that of all the planets in the Solar System, Earth is the only one with active volcanoes, and that the next-best candidate for sustaining life, Mars, happens to be the only other planet to have experienced volcanic activity in the past billion years. —JUDSON KNIGHT

Viewpoint:

Yes, the timing of the rise in Earth's atmospheric oxygen was triggered not by biological processes but by geological processes such as volcanic eruption, which transported elements (among them oxygen) from Earth's interior to its atmosphere.

The geophysical and fossil record presents abundant, clear, and convincing evidence that approximately 2.2 to 2.4 billion years ago, there was a dramatic increase in the oxygen content of Earth's atmosphere. The atmospheric changes correspond to the appearance of complex multicellular life forms in the fossil record. Scientists have argued for decades whether the increase in atmospheric oxygen resulted primarily from geophysical processes that then facilitated the evolution of complex multicellular life, or whether evolutionary processes produced organisms capable of biogenically altering Earth's atmosphere.

The scientific argument involves classic arguments of correspondence (the simultaneous occurrence of phenomena) versus causation (a dependence or linking of events where the occurrence of one event depends upon the prior occurrence of another). Although the extent (visible and existing) fossil record establishes correspondence between the rise of complex life-forms and higher oxygen content in the atmosphere, it does not conclusively establish a causative correlation between the two events.

In addition to providing evidence of the rise in atmospheric oxygen, the geological and fossil records also provide evidence of oxygen in the more primitive atmosphere prior to the precipitous rise in oxygen content levels. Sedimentary rocks believed to be approximately 2.2 billion years old contain evidence of changes in the oxygen and sulfur chemistry that are consistent with the rise in atmospheric oxygen. The global distribution of this evidence is consistent with a global atmospheric change. However, these chemical indicators of primitive atmospheric oxygen also extend farther back into the geological record to a time well before the established rise of oxygen content levels.

Because evidence exists that reactions creating free atmospheric oxygen operated before the global accumulation of atmospheric oxygen, it becomes essential to address the rate of oxygen and associated ozone build-up when assessing the relative contributions of biogenic and geophysical processes. Moreover, it can be fairly argued that biogenic contributions to atmospheric oxygen depended upon the development of a sufficient ozone layer to provide adequate protection from ultraviolet radiation and thus allow the development of more advanced organisms.

Admittedly, there is strong evidence of biogenic alteration of the oxygen content of Earth's atmosphere. Recent issues of *Science* and *Nature* contained articles presenting evidence for two different mechanisms of biogenic alteration. Arguing in *Science*, David Catling and colleagues presented evidence that a loss of hydrogen and the consequential enrichment of Earth's atmospheric oxygen content was accomplished through the microbial splitting of hydrogen and oxygen in water molecules, with the subsequent loss of hydrogen in biogenically produced methane gas. Methane is made of carbon and hydrogen (CH_4). So when the H_2O is split, the hydrogen combines with carbon to form methane. In *Nature*, Terri Hoehler and colleagues presented a different mechanism of atmospheric oxidation through the action of the direct loss of hydrogen to space (hydrogen molecules have a mass too low to be held in Earth's atmosphere by Earth's gravitational field) and the physiochemistry of "microbial mats" that utilize hydrogen to produce methane. Importantly, however, both of these biogenic mechanisms, in part, rely on geophysical and inorganic chemical reactions.

Although most arguments assert that biogenic processes—especially photosynthesis—played a far more significant role in producing oxygen in Earth's primitive atmosphere than did the geophysical process associated with volcanic action and hydrothermal venting, geophysical processes likely contributed significant amounts of atmospheric oxygen.

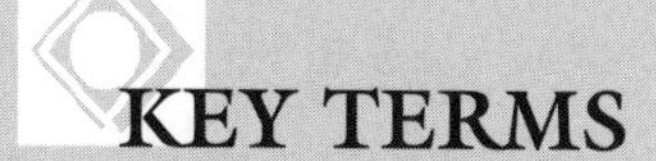

KEY TERMS

Key Terms

BASALTS: A type of volcanic rock that is composed of Fe- and Mg-rich silicates.

CYANOBACTERIA: Photosynthetic bacteria. They are prokaryotic and represent the earliest known form of life on Earth.

KOMATIITES: A very magnesium-rich volcanic rock.

METHANE: The primary component of natural gas and the simplest member of the alkane series of hydrocarbons.

OXIDATION: A chemical reaction that combines oxygen with other chemical species. Also, used to describe the loss of electrons from an atom or ion.

OXIDATION-REDUCTION REACTIONS: Termed "redox" reactions, these are chemical reactions involving atoms or ions that result in changes in electron configurations. During oxidation-reduction reactions the oxidation state of the atoms (and of ions subsequently produced) changes and electrons are lost or gained. Electron loss increases oxidation (i.e., the oxidation state) and produces a positive oxidation state. In contrast, the capture or addition of electrons results in reduced or negative oxidation state. As the electrons are not destroyed in chemical reactions (i.e., the net number of electrons must remain the same) oxidation-reduction reactions are coupled so that oxidation cannot occur without a corresponding reduction.

PALEOSOL: A stratum or soil horizon formed as a soil in a past geological age.

REDUCTION: A term used variously to describe the transformation of a compound into component molecular or atomic species. Also used to describe the gain of electrons by an atom or ion during a chemical reaction.

SILICATES: Any of a number of minerals consisting of silica combined with metal oxides; they form a chief component of the rocks in volcanic sinks, and are the minerals that comprise most of Earth.

TECTONIC PLATES: A massive, irregularly shaped slab of solid rock, generally composed of both continental and oceanic lithosphere (the rigid outer part of Earth, consisting of the crust and upper mantle). Plate size can vary greatly, from a few hundred to thousands of kilometers across.

Photodissociation One such geophysical or chemical process is that of photodissociation. Photodissociation is a process driven by the influx of ultraviolet radiation—especially in Earth's upper atmosphere—that results in the liberation of oxygen from water molecules. Just as oxygen-oxygen bonds in ozone molecules are shattered, the oxygen-hydrogen bonds in water molecules are broken by exposure to intense ultraviolet radiation. A portion of the hydrogen ions produced from the breakage then combined to form hydrogen gas that, because of its low mass, can escape Earth's atmosphere into space. Hydroxyl ions (OH^-)—produced by the breaking of an oxygen-hydrogen bond and the accompanying loss of a proton from the water molecule—react as part of the photodissociation reaction to produce water and atomic oxygen that, because of the net loss of hydrogen, can then react with other atomic oxygen atoms to produce molecular oxygen gas (O_2). The mass of oxygen molecules allows sufficient gravitational attraction so that the molecules do not escape into space and accumulate in the atmosphere.

Given that water vapor was present in Earth's primitive atmosphere, this process and the net gain of oxygen began long before the evolution of microbes capable of photosynthesis. Accordingly, the mechanism for the slow accumulation of oxygen was in place before the precipitous rise in oxygen levels. That such photodissociation mechanisms have been observed to occur in the upper atmosphere of Venus—a presumably abiotic environment—is often offered as evidence of the ubiquity of such reactions in a primitive atmosphere.

Why a Delay? If geophysical mechanisms that created free atmospheric oxygen existed, it is fair then to ask why there was a delay in the accumulation of atmospheric oxygen until approximately 2.2 billion years ago. One answer may lie in the fact that oxygen is a highly reactive gas and vigorously enters into a number of reactive processes. Throughout Earth's early history, the existence of highly active oxygen sinks (reactions that utilize oxygen and therefore remove it from the atmosphere) prevented any significant accumulation of oxygen. Moreover, the ubiquitous existence of organisms interpreted to be anaerobic in the fossil record provide abundant evidence that some mechanism or mechanisms operated to remove free oxygen from the primitive atmosphere.

Analysis of carbon isotopes in geologic formations and microfossils provides evidence that cyanobacteria capable of photosynthesis reactions evolved long before the abrupt rise in atmospheric oxygen. More importantly to the assessment of the role of non-biogenic process in oxygen production, the same evidence argues that atmospheric oxygen production may not have greatly fluctuated over time. Accordingly, an alternative model to explain a rise in atmospheric oxygen (i.e., an alternative to cyanobacteria producing greater quantities of oxygen which then accumulated in the atmosphere) rests on rate variable geophysical processes that altered aspects of oxygen utilization, which, in turn, allowed for accumulation of atmospheric oxygen.

For example, if geophysical processes that utilized, and thus removed, atmospheric oxygen occurred at higher rates earlier in Earth's history, the slowing of these processes more than two billion years ago would then allow and account for the accumulation of atmospheric oxygen.

Other reactions also account for the lack of free oxygen in Earth's primitive atmosphere. For example, methane—a primitive atmospheric component—reacts with available atmospheric oxygen to form water and carbon dioxide. Ammonia gases react with oxygen to form water and nitrogen gas. Until the levels of methane and ammonia gases lessened to negligible rates, these reactions would have prevented any significant accumulation of atmospheric oxygen.

Atmospheric oxygen is also consumed during oxidative reduction of minerals and in the oxidative reduction of gases associated with volcanic activity (e.g., carbon monoxide, hydrogen, sulfur dioxide, etc.). Oxygen is also ultimately utilized in reducing the gases produced in hydrothermal vents (e.g., volcanic vents associated with the Mid-Atlantic Ridge). The vent gases include reduced species of iron and sulfur (e.g., FeII and SII-). Currently, the majority of free or net oxygen usage occurs in the weathering of rock via reduction of carbon, sulfur, and iron.

At current levels of volcanic activity, estimates of oxygen utilization indicate that oxygen not recycled in the respiration/photosynthesis cycle is consumed by these geophysical reactions and so atmospheric oxygen levels remain stable. Earlier in Earth's history, however, a vastly greater percentage of oxygen utilization existed, exceeding that used in current weathering reactions. Combined with evidence of increased levels of volcanic activity earlier in Earth's history, such an oxygen sink could have been sufficient to prevent oxygen accumulation.

Some scientists, including L. R. Kump and others, have recently suggested that a quantitative change in volcanic activity (including hydrothermal venting) alone would not account for increased oxygen consumption (i.e., a more vigorous oxygen sink) because the concurrent production of other gases, including carbon dioxide, would also have stimulated increased photosynthesis. Instead, they argued that possible qualitative changes in the reduction state of materials increased the required uptake of oxygen sufficiently to prevent atmospheric oxygen accumulation.

Oxidation-Reduction Reactions The postulated qualitative shift is based upon evidence that indicates that the earth's mantle was far more reduced in Earth's early history. Accordingly, more oxygen would be required to complete oxidation-reduction reactions. The increased reduction in upper mantle materials would have also meant that volcanic gases arising from the upper mantle would have been more reduced and therefore utilized more oxygen in oxidative reaction processes. Kump and his colleagues estimate that "the potential O_2 sink from surface volcanism could therefore have been higher than today by a factor of 40 or more, allowing it to easily overwhelm the photosynthetic O_2 source."

One possible explanation for differential states of reduction involves changing states of mantle materials that contributed the reduced gases. Moreover, there is geological evidence that mantle materials that contribute to volcanic and hydrothermal gases have not remained constant in terms of their state of reduction. The geologic record provides direct evidence that prior to the rise in atmospheric oxygen, mantle materials were more reduced and therefore would have utilized more oxygen in oxidation-reduction reactions. This oxygen sink then acted to prevent the accumulation of atmospheric oxygen. In addition, the geologic record provides evidence that oxidized material was transported (subducted) from the crustal surface to the lower lithosphere (a region of the crust and upper mantle). This movement of oxidized materials away from the crust-atmospheric interface only increased the exposure of more reduced materials to whatever atmospheric oxygen existed and thus increased the rate of atmospheric oxygen utilization.

If the exposure of oxidized materials changed—as the geologic record indicates happened near the rise in atmospheric oxygen—then once-buried, more oxidized materials would have became dominant in volcanic gases. The increased oxidation state of the volcanic and hydrothermal gases then reduced the utilization of atmospheric oxygen and allowed for the subsequent accumulation of atmospheric oxygen.

The existence of geophysical processes to produce atmospheric oxygen and evidence of variability in the vigor (reflected in oxygen consumption) and abundance of these reactions provide evidence of a substantially non-biogenic transition of Earth's primordial atmosphere that combined hydrogen, ammonia, methane, and water vapor to one dominated by nitrogen and oxygen (approximately 21% oxygen).

Although the current evolutionary record reveals that organisms capable of photosynthesis existed at least a half billion years before the rise in Earth's atmospheric oxygen content, the contributions of cyanobacteria and botanical photosynthesis to free net atmospheric oxygen do not stand up to quantitative scrutiny. A review of the biogenic oxygen cycle reveals that more than 98% of the oxygen produced biogenically is reused in photosynthesis/respiration process. This strongly argues against a purely biogenic source of rapid oxygen accumulation approximately 2.2 billion years ago. —K. LEE LERNER

Viewpoint:

No, the theories based on geological principles accounting for the timing of the rise in Earth's atmospheric oxygen have insufficient data to supplant biological processes as the cause.

Most of us love a good mystery, especially scientists. The rise of oxygen in Earth's atmosphere approximately 2,200 to 2,400 million years ago engenders a mystery worthy of Sherlock Holmes. Cyanobacteria are simple organisms that produce oxygen when they harnesses the power of the Sun, which is called photosynthesis. They have been around for at least 2,700 million years; some scientists estimate that cyanobacteria have thrived for nearly 3,500 million years. Nevertheless, even at the younger age estimate, it took at least 300 to 500 million years for Earth's atmospheric oxygen levels to rise significantly and enable more complex life to evolve. The mystery lies in why there is such a big gap between the emergence of cyanobacteria and the rise of atmospheric oxygen.

Although biological processes have long been considered the primary source for increased oxygen in the atmosphere, some scientists propose that geological processes, such as volcanic gases, initiated the timing of the precipitous rise in oxygen levels. However, the evidence to support a geological theory is minimal at best, and biological processes remain the best possible explanation for controlling the evolution of the atmosphere's oxygen content.

One of the most popular geological theories behind the rise in oxygen was proposed by James F. Kasting and colleagues at the Department of Geosciences and Astrobiology Research Center at the Pennsylvania State University. According to these scientists, it was escaping volcanic gases that most influenced the sharp rise in atmospheric oxygen. This theory is based on composition aspects of the earth's mantle (the region below the crust and above the core), where some volcanic gases originate. Before oxygen, the mantle contained an abundance of "reducing" components like iron silicates. These components, when brought to, or near, the surface by volcanic processes react with oxygen and extract most of the oxygen produced by photosynthesis. However, according to this theory, about 2.7 billion years ago the mantle underwent a change in which oxidized material that had "sunk" to the base of the mantle was pushed to the top, resulting in the volcanic gases emitted becoming less reducing. The bottom line is that as volcanic gases became more oxidized over time, their ability to "mop up" atmospheric oxygen decreased. As a result, the net photosynthetic production of oxygen was greater than the volcanic gases' ability to react with the oxygen. This resulted in an oxygen-rich atmosphere. However, studies of ancient basalts and komatiites (rocks made of Fe- and Mg-rich silicate) show that the mantle's oxidation state did not change appreciably with time, indicating that the volcanic gases theory is not correct.

New Theories Bolster Biological Causes

NASA scientists David Catling and colleagues recently proposed that because the early atmosphere had low oxygen content, methane (which is normally broken down in high oxygen environments) accumulated in the atmosphere. Made up of one carbon atom and four hydrogen atoms, the methane eventually ascended to the upper atmosphere, where intense ultraviolet rays broke it down. The extremely light hydrogen atoms drifted from Earth's gravitational field and off into space. The basic result was that the hydrogen atoms and oxygen atoms (or the organic matter made from hydrogen) were kept apart, thus creating more oxygen. Because there were less hydrogen atoms to combine with oxygen and make it disappear, Earth's oceans and rocks became saturated with oxygen, which then could accumulate in larger amounts in the atmosphere.

Recent research by another NASA scientist focuses on the methane production theory and attempts to answer the primary question inherent in the theory: Where did the methane come from? Terri Hoehler and colleagues measured the gases released from microbial mats in Baja, Mexico. These mats, which are aggregates of microorganisms composed mainly of bacteria and algae, are closely related to the ones that covered much of Earth's early biosphere and were found living in similar conditions. The group discovered that the mats released large quantities of hydrogen during the night. "If the Earth's early microbial mats acted similarly to the modern ones we studied," said Hoehler in a NASA news release, "they may have pumped a thousand times more hydrogen into the atmosphere than did volcanoes and hydrothermal vents, the other main sources."

Some of this hydrogen might have escaped directly to space, and the remainder could have provided "food" for microbes producing methane. The elevated levels of hydrogen within the mats favors the biological production and release of methane and supports the premise of Catling's work. "The bizarre implication," said Catling, "is that we're here as a result of these gases that came from microbial scum, if you like, back on early earth."

Although microbial mats can grow in various environments, including shallow seas, lakes, and rivers, the discovery of microbial mats intermixed with soil indicates that life possibly emerged from the sea to land as far back as 2.6 billion years ago, which contradicts what many scientists think about the timing when atmospheric conditions were favorable to life on land. Since the 2.6 billion year date is 200 to 400 million years earlier than previously estimated for the rise in atmospheric oxygen, the finding suggests that the estimates were possibly wrong and that life was already flourishing on land in an oxygen-rich atmosphere. It also indirectly supports the theory that these microbial mats were prolific enough to produce the gases needed to cause an increase in atmospheric oxygen.

New Information on the Age of Oxygen-Rich Atmosphere One way to resolve the mystery of why it took oxygen so long to accumulate after the appearance of cyanobacteria is to say that the paradox does not actually exist. A Pennsylvania State University geochemist, Hiroshi Ohmoto, says that his research shows that an oxygen-rich atmosphere did not emerge as late as evidence suggests. One of the pieces of evidence about the timing of the rise in atmospheric oxygen is based on iron levels found in paleosols (ancient soils that are now rock). The evidence supporting the theory about when the rise in atmospheric oxygen occurred centers around insoluble iron hydroxides, which results when oxygen converts iron silicates. Iron hydroxides have only been found in paleosols less than 2.3 billion years old, indicating that the atmosphere did not have high oxygen levels before this time. However, according to Ohmoto and colleagues, just because the iron is not present in the paleosols doesn't mean that it was never there. There are a number of explanations for its disappearance. It could have been removed by hot water from volcanoes or by organic acids produced by cyanobacteria. As a result, the microorganisms might have confused the biological record by both creating oxygen that placed iron in the soils while producing the very acids that worked to remove it. As a result, it makes the supposed timing of the jump to high oxygen levels less certain.

Most of the early oxygen produced by cyanobacteria is still "imprisoned" in massive banded iron formations (sedimentary deposits of iron oxides), which were created on ocean floors in a process involving early reducing iron ions (Fe^{2+}) that took up the oxygen produced by shallow water cyanobacteria during oxidizing reactions. Abundant oceanic Fe^{2+} suggests a lack of atmospheric oxygen, and banded-iron formations of ferric oxide would have formed an oxygen sink for any oxygen in the environment. So far, the vast majority of these banded iron formation have been found in rocks more than 2.3 billion years old, again indicating that oxygen levels were extremely low. In answer to this evidence, Ohmoto points out that similar banded iron formations have been found dating back to only 1.8 billion years, indicating that these structures can also form in the presence of high oxygen levels.

The idea that oxygen may have been around a lot longer than originally thought is further supported by research performed at Yale University by geochemist Antonio Lasaga. He has modeled the cycles of oxygen, carbon, sulfur, and iron from the early Earth and determined that within 30 million years after the appearance of cyanobacteria, the atmosphere would have contained very high oxygen content and that the atmosphere would have maintained this basic content.

Another group of scientists have also challenged the long-held theory of when Earth's atmosphere became enriched with oxygen. According to iron-rich nodules found in the deep strata of South Africa's Witwatersrand, Earth's atmosphere may have been rich in oxygen nearly 3 billion years ago. The researchers say the iron-rich nodules contain ferric iron produced by exposure to an oxygen-rich atmosphere. The nodules were dated at 2.7 to 2.8 billion years old. This theory is bolstered by evidence found in Western Australia's Pilbara region that found the presence of sulfates in rocks up to 3.5 billion years old. These sulfates also could not have formed without an oxygen-rich atmosphere.

If proven, these theories about a much earlier origin for an oxygen-rich atmosphere would make the debate over the geological versus the biological processes involved in the rise of an oxygen-rich atmosphere moot. If the higher oxygen levels existed hundreds of millions of years earlier, then the precipitous rise in atmospheric oxygen content did not even occur.

The Final Analysis Scientists have long been tantalized by the great mystery of the rise in Earth's atmospheric oxygen levels between 2,200 and 2,400 million years ago. Many theories have been suggested, but none have become widely accepted. In fact, the geochemical and biological data can be interpreted in many ways. Some scientists are beginning to propose that the quick (in geological terms) transition from an atmosphere with little or no oxygen to an oxygen-rich atmosphere never occurred. Rather, they say that evidence shows that Earth had higher oxygen content far longer than scientists have thought. If this is the case, then the geological hypotheses proposed are wrong, and the biological processes remain the most likely reason for the rise in oxygen.

Barring a momentous discovery, the debate over whether the timing of the rise in atmospheric oxygen was triggered by geological or biological processes will continue for some time to come. Most likely, ancient geological processes have played some part in determining the amount of oxygen in the atmosphere. For example, tectonic plates may have opened up deep basins on the ocean floor that were free of oxygen and, as a result, served as oxygen sinks where organic material could have settled. Once these sinks were filled, oxygen began to accumulate rapidly. Nevertheless, there is not enough evidence to supplant biological processes as the dominating influence in the evolution of Earth's atmosphere. For now, the generally accepted theory is that cyanobacteria, as the only life on Earth for about two billion years, single-handedly saturated the atmosphere with oxygen. —DAVID PETECHUK

Further Reading

Arculus, R. J. "Oxidation Status of the Mantle: Past and Present. Annual Review, Earth Planet." *Science* 13 (1985): 75–95.

Brocks, J. J., G. A. Logan, R. Buick, and R. E. Summons. "Archean Molecular Fossils and the Early Rise of Eukaryotes." *Science* 285 (1999): 1033–36.

Castresana, J., and D, Moreira. "Respiratory Chains in the Last Common Ancestor of Living Organisms." *Journal of Molecular Evolution* 49 (1999): 453–60.

Catling, David, et al. "Biogenic Methane, Hydrogen Escape, and the Irreversible Oxidation of Early Earth." *Science* 293 (August 3, 2001): 839–43.

Cloud, P. E. "A Working Model of the Primitive Earth." *Am. J. Sci.* 272 (1972): 537–48.

———. *Oasis in Space: Earth History from the Beginning*. New York: W.W. Norton, 1988.

Darwin, Charles. *On the Origin of Species by Means of Natural Selection, or the Preservation of Favored Races in the Struggle for Life.* Cambridge: Cambridge University Press, 1859.

Delano, J. W. "Oxidation State of the Earth's Upper Mantle During the Last 3800 Million Years: Implications for the Origin of

Life, Lunar Planet." *Science* 24 (1993): 395–96.

Grotzinger, J. P., and A. H. Knoll. "Stromatolites in Precambrian Carbonates: Evolutionary Mileposts or Environmental Dipsticks?" *Annual Review of Earth and Planetary Sciences* 27 (1999): 313–58.

Hoehler, Terri, et al. "The Role of Microbial Mats in the Production of Reduced Gases on Early Earth." *Nature* (July 19, 2001): 324–27.

Jorgensen, Bo Barker. "Biogeochemistry: Space for Hydrogen." *Nature* (July 19, 2001): 286–88.

Kasting, J. F. "Earth's Early Atmosphere." *Science* 259 (1993): 920–26.

Kasting, J. F., J. F. Kump, and H. D. Holland. "Atmospheric Evolution: The Rise of Oxygen." In *The Proterozoic Biosphere: A Multidisciplinary Study*, edited by J. W. Schopf. New York: Cambridge University Press, 1992: 159–63.

Kasting, J. F., D. H. Eggler, and S. P. Raeburn. "Mantle Redox Evolution and the Oxidation State of the Archean Atmosphere." *J. Geol.* 101 (1993): 245–57.

Kirschvink, J. L., E. J. Gaidos, L. E. Bertani, N. J. Beukes, J. Gutzmer, L. N. Maepa, and R. E. Steinberger. "Paleoproterozoic Snowball Earth: Extreme Climatic and Geochemical Global Change and Its Biological Consequences." *Proceedings of the National Academy of Sciences* 97 (2000): 1400–05.

Kump, L. R., J. F. Kasting, and M. E. Barley. "Rise of Atmospheric Oxygen and the 'Upside-down' Archean Mantle." *Geochemistry Geophysics Geosystems* (2001).

Lecuyer, C., and Y. Ricard. "Long-term Fluxes and Budget of Ferric Iron: Implication for the Redox States of the Earth's Mantle and Atmosphere, Earth Planet." *Sci. Lett.* 165 (1999): 197–211.

Mojzsis, S. J., G. Arrhenius, K. D. Mckeegan, T. M. Harrison, A. P. Nutman, and C. R. L. Friend. "Evidence for Life on Earth Before 3,800 Million Years Ago." *Nature* 384 (1996): 55–59.

Ohmoto, H. "Evidence in Pre-2.2 Ga Paleosols for the Early Evolution of Atmospheric Oxygen and Terrestrial Biota." *Geology* 24 (1996): 1135–38.

Should energy resources in the Arctic National Wildlife Refuge be developed?

Viewpoint: Yes, energy resources in the Arctic National Wildlife Refuge should be developed because dependence on foreign resources poses a serious threat to the security of the United States.

Viewpoint: No, energy resources in the Arctic National Wildlife Refuge should not be developed because the gain would be minimal and the cost to the environment unacceptable.

When the first major oil strike occurred, at Titusville, Pennsylvania, in 1859, the momentous character of the event could hardly have been appreciated. There were no automobiles then, nor could anyone have imagined the importance that petroleum would acquire within three-quarters of a century. Even if they had, most observers would have presumed that America had all the oil supplies it needed.

At that time, paleontologists were just beginning to understand the fact that dinosaurs had once ruled the planet. It would be years before scientists understood that the deposition of their bodies in the earth—the result of a great cataclysm some 65 million years ago—had provided Earth's "new" human rulers with a vast but nonetheless limited supply of energy in the form of petroleum.

Likewise it would be years before prospectors began to discover ever more abundant supplies of oil on U.S. soil, first in Texas and surrounding states, and later in Alaska. Also in the future lay the discovery of even more impressive reserves in the Middle East, a region that until the mid-twentieth century was economically impoverished and geopolitically insignificant.

By 1973, much had changed. America and other industrialized nations had millions of cars on the road and regularly imported millions of barrels of crude oil from Saudi Arabia and other nations, which—because of their vast oil wealth—had become powerful players on the international scene. In 1973, Americans would be forced to confront their dependence on foreign oil. That was the year when the Organization of Oil-Exporting Countries (OPEC), composed primarily of Arab nations, placed an embargo on oil sales to the United States in retaliation for U.S. support of Israel. The results were disastrous: suddenly scarce, petroleum was in high demand, and anyone who lived through that time can recall the long lines at the gas pumps, the frustration of cancelled family trips, the exigencies of enforced conservation, and the fears raised by dependency on hostile foreign powers for the fuel that literally runs our nation.

Those fears subsided somewhat with the conclusion of the embargo—unsuccessful inasmuch as it failed to sway U.S. policy in the Middle East—but they have resurfaced again and again over the years. One of the most dramatic instances of such resurfacing occurred after the September 11, 2001, attacks on the World Trade Center towers and the Pentagon. Though a small band of Muslim extremists actually carried out these acts, the subsequent response by a large segment of the Arab world—most notably, a refusal to wholeheartedly condemn the terrorists' actions—once again gave notice to Americans that their trade agreements with Saudi Arabia and other oil-exporting Arab nations could break down at any time.

In the eyes of many Americans, three decades of strained relations with the Arab world only served to reinforce the need to develop oil resources on U.S. soil. By that time, most economically viable reserves in the continental United States had been tapped, but Alaska still presented a potentially rewarding frontier for future exploration. In response to the embargo of 1973, America stepped up its efforts to extract oil from Alaska.

Significant oil reserves had first been discovered along the shores of the Arctic Ocean in 1968, but the lack of infrastructure (including roads) in that largely uninhabited region had presented an impediment to exploitation of those reserves. Then, in 1969, a group of petroleum companies paid the state $1 billion for drilling rights and proposed the building of a pipeline. Opposed by environmentalists, the proposed pipeline had gone unbuilt, however, until November 1973, a month after the Arabs imposed the embargo, at which point Congress authorized the construction of a 789-mi (1,262-km) pipeline.

The pipeline made possible the transport of oil from the Prudhoe Bay oil field on the Arctic coast to the harbor at Valdez, from whence oil tankers transported it to the ports on the West Coast of the United States. However, environmental concerns remained, and loomed greater in the years that followed. In part, this was due to the growth of an increasingly powerful and politically forceful environmental movement, which was still in its infancy at the time the original pipeline was proposed. But concerns over the threat posed to the Alaskan wilderness by the extraction and transport of oil had also spread far beyond the limits of the environmentalist movement to the population as a whole. This was particularly the case after the 1989 disaster of the Exxon *Valdez,* an oil tanker that ran aground off of Prince William Sound, causing the worst oil spill in the history of North America and doing immeasurable damage to the environment.

Such was the state of affairs at the end of the twentieth century and the beginning of the twenty-first, the setting for a renewed debate—involving scientists, politicians, oil companies, state officials, native groups, environmentalists, and the public at large—concerning the exploitation of oil resources in Alaska. At issue was the Arctic National Wildlife Refuge (ANWR), a wilderness of some 19 million acres (29,688 square mi or 76,891 square km—about the size of Maine) protected since 1960 by an act of Congress. Within ANWR is a shoreline region designated as 1002 Area, estimated to contain between 16 billion and 32 billion barrels of oil. (A barrel is equal to 23.1 gallons, or 159 liters.)

In the essays that follow, arguments are made, based on scientific speculation regarding the recoverability of resources, both for and against the development of energy resources in ANWR. In other words, not all arguments against drilling necessarily revolve around environmental concerns alone, and not all arguments in favor of it are based purely on concerns for national security. Obviously, those two positions—the environment on one hand, and national security on the other—play a major role in governing one's views on the advisability of exploiting the resources at ANWR. However, at heart the question is one of costs versus benefits.

On the one hand, there are the costs involved in the impact on the environment, as well as on the lives of native peoples. Furthermore, there is the sheer economic cost of extraction and development of ANWR reserves against the economic and political benefits to be accrued from those actions. On the other hand, there is the political cost of dependence on oil exported by increasingly hostile nations, combined with the potential economic benefits, both to the state of Alaska and to the nation as a whole. —JUDSON KNIGHT

Viewpoint:

Yes, energy resources in the Arctic National Wildlife Refuge should be developed, because dependence on foreign resources poses a serious threat to the security of the United States.

For the last 40 years of the twentieth century and into the first decade of the twenty-first century, drilling for oil in the Coastal Plain of the 19 million-acre Arctic National Wildlife Refuge (ANWR) has been strongly debated. The ANWR, located in northeastern Alaska, is an area abundant with fauna and flora, and rich with oil potential. Energy experts agree that the Coastal Plain (commonly called 1002 Area) is currently the most promising domestic onshore oil and gas prospect. The consensus of most petroleum geologists is that its potential is on the order of billions of barrels of recoverable oil and trillions of cubic feet of recoverable gas.

A view of the Jago River, which runs along an arctic plain in the Arctic National Wildlife Refuge. *(© Tom Bean/Corbis. Reproduced by permission.)*

Opponents, Proponents, and the Government Since 1985 oil and gas leasing on this federally owned land, administered by the U.S. Fish and Wildlife Service (within the Department of the Interior), has been banned. Exploration and development of oil and gas are supported by the oil companies and by the majority of Alaskan officials and citizens, but has been opposed by many environmental organizations. The inhabitants of the area, the Inupiat Eskimos and Gwich'in Indians, are also actively involved since they have a direct economic relationship to the land, and to its wildlife and other natural resources. The U.S. Congress will ultimately decide whether to open up the ANWR for oil exploration and development.

ANWR Development Domestic oil production, which has declined from nearly 9 million barrels per day in 1985 to about 5.9 million barrels per day in 2001, is projected to decline to less than 5 million barrels per day in 2010. Even with only a modest growth in demand, the deficit at that time in U.S. supplies (the difference between domestic supply and demand) is estimated by the National Association of State Energy Officials to be around 10 million barrels per day. The deficit can only be made up by new discoveries or imports. As a result, it is the contention of this article that the energy resources of the ANWR be developed. The basic arguments for developing the ANWR energy reserves are as follows:

- Small fractional area affected: less than 8% of the ANWR will be open for oil drilling and production
- Security and energy independence: increasing oil imports present a threat to U.S. security, so increased domestic supplies help

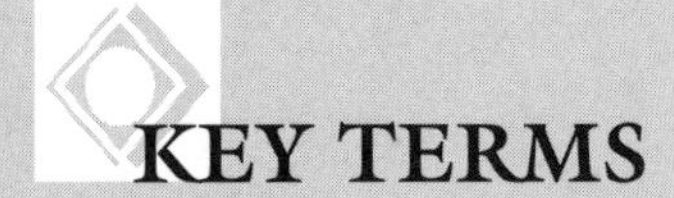

KEY TERMS

FAUNA: The animals of a particular habitat, region, or geological period.

FLORA: The plants of a particular habitat, region, or geological period.

SEISMIC: Relating to or denoting geological surveying methods involving vibrations produced artificially, such as by explosions.

minimize that security threat and increase self-reliance

- Safe technology: advances in energy exploration could be safely applied, greatly reducing the environmental impact caused by arctic oil development
- Economic advantages: development will increase both Alaskan and U.S. oil production, add usability to the Trans-Alaska Pipeline System, create Alaskan jobs and employment nationwide, improve the Alaskan economy, and raise the U.S. gross national product by billions of dollars
- Best chance for discovery: contains the best prospect for significant U.S. oil discovery
- Major support in Alaska: the majority of its citizens favor development

Small Fractional Area Affected Under federal law, 17.5 million acres (about 92%) of ANWR will remain permanently closed to all development. The remaining 1.5 million acres (about 8%) on the northern Coastal Plain of ANWR is the only area being considered for development. If oil drilling commenced, less than one half of one percent, or about 2,000–5,000 acres, of the 1.5 million acres of the Coastal Plain would be directly affected. According to Senator Frank Murkowski (R-AK), the actual area in the ANWR covered by such structures as drilling rigs, buildings, and roads would be less than the size of the Anchorage International Airport.

Security and Energy Independence The 11 countries of OPEC (Organization of the Petroleum Exporting Countries) account for more than 40% of the world's annual oil production and possess about 75% of the proven reserves. At the time of the 1973 Arab oil embargo, the United Stated imported only 36% of its oil. As of 2001, the U.S. imported, according to the American Petroleum Institute, a little over 55% of the oil needed domestically. Experts at the Department of Energy are expecting the import percentage to increase from 55% to 65% by the year 2020.

Oil Prices. Little interest was given to oil potential in ANWR when oil prices were low. However, in 1999, OPEC and other oil-exporting nations agreed to cut oil production. Within five months world oil prices went from $10 to $33 per barrel; but leveled off in 2001 to about $26. Consequently, interest in Alaskan oil, and especially in ANWR development, has drastically increased. Recent OPEC production agreements have renewed calls for U.S. policies that might reduce dependence on foreign oil and help with uncertainties in energy markets due to continuing Middle East crises.

Declining Domestic Production. United States oil production has steadily declined since its peak in 1973. ANWR development would alleviate current oil supply problems. According to the Energy Information Administration (EIA), peak annual ANWR production—assuming that 9.4 billion barrels of petroleum are recoverable at a market price of $24 per barrel—would be approximately 1.4 million barrels per day, compared with projected U.S. oil imports of 15.3 million barrels per day and total oil use of 24.3 million barrels per day in 2015. (The level and timing of peak production would depend upon the rate at which the ANWR oil fields are developed; according to the EIA, from the time of approval to first production would be from 7 to 12 years.) Supporters of development argue that ANWR oil would reduce dependence on foreign sources, and that the additional incremental supply could be crucial in determining oil prices. At the upper estimate, the U.S. Geological Survey projects that ANWR reserves could total as much as 16 billion barrels of recoverable oil, an amount that represents all the oil currently imported from Saudi Arabia for the last 30 years.

Safe Arctic Technology New petroleum technology, including advances in computing and exploration methods, developed since the early 1970s will allow companies to tap underground reservoirs with a much smaller surface impact. Senator Murkowski recently said that through such technologies oil companies could safely develop a manageable "footprint" (the area occupied by infrastructure) in order to protect the surrounding environment and the ecology. If the Prudhoe Bay oil fields were built today using the latest techniques, the footprint would be 1,526 acres, 64% smaller than when actually built. In addition, millions of dollars of research on wildlife resources and their habitat on Alaska's North Slope have greatly increased the scientific understanding of arctic ecosystems

Oil drilling platforms off the coast of Texas.
(© Jay Dickman/Corbis. Reproduced by permission.)

and have also shown that wildlife and petroleum extraction can coexist.

The indigenous Gwich'in Indians are most apprehensive about oil development in the Refuge because they depend on the barren-ground caribou of the Porcupine herd for food. They fear that ANWR development would disrupt the caribou's calving grounds and diminish their food supplies. However, Department of Interior Secretary Gale Norton stated in 2002 that the wildlife of the ANWR can be adequately protected and America's treaty obligations with native tribes would not be violated by oil exploration in the ANWR chiefly due to improvements in oil-drilling technology.

Proponents of drilling point to the oil fields at Prudhoe Bay and Kuparuk, both about 60 mi (100 km) west of the Refuge, and the central Arctic caribou herd. According to Pat Valkenberg, a research coordinator with the Alaska Department of Fish and Game, the herd has increased its numbers from 15,000 in 1985 to 27,100 in 2000 despite several hundred miles of roads and more than a thousand miles of elevated pipe.

As mentioned previously, drilling in the ANWR can now be accomplished with far less impact than could have been possible using older technology. Among the improvements are smaller gravel pads to support the wells, greater distances that can be reached from a single (draining) pad, the re-injection of drilling wastes (eliminating the need for large waste pits), and a reduction in the need for gravel roads. In addition, other improvements include three-dimensional seismic exploration, four-dimensional time-lapse imaging, ground-penetrating radar, and enhanced computer processing of resulting data on geological structures. According to the

A polar bear guards a whale carcass in the Arctic National Wildlife Refuge.
(© Lowell Georgia/Corbis. Reproduced by permission.)

Energy Department, these latter improvements have increased the number of successful wells from about 10% to as much as 50%, resulting in a decrease in the number of pads and exploration costs. Together, these advances decrease the developmental impact.

Economic Advantages Between 1980 and 1994, Alaska's North Slope (ANS) oil field development and production activity contributed over $50 billion to the U.S. economy. The oil-rich region extended eastward from the 2-billion-barrel Kuparuk River field, past the Prudhoe Bay field (originally 11 billion barrels, now down to about 4 billion barrels), and a few smaller fields (such as Lisburne and Endicott), and in all likelihood extends through ANWR's 1002 Area. With the close proximity of these current oil resources, it is believed that the ANWR could easily contain the largest undrilled, petroleum potential known in the United States.

The state of Alaska is already strongly tied to oil with regards to its revenue base. The Prudhoe Bay field, the largest single accumulation of oil ever discovered in North America, is located on lands owned by Alaska. The proportion of the state budget utilizing petroleum revenues has risen from an annual average of 12%, initially in 1968, to 90% 15 years later, and currently holds at 80%. According to Arctic Power, an Alaskan nonprofit organization, since 1987 the North Slope oil fields have provided the country with nearly 20% of its domestic production. But its actual production has been on the decline during that same period. Peak production was reached in 1980 with 2 million barrels per day, but is currently at nearly 1.2 million barrels per day. By 2010 oil flow is predicted to drop to just 315,000 barrels per day, and no new fields have been identified to compensate for the decline.

Industry interest in the 1002 Area is based in part in keeping the Trans-Alaska Pipeline operating efficiently. Oil is transported from the North Slope by the 800-mi (1,290-km) system, from Prudhoe Bay to Valdez in south-central Alaska, where it is then transferred to tankers. Pipeline costs are largely fixed; a smaller flow of oil means higher pipeline rates per barrel.

Both the national and Alaskan state governments have important political and economic reasons to develop natural resources and assist those corporations extracting them. Revenues (estimated in the billions of dollars in the form of bonus bids, lease rentals, royalties, and taxes) and employment (estimated by Senator Murkowski to create between 250,000 and 735,000 jobs) from such companies are central to their financial welfare. Backed by the majority of Alaskans, two U.S. senators and one house representative from Alaska are strong advocates of oil development in the ANWR in order to meet the state's economic needs.

Best Chance for Discovery Mark Rubin of the American Petroleum Institute says that ANWR is the best place to look for domestic oil. The Coastal Plain lies between two known major discovery areas. To the west, the Prudhoe Bay, Lisburne, Endicott, Milne Point, and Kuparuk oil fields are currently in production. To the east, major discoveries have been made in Canada, near the Mackenzie River delta, and in the Beaufort Sea. Rubin adds that if current predictions hold, then there is more oil in ANWR than any other place in the United States. Petroleum geologists agree that the Coastal Plain is America's best possibility for a major discovery like the giant "Prudhoe Bay–sized" discovery. Estimates from the Department of Interior range from just under 6 to upwards of 16 billion barrels of recoverable oil.

Majority Support in Alaska According to a February 2000 poll performed by the Dittman Research Company, more than 75% of Alaskans favor exploration and production of the oil reserves in ANWR as long as it is done in an environmentally sound way that keeps land disturbances to a minimum. According to a poll taken by the *Los Angeles Times* between January 31 and February 3, 2002 (and backed by earlier polls with similar results), 48% of all Americans favor opening a part of the ANWR to energy exploration. Alaska Democratic Party Chairman Scott Sterling said that he supports oil drilling in ANWR as long as wildlife is protected.

The Inupiat people of the North Slope have called the Arctic their home for thousands of years. They have depended on Alaska's North Slope and the area within the ANWR for their very lives. The majority of Inupiat Eskimos support onshore oil development on the Coastal Plain. These partial-subsistence hunters will gain financially (by providing jobs, schools, and a better economy) from leasing ANWR's potentially rich land to the oil companies. On the other hand, the Inupiat people fear the potential disruption of wildlife and their way of life. However, former Inupiat Mayor Benjamin P. Nageak said that since the first discovery of oil in 1969 the oil companies have consistently met the strict standards and regulations that were imposed on them in order to protect the land. Nageak went on to say that his people have the greatest stake possible in seeing that all development is performed in an environmentally safe way.

Oil-Dependent United States Oil development in the ANWR could yield up to 16 billion barrels of oil. The U.S. economy is highly dependent on oil and because of this the country is currently importing over half its oil, much from the Middle East. Conservation and renewable energy have great potential, but for today and for many years to come, oil will be a major source of energy. While it may be gratifying for some to strongly resist drilling in sensitive areas, it is naïve to ignore the national security and economic benefits that ANWR oil extraction can provide. Rather than trying to prevent the inevitable, ecologists (and others sensitive to the environment) should work hard to design and implement enforceable environmental standards for drilling in sensitive areas such as ANWR. Because the United States—at least for the foreseeable future—is so dependent on petroleum, the energy resources of the ANWR should be developed. —WILLIAM ARTHUR ATKINS

Viewpoint:

No, energy resources in the Arctic National Wildlife Refuge should not be developed, because the gain would be minimal and the cost to the environment unacceptable.

Background Early explorers to the North Slope of Alaska often noted oil seeps and oil-stained sands on the surface. Yet it was not until 1944 that oil exploration began in earnest on 23 million acres of what was then called Naval Petroleum Reserve Number 4 in northwestern Alaska.

In 1968, North America's largest oil field was discovered near Prudhoe Bay, Alaska. The Trans-Alaska pipeline was finished in 1977. By

A caribou herd crosses the frozen Kongakut River in the Arctic National Wildlife Refuge. *(© Steve Kaufman/Corbis. Reproduced by permission.)*

1981, the oil fields around Prudhoe Bay, Endicott, and offshore in the Beaufort Sea, yielded about 1.7 million barrels of oil per day—about 25% of U.S. domestic production. By 1988, these oil fields showed signs of exhaustion and oil yields declined. No new oil fields have been discovered in this area.

In the late 1980s and early 1990s, oil companies leasing off- and onshore sites farther east announced the potential for oil drilling on these holdings. These sites are situated in or just offshore of the Arctic National Wildlife Refuge (ANWR), designated by Congress in 1960 as an 8.9-million acre protected wilderness. The Alaska National Interest Lands Conservation Act (ANILCA, 1980) doubled ANWR's size, but omitted protection for the shoreline region known as 1002 Area, pending further research into its oil-production potential and its effects on wildlife. This region is the main focus of interest for oil companies and conservationists. The U.S. Geological Survey (USGS) has analyzed the region's energy potential, and found that ANWR and 1002 Area do contain oil. The questions remain, though, about whether it is economically feasible to recover it.

The U.S. Fish and Wildlife Service (FWS) meanwhile conducted numerous studies to assess the effects of oil exploration and extraction on ANWR's wildlife. They concluded that resource exploitation would seriously affect the Porcupine herd of caribou, destroy polar bear den areas, and disrupt key breeding bird sites.

How Much Oil Is There? There is oil under ANWR. But the first two questions that need to be answered are: "How much is there?" and "Is it economically and technically feasible to extract it?"

The most reliable studies of potential oil deposits under ANWR have been conducted by the USGS. The agency assessed the potential for technically recoverable oil in ANWR and in 1002 Area. The agency has reported that there are three likely scenarios surrounding the amount and recoverability of the oil.

Scenario 1: Most Likely, Most Recoverable, Least Amount. There is a 95% probability of finding 5.7 billion barrels of recoverable oil from ANWR's Coastal Plain. Peak production rates for this scenario are estimated between 250 and 400 million barrels per year.

Scenario 2: The Mean. There is a 50% probability of finding 10.3 billion barrels of oil that are recoverable from ANWR's Coastal Plain. Peak production rates for this scenario are estimated between 400 million and 600 million barrels per year.

Scenario 3: Least Likely, Least Recoverable, Greatest Amount. There is a 5% probability of finding 16 billion barrels of recoverable oil from ANWR's Coastal Plain. Peak production rates for this scenario are estimated between 600 and 800 million barrels per year.

Time No matter how much oil is under ANWR, it will take time to develop the area and extract it. For all scenarios, and at the highest extraction rate, significant amounts of oil will not be produced until about 10 years after development begins, with the greatest yield after 15–20 years. Maximum yield will be recoverable for less than five years, before it begins declining rapidly. Thus, any amount of oil recovered from ANWR will not affect immediate energy needs and will contribute most to our energy needs for a period of only a few years. As discussed below, even at its greatest, this contribution is insignificant compared with U.S. energy consumption.

Cost Not all of the oil believed to underlie ANWR is technically or economically recoverable. The FWS reports that at prices lower than $16 per barrel, there is no economically recoverable oil in 1002 Area of ANWR. The amounts and rates of oil extraction set out in the above scenarios assume an oil price of $24 per barrel to make extraction economically feasible. It is clear that the most recoverable oil (at 95%) yields the least economic return in terms of oil production.

Much of the oil under ANWR and 1002 Area occurs as small, widely dispersed deposits. The USGS, with concurrence from oil companies, states that accumulations of less than 100 million barrels of oil are uneconomic for mining. Other economic factors that influence the cost of oil extraction are the current state of oil-extraction technology and the field's proximity to infrastructure.

The region known as 1002 Area is 100 mi (161 km) wide and located more than 30 mi (48 km) from the tail end of the nearest pipeline and more than 50 mi (80 km) from the nearest gravel road and oil support facility. To be fully developed, 1002 Area would require the construction of all the above, plus production sites, power plants, processing facilities, loading docks, living quarters, airstrips, gravel pits, utility lines, and landfills. Thus, its isolation adds significantly to the costs of developing its oil. The USGS report shows that far less oil can be economically recovered than actually occurs, or even than is technically recoverable. (Technically recoverable oil is defined as the in-place oil that is available for extraction based on geology and the state of current oil-extraction technology, regardless of cost.) Further, when oil prices decline, some oil companies may simply abandon their oil fields, leaving the infrastructure and waste lagoons intact.

ANWR and Energy Independence The crux of the argument in favor of oil drilling in ANWR and 1002 Area is that it will significantly reduce U.S. dependence on foreign oil; that it will make the United States more "energy independent." That argument is patently untrue.

As of 2000, the United States consumes about 19.6 million barrels of oil each day, or 7.154 billion barrels per year. If development of ANWR's oil fields began tomorrow and oil began to flow by 2010, the national rate of oil consumption would by then be more than 8 billion barrels per year. By the time of ANWR's maximum output, in 2020, U.S. oil consumption is projected to be about 9.5 billion barrels per year. Yet at its maximum output, ANWR would yield merely 800 million or so barrels of oil per year. And this high yield would last for only about five years, before declining drastically.

In short, even if ANWR could produce the maximum amount of oil (16 billion barrels, at a 5% chance of recovery), it would at best contribute less than 10% of U.S. oil needs. If the mean amount of oil was recoverable (10.3 million barrels, at 50% chance of recovery), at best about 5% of U.S. oil needs would be met. For the most recoverable oil (5.7 billion barrels, at 95% chance of recovery), only about 2% of our oil needs would be met by ANWR.

Energy Independence and Fuel Efficiency Those who are concerned with U.S. dependence on foreign oil—which should be everyone—obviously cannot find relief in ANWR's oil. However, raising fuel-efficiency standards in American cars and trucks—or even using better tires—will significantly reduce our oil consumption and thus our need for foreign oil. Raising the fuel-efficiency of cars, SUVs, and trucks to

PRUDHOE BAY

Prudhoe Bay, on the North Slope west of ANWR, has been in oil development since the early 1980s. Since few people ever go there and see for themselves, not many people know what the impact of oil exploitation has been. Below are a few facts that should be considered before developing ANWR for oil extraction.

1. More "Footprint" Myths. More than 1,000 square miles of tundra are covered with gravel (an area greater than Rhode Island); 23 additional oil fields cover another 2,000 acres; the Trans-Alaska pipeline and haul road eat up another 10,000 acres; there are more than 1,123 mi (1,807 km) of subsidiary roads and pipelines connected to oil wells, with an additional 500 mi (804 km) of roads along these pipelines; there are two 6,500-ft (1981-m) airstrips.

2. Gravel. More than 10,000 acres of wetlands have been filled and covered by gravel for roads, airstrips, drill pads, and other facilities. All told, more than 22,000 acres of the North Slope is covered in gravel; there are 350 mi (563 km) of gravel roads; one mile of road requires 50,000 cubic yards of gravel. (Phillips Petroleum used more than 1 million cubic yards of gravel to fill in a 100-acre wetland for just one well.) Gravel is mined from open pits along floodplains, river deltas, and riverbanks; gravel mines covered the entire floodplain of the Put River. The FWS estimated that over 60-million cubic yards of gravel have been mined, enough to cover Rhode Island with a 1-in (2.5-cm) layer of gravel.

3. Water Use. The Arctic is arid, getting a mere 3–12 in (8–30 cm) of precipitation per year. Oil exploitation is water-intensive: each year 27 billion gallons of water are used for oil extraction; a single well requires 1.5 million gallons for drilling and another 360,000 gallons for "camp" use. Winter roads are "paved" with 6-in (15-cm) thick ice: one mile of road requires 1–1.5 million gallons of water for "paving," ice helipads need 2–3.6 million gallons; one airstrip gets 8 million gallons of water for ice-paving.

4. Water Pollution. Between 1991 and 1997, 25 billion gallons of toxic pollutants were discharged into surface waters of the North Slope. Saline water (estimated at 16.4 million gallons) from the Prudhoe Bay is drawn into wells, contaminating freshwater resources.

5. Oil Spills. Oil and other contaminant spills are a chronic problem: there are about 400 spills a year of various toxic substances onto the surface; between 1984 and 1993, there were 1,955 crude oil spills (376,321 gallons), 2,390 diesel spills (464,856 gallons), 977 gasoline spills (13,382 gallons), and 1,360 hydraulic fluid spills (77,301 gallons). From 1996 to 2000, more than 1.3 million gallons of oil (crude, diesel) were spilled. In 2000 alone, 18,000 gallons of drilling mud spilled at a BP facility, 9,000 gallons of crude gushed from a ruptured pipeline, destroying a wetland, 92,000 gallons of a salt water/crude oil mixture covered the tundra near a Phillips Petroleum site.

As of 1990, more than 4 million gallons of crude, gasoline, diesel, hydraulic fluid, acids, corrosives, heavy metals, lead, and other toxic chemicals had been spilled on the North Slope. Oil companies admit that they are unable to control or to clean up spills, citing the inclement Arctic weather. Their "spill drills," or clean-up tests, fail repeatedly.

6. Drilling Wastes. Before 1988, about 6 billion gallons of drilling wastes were dumped into 450 unlined reserve pits. After mixing with snow and freezing, during spring melt, the wastewater flows over the tundra as toxic runoff and accumulates in wetlands and ponds. Lawsuits have forced oil companies to abandon surface pits. Today, the waste is injected under the permafrost. To date, less than half the surface reservoirs have been cleaned up. In March 2001, more than 5,000 gallons of drilling waste spilled onto the tundra from a waste injection facility.

—Natalie Goldstein

39 mi (63 km) per gallon (mpg) would save vastly more oil than occurs in all of ANWR. At 39 mpg, we would save a whopping 51 billion barrels by 2050. Just using more efficient tires would save more oil than ANWR can produce. On an annual basis, by 2020 (the time of ANWR's maximum output), if vehicles are getting 39 mi (63 km) to the gallon, we would be saving more than 1.2 billion gallons of oil per year—far more than ANWR yields.

What We Would Lose ANWR is one of our last, great undisturbed wildernesses and the last U.S.-owned remaining intact arctic-subarctic

Senator Joe Lieberman (left), actor Robert Redford (center), and Senator Russ Feingold (right) following a November 2001 press conference in which the men voiced their opposition to oil drilling in the Arctic National Wildlife Refuge.

ecosystem. It is recognized as a key component in an international arctic-subarctic refuge network (including neighboring preserves in Canada, where resource exploitation is forbidden). The FWS studies have shown that oil extraction in ANWR, and particularly in 1002 Area, will have severely negative effects on wildlife.

Caribou. ANWR is home to the Porcupine herd of caribou, which migrates through the refuge and into and out of adjacent preserves in Canada. However, the largest concentration of caribou calving locations lies within 1002 Area. The FWS has shown that oil extraction in this area would reduce suitable calving locations, reduce available forage for caribou, restrict their access to insect-relief habitat on the coast, expose the herd to higher predation, and alter ancient migratory pathways, with unknown consequences for the herd. The FWS concludes that there is no doubt that oil drilling will damage, perhaps irremediably, the Porcupine herd of caribou. A 1987 FWS report also indicates a similar "major" effect (defined as "widespread, long-term change in habitat availability or quality" that would negatively affect wildlife) for ANWR's muskoxen.

Polar Bears. The FWS reports that ANWR and 1002 Area are "the most important land denning habitat for the Beaufort Sea polar bear population." Denning bears are female bears with cubs. Without these denning areas, this population of polar bears would decline, due to increased cub mortality. Development might also result in more deadly bear-human confrontations.

Birds. More than 135 species of birds are known to nest in 1002 Area, including many shorebirds, waterfowl, loons, songbirds, and raptors. Oil drilling would result, at the least, in disturbance of nests, at worst, in destruction of the habitat required by breeding birds for nesting and feeding. Most affected birds include snow geese, seabirds, and shorebirds.

The Land. ANWR is tundra. Permafrost lies beneath it. This land and its vegetation are extremely vulnerable to disturbance and mend slowly or not at all. The land is still scarred, and the vegetation still damaged, where simple seismic studies were conducted more than 15 years ago. Tire tracks leave scars and ruined vegetation that takes decades to recover, if they recover at all. Wildlife cannot survive in this harsh place while waiting for vegetation to recover. And if simple tire tracks leave permanent scars, the damage inflicted by the infrastructure built to sustain oil extraction will do irremediable damage.

Native Peoples. The Inupiat and Gwich'in peoples of northeast Alaska depend on the caribou for survival. (Gwich'in means "people of the caribou.") Not only do these native people have a deep spiritual and cultural attachment to the caribou, but they rely on them for much of their subsistence, from meat to skins for clothes and shelter to tools and trade. The Gwich'in are most closely associated with the Porcupine caribou herd, which is most threatened by oil development. Both the Gwich'in and the Inupiat continue to lobby the U.S. government to abandon plans for oil development in ANWR, and to pressure the Canadian government to oppose it.

The Myth of Footprints To make oil drilling in ANWR more palatable, Congress nominally reduced the size of the area open to "production and support facilities" to 2,000 acres on the Coastal Plain (1002 Area), or a 2,000-acre "footprint." "Footprint" is an oil-industry term used to describe the number of acres actually covered with a layer of gravel to support oil-field infrastructure.

However, this provision excludes "leasing and exploration," which will be permitted on the entire 1.5-million-acre Coastal Plain. Exploration means drilling. Oil companies do not drill anywhere they will be prohibited from extracting profitable oil.

Further, the 2,000-acre limit applies only to "surface" acreage. It does not cover "seismic or other exploratory" activities. Seismic research is done with convoys of bulldozers and "thumper trucks." Oil exploration entails erecting large oil rigs and airstrips for aircraft.

The limitation also does not apply to aboveground pipelines, which, because they do not touch the ground, are not considered as part of the acreage developed! FWS studies show that caribou avoid calving within 2.7 mi (4.4 km) of pipelines and roads, thus greatly expanding the "footprint" impact of development on wildlife.

The impression given by the limitation is that the 2,000 acres will be contiguous and compact. That is false. Nothing prevents smaller, intermittent developments over a vast area of the Coastal Plain. In fact, the USGS has reported that "oil under the coastal plain is not concentrated in one large reservoir, but is spread underneath [it] in numerous small deposits." Developing these widely dispersed deposits would damage an area far greater than 2,000 acres—and would affect wildlife throughout the region.

Conclusion There is oil under ANWR. However, the amount of oil that can be extracted technically and profitably does not begin to approach the amount of oil the U.S. needs to reduce its dependence on imports. Improving fuel efficiency will have a far greater impact on reducing our oil consumption and dependence on foreign oil. Furthermore, it will save a priceless and irreplaceable wilderness that a secretary of the interior described this way: "[ANWR's] wildlife and natural [wilderness] values are so magnificent and so enduring that they transcend the value of any mineral that may lie beneath the surface. Such minerals are finite. Production inevitably means changes whose impacts will be measured in geologic time in order to gain marginal benefit that lasts only a few years." —NATALIE GOLDSTEIN

Further Reading

Arctic National Wildlife Refuge. U.S. Fish and Wildlife Service [cited July 12, 2002]. <http://www.r7.fws.gov/nwr/arctic/index.html>.

"Arctic National Wildlife Refuge: Potential Impacts of Proposed Oil and Gas Development on the Arctic Refuge's Coastal Plain: Historical Overview and Issues of Concern." U.S. Fish and Wildlife Service Report [cited July 12, 2002]. <www.fws.gov>.

"Arctic National Wildlife Refuge, 1002 Area, Petroleum Assessment, 1998." USGS Report [cited July 12, 2002]. </geology.cr.usgs.gov/pub/fact-sheets/fs-0028-01/fs-0028-01.htm>.

Baden, John A. "Drill in the Arctic National Wildlife Refuge?" *Bozeman Daily Chronicle* (31 October 2001).

Chance, Norman. "The Arctic National Wildlife Refuge: A Special Report." *Arctic Circle.* University of Connecticut [cited July 12, 2002]. <http://arcticcircle.uconn.edu/ANWR/>.

Corn, M. Lynne, Lawrence C. Kumins, and Pamela Baldwin. "The National Arctic Wildlife Refuge." *The CRS Issue Brief for Congress.* National Council for Science and the Environment [cited July 12, 2002]. <http://cnie.org/NLE/CRSreports/Biodiversity/biodv-14.cfm>.

Doyle, Jack. *Crude Awakening: The Oil Mess in America: Wasting Energy, Jobs, and the Environment.* Friends of the Earth, 1994.

Lentfer, Hank, ed. *Arctic Refuge: A Circle of Testimony (Literature for a Land Ethic).* Milkweed Editions, 2001.

Mitchell, John G. "Oil Field or Sanctuary?" National Geographic Society [cited July 12, 2002]. <http://magma.nationalgeographic.com/ngm/data/2001/08/01/html/ft_20010801.3.html>.

Natural Resources Defense Council (NRDC) [cited July 12, 2002]. <www.nrdc.org/land/wilderness/anwr>.

NWR News [cited July 12, 2002]. <http://www.anwr.org/index.html>.

The Oil and Gas Resource Potential of the Arctic National Wildlife Refuge 1002 Area,

Alaska. Open File Report 98-34. Staff of the U.S. Geological Survey. Reston, VA: U.S. Geological Survey, 1999.

Petroleum Supply, Consumption, and Imports, 1970–2020 (million barrels per day). National Association of State Energy Officials [cited July 12, 2002]. <http://www.naseo.org/events/winterfuels/2001/presentations/Blake1.pdf>.

Potential Oil Production from the Coastal Plain of the Arctic National Wildlife Refuge: Updated Assessment. Report # SR/O&G/2000-02, U.S. Department of Energy [cited July 12, 2002]. <http://www.eia.doe.gov/pub/oil_ gas/petroleum/analysis_publications/arctic_national_wildlife_refuge/html/execsummary.html>.

"Senator Launches Battle over Drilling in Arctic Refuge." *Inside Politics*. February 21, 2001 [cited July 12, 2002]. <http://www.cnn.com/2001/ALLPOLITICS/02/26/arctic.refuge.drilling/>.

Strohmeyer, John. *Extreme Conditions: Big Oil and the Transformation of Alaska*. New York: Simon & Schuster, 1993.

Trustees for Alaska [cited July 12, 2002]. <www.trustees.org>.

Union of Concerned Scientists [cited July 12, 2002]. <www.ucsusa.org/energy/brf_anwr.html>.

Ward, Kennan. *The Last Wilderness: Arctic National Wildlife Refuge*. Wildlight Press, 2001.

Does rock varnish accurately record ancient desert wetness?

Viewpoint: Yes, desert varnish (rock varnish) may be an accurate indicator of ancient desert wetness.

Viewpoint: No, rock varnish does not accurately record ancient desert wetness because it cannot be dated effectively and its mineral composition cannot exclusively be attributed to climate change.

The subject of desert varnish, or rock varnish, as we shall see, carries with it a controversy concerning the relative levels of wetness that once existed in what are today desert environments. But this particular topic also illustrates several persistent themes in the earth sciences as well. Among these are the long periods of time usually required to bring about any noticeable change in geologic features (that is, aside from dramatic instances of tectonic activity, such as volcanism). There is also the matter of the gradualism that characterizes the transport of solid-earth material from one place to another, and its accretion in one place. Also interesting, from the perspective of the geologic sciences as a whole, is the combination of physical, chemical, and biological processes, discussed in the essays that follow, which bring about the "varnishing" of rocks.

Large outcroppings of rock do not tend to be all of one color, even if the rock itself is of the same mineral makeup. The reason relates to rock varnish, a type of coating that accumulates when rock is exposed to the elements for millions of years. The term *rock varnish* applies to the dark coloration that coats rocks on wet cliffs, on the walls of caves, and in a variety of other locales. One particular kind of rock varnish is known as desert varnish, a term for a dark coating that gathers on the surfaces of rocks in arid regions.

The layer of varnish on desert rocks is extremely thin: usually no more than 200 microns, or 0.00787 inches. Also known as a micrometer, a micron (represented by the Greek letter μ) is equal to one-millionth of a meter, or 0.00004 inches. Despite the fact that this constitutes a membrane-thin coating, desert varnish is easily visible to the naked eye, in part because of the minerals of which it is made.

Desert varnish is composed of various iron oxides, which are reddish, and varieties of manganese oxide, which tends toward slate-gray or black. Also contributing to the coating are various clays, which further the darkening of the rock surface, assuming that the surface remains undisturbed. Naturally, not all surfaces of a rock are exposed for the same amounts of time: one particular face may be much more susceptible to weathering than another, for instance, or heavy winds and rains may remove a chunk of rock from one side of an outcropping, exposing virgin material to new weathering.

Ancient Native Americans and others in dry regions around the planet used desert and rock varnishes as a surface in which to carve out images known as petroglyphs, or rock carvings. Using sharp stones to cut into the darkened rock surface, they carved out light-colored images—using the type of contrast seen in a photographic negative today—to represent hunters, animals, and other scenes. But these man-made pictures may not be the only stories contained in the mineral-varnished surfaces of desert rocks.

There is a body of scientific opinion to the effect that desert varnish on rocks provides evidence of wetness in deserts prior to the beginning of the present geologic epoch, the Holocene, which began at the end of the last ice age about 11,000 years ago. As shown by the essays that follow, there is considerable evidence on either side of the argument, and indeed, the same experts can be cited as proponents of either theory. Thus, both essays make favorable note of a study published in the August 2001 *GSA Today* by Wallace S. Broecker and Tanzhou Liu of Columbia University's Lamont Doherty Earth Observatory.

Critical to understanding the information contained in the desert-varnish record—and the information contained in these essays—is an appreciation of radiometric and other forms of dating used by geologists. The degree to which a given rock is varnished by iron oxides, manganese oxide, clays, and other substances is a form of relative dating, which indicates the age of a rock or other object in relation to other items. Absolute dating, by contrast, involves the determination of age in actual years or millions of years.

Absolute dating methods usually center around the idea that over time, a particular substance converts to another, mirror substance. By comparing the ratios between them, it is possible to arrive at some estimate as to the amount of time that has elapsed since the organism died. Many varieties of absolute dating are radiometric, involving the decay of radioactive isotopes, or radioisotopes, which eventually become stable isotopes.

Each chemical element is distinguished by the number of protons (positively charged particles) in its atomic nucleus, but atoms of a particular element may have differing numbers of neutrons, or neutrally charged particles, in their nuclei. Such atoms are referred to as isotopes. Certain isotopes are stable, whereas others are radioactive, meaning that they are likely to eject high-energy particles from the nucleus over time, until they eventually become stable.

The amount of time it takes for half the isotopes in a sample to stabilize is called its half-life. This varies greatly between isotopes, some of which have a half-life that runs into the billions of years. By analyzing the quantity of radioactive isotopes in a given sample that have converted to stable isotopes, it is possible to determine the age of the sample.

Isotopes are usually identified by the chemical or element symbol, with a superscript number before or after the symbol indicating the combined number of protons and neutrons. For example, one isotope mentioned below as an useful indicator for radiometric dating is beryllium-7, designated as ^{7}Be. A quick glance at the periodic table of elements reveals that beryllium has an atomic number of 4, meaning that there are four protons in the nucleus; thus, the superscript number 7 indicates the presence of three neutrons in the nucleus of this radioisotope. —JUDSON KNIGHT

Viewpoint:

Yes, desert varnish (rock varnish) may be an accurate indicator of ancient desert wetness.

All desert varnish is rock varnish, but not all rock varnish is desert varnish. The very thin (less than 200 microns-thick), but very visible, coating found so commonly in desert regions is called desert varnish. Similar coating is found on wet cliffs and the walls of some caves that are not all in desert regions. The term rock varnish applies to the dark patina found on some rocks, anywhere.

Desert varnish is of particular interest because it may provide an accurate record of ancient desert wetness. Admittedly the adjective "accurate" seems to be too optimistic a term to describe records of climatic events that occurred more than 30,000 years ago. But, it is used here in the context of being as accurate as the dating of the interglacial periods, the time frame in paleoclimatology where records of events are not readily available from other sources.

Researchers Wallace S. Broecker and Tanzhuo Liu at the Lamont Doherty Earth Observatory of Columbia University spent five years studying desert varnish to evaluate its uses as a recorder of paleo-wetness. The results of their U. S. Department of Energy–funded efforts convinced them that "rock varnish does indeed bear an amazing record of wetness in Earth's desert regions."

The researchers examined tens of thousands of varnished rocks in the field and thousands of thin sections in the lab. What they found explains why others before them were unsuccessful in relating desert varnish to ancient wetness. Random sampling rarely provided a reliable full sequence of chemical events over the time the varnish formed. They got consistent results by using extreme care to select samples

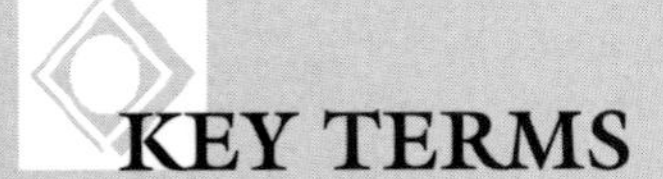

KEY TERMS

ACCRETION: Process of growth or enlargement by gradual build-up.

CATION: A positively charged ion, an atom with one more proton in the nucleus than electrons in the region around the nucleus.

DETRITUS: Loose, organic particles that result directly from disintegration.

GEOMORPHIC: Of or relating to the form of the earth or a celestial body, or its solid surface features.

ISOTOPE: Any of several atoms of an element having different atomic weights as a result of having different numbers of neutrons in the nucleus; radioisotopes are radioactive isotopes.

NUCLIDE: A particular isotope with a specific number of neutrons; identified with the atomic weight as a superscript; example ^{26}Al for aluminum, element 13 which has 13 protons in the nucleus as all aluminum atoms do, but also 13 neutrons which not all aluminum atoms have.

PALEO: Prefix meaning ancient.

PYROLYZED: A substance in which chemical change occurs due to the action of heat.

STRATIGRAPHY: Geology that deals with the arrangement of strata (layers).

SUBSTRATE: Base on which an organism lives; a substance acted upon.

from the most stable rock surfaces and looking only at the varnish-filled microbasins that have a width-to-depth ratio of approximately 10. These selected samples contain the most complete varnish microstratigraphy. Element maps and element line profiles were developed using an electron microprobe to establish patterns of concentration relative to depth as indicators of past regional environmental conditions.

Corroborating Evidence Where the Columbia team used an electron microprobe to study the sedimentary bands in desert varnish that represent the records of micrometers of accumulated elements per millennium, another team used scanning electron microscopy (SEM) to study them. Under the direction of Richard L. Orndorff, Assistant Professor at the University of Nevada, Las Vegas, a team of graduate students used SEM to study laminations that can be identified and energy dispersion spectrometry (EDS) to determine differences in relative abundances of elements which they hypothesized reflect changes in climate. Their samples were taken from boulders along the shorelines left by a paleolake near Fallon, Nevada. All the samples studied had well-developed desert varnish, some with thickness of 200 microns. The team found layers of greater quantities of elements consistent with wind-blown clay that appear to correspond to dry periods, and layers of dark manganese oxide that are believed correspond to humid periods. The conclusion of the team is that sublayers in the microlaminations reflect climate oscillations in the region studied. Their work was presented at an Annual Meeting of the Geological Society of America in May 2002.

Composition of Rock Varnish Clay minerals dominate desert varnish. These are composed of Mg-Al-Si (magnesium-aluminum-silicon) oxides, representative of the weathered rock in the local region. The trapped fine dust particles are submicrometer in size and are not resolved in electron probe maps. Oxides of Mn (manganese), Fe (iron), Ba (barium), and Ca (calcium) are also found in the chemical makeup. The calcium and manganese interrelate, but the iron oxide content of varnish is unchanged through the entire record of the varnish chemistry. The color in rock varnish comes from black manganese oxide (the mineral birnessite) and red iron oxide (the mineral hematite), according to researchers at Caltech's Geological and Planetary Sciences division. NASA scientists at the Ames Research Center (ARC) have also recently documented magnetite, another iron oxide but one that is not as completely oxidized as hematite, a curious discovery in that environment.

The record of manganese oxide, MnO_2, content relates to wetness periods. The Columbia researchers note there is a trend toward higher content during periods of increased precipitation, although they admit "the Mn content in varnish cannot be used as an absolute paleorain gauge." A plot of manganese oxide against annual rainfall for Holocene varnish produces a nonzero intercept. This suggests that manganese enters the varnish from some other source—possibly with dust, aerosols, and dew. Holocene refers to the recent epoch that covers from 11,000 years BP (before the present) to today.

To test their study of manganese oxide as indicators of ancient wetness, in spite of its nonzero intercept, the Columbia research team set up an eight-month experiment in Columbia's Biosphere 2 campus where eight rock samples were studied, four were protected from environmental variables and four were exposed. The varnish was analyzed for the presence of ^{7}Be, a 53-day half life radioisotope produced in the atmosphere by cosmic radiation. The rocks exposed to the Biosphere's environment had marked accumulation of the isotope compared to the sheltered rocks. The test suggested to the team that the nonzero intercept for manganese is the result of immediate environmental conditions.

Barium oxide is in much lower concentration in desert varnish than manganese oxide, but it correlates well to the rise and fall of manganese oxide and produces more conspicuous dark layers in the thin section probes that are interpreted by the team to correspond with glacial time. The Columbia team sees this as reinforcing the idea that desert varnish can be considered as an indicator of ancient wetness.

Origin of Rock Varnish Whereas clay makes up about 70% of desert varnish, it is generally thought the process starts with fine windblown soil deposited as a thin film on the rock, possibly sticking because there is a little dew on it. The composition of the varnish is totally different from the rock under it. Gradually the oxides of manganese and iron that were transported with the soil bind it to the surface so layer upon layer of varnish evolves, and there is evidence that it continues to evolve.

The Columbia team also investigated varnish samples for such things as lead and zinc that have been added to the atmosphere in the last hundred years from the use of leaded gasoline, and from smelters releasing zinc into the environment. Both were found in the outer desert varnish layers. They also tested for nuclides produced during nuclear bomb tests. They were found, too. Lead (a decay product of heavy radioactive elements) was the most convincing. It was found in a concentration ten times higher in outer layers than in records of earlier periods.

There are two scenarios proposed for how rock varnish is formed. The older hypothesis suggests a chemical origin. A newer hypothesis suggested by scientists in the Department of Geosciences at the University of Arizona proposes varnishes are the result of a biogenic deposit produced by mixotrophic bacteria living on rock surfaces. A plot of the rate of varnish production versus moisture, published in 1982 by Dorn and Osborne of the University of Arizona group, indicates semiarid environments are optimal for growth of varnish. Hyper-arid environments lack water for bacterial growth, and varnish growth is indicated as extremely slow or nonexistent on the graph. It has also been suggested that the bacteria live on the energy produced from the oxidation of manganese. The authors of the Columbia University study on the correlation of desert varnish layering features to wet events in ancient times think that enough evidence is not available for them to support the biogenic hypothesis.

There are convincing studies on the presence of bacteria in layered rock varnishes but whether they are the reason for the varnish being there is still in dispute. Studies using transmission electron microscopy (TEM) conducted by geologists David H. Krinsley and Brian G. Rusk at the University of Oregon suggest the presence of bacteria in varnish from images of the proper size and shape for bacteria that are apparently impregnated with Mn and Fe oxides. Their measurements of varnish pH indicate conditions unsuitable for physical and chemical oxidation of manganese, suggesting the bacteria as the cause. Their work was presented at the Second International Conference on Mars Polar Science and Exploration 2000, held August 2000 in Reykjavik, Iceland. Their study also speculated on the possible presence of bacteria in layered rock varnish on Mars.

The Dating Dispute There has been considerable controversy on the direct dating of rock varnish. Interest in direct dating has been intense among those involved with dating ancient rock art, also called petroglyphs, although dating is also very important to paleoclimatology. Petroglyphs represent a record of the very earliest human activities. Pictures were cut into the dark varnish exposing the lighter underlying rock. Paleoclimatology deals with climate before humans began collecting instrumental measurements.

Since the 1980s there have been attempts to directly date rock varnish using what is called cation-ratio dating. For several years this method was hailed as a great breakthrough in dating rock varnish, but as more research was done by a number of different geochemists the method gave inconsistent results. Rock varnish contains very little carbon, and the source of any carbon that is found is uncertain.

Eventually the first scientists to use cation-ratio dating chose to withdraw their conclusions. The end result of the controversy is that no method of direct dating of rock varnish is presently recognized as having much merit by many leading scientists in the field.

The problems of dating came about because the source of carbon used in the cation-ratio dating method (accelerator mass spectrometry dating, AMS ^{14}C) came from carbon compounds and sources that were inadequately identified. The results first produced with this method could not reliably be reproduced by other researchers. The use of $^{232}Th/^{238}U$, a common isotope pair suitable for radiometric dating, is not possible because the ratio of these two isotopes in rock varnish is near unity.

With the problem that rock varnish cannot be directly dated, how can it provide a useful record of ancient desert wetness? The dates are determined by other features in the environment such as raised shorelines, basalt flows, alluvial fans, and moraines. In some places algal deposits from periods of ancient wetness can be dated using radiocarbon methods. Cosmogenic isotopes may also be useful. Some high energy

THE MYSTERIES OF PETROGLYPHS

Just about everything related to petroglyphs is a mystery: Are they symbols representing individual accomplishments, or the beliefs of the carver? Or are they merely idle doodles amounting to nothing more than ancient graffiti?

Carvings are found pecked and chiseled into the varnish coating of desert rocks, on the walls of cliffs, and in caves around the world, wherever ancient people lived. Many petroglyphs have been found in the American Southwest. More than 7,000 have been cataloged in Utah alone. Many of the petroglyphs in the Southwest are thought to have been created by the people of three distinct, though related, cultures which emerged between 100 and 500 A.D.: the Hohokam, the Mogollon, and the Anasazi. These ancient people were thought to have disappeared around 1500 A.D., but it is now believed they are the ancestors of present-day Pueblo Indians.

Rock carvings fall into three general classifications. Any type of human figure is known as anthropomorphic, animal figures are called zoomorphics, and all the rest are described as geometric. Carvings of handprints are common in the Southwest. Are these doodles, or do they stake a claim to a territory? Scenes of hunters pursuing animals are great finds. If the hunter has a spear, the petroglyphs were probably drawn prior to 200 A.D., since in Arizona bows used for hunting have been found around that time. But dating petroglyphs is uncertain at best and depends greatly on where the drawings are found.

Most carvings of zigzags, spirals, and dots are a mystery, although some appear to have celestial meanings. The Anasazi positioned large stone slabs so sunlight would fall on spiral-shaped petroglyphs. Could this be an ancient calendar? Archeoastronomers have found rock carvings that seem to indicate solstices, and some that seem to record planetary movements. The most unusual theory suggests that some petroglyphs are pictures of extraterrestrial visitors.

—M. C. Nagel

cosmic radiation is always bombarding the Earth. It can penetrate meters into rock and produce long-lived radionuclides. The concentration of isotopes is extremely small but the build-up of isotopes over ages can provide a way to date the surface of such features as lava flows.

Rock Varnish Microbes on Mars? The rock varnish record may go beyond the deserts on Earth. No one expects to find petroglyphs carved into rock varnish on Mars, but there is considerable evidence that there are some rocks on Mars that have shiny dark coatings consistent with desert varnish. If NASA's future rovers on Mars corroborate what has been reported by CNN.com (July 2, 2001) that Martian meteorites have high concentrations of manganese, and the information is combined with the known presence of red iron oxides on the surface of Mars (which is why it is called the Red Planet), then there is a real possibility that what appears to be desert varnish *is* desert varnish. The next speculation is that Martian desert varnish could be formed by microbial actions and that would be firm evidence that there is, or was, life on Mars!

Researchers at NASA-ARC report that magnetite has been found on the surface of Mars, as unfavorable a location to find it as the Earth-bound desert varnish where they also found magnetite. This discovery adds credibility to the idea that there is desert varnish on Martian boulders. Plus, there is the information from the Krinsley team presented at the Second International Conference on Mars Polar Science and Exploration on their observation of bacterial presence in layered rock varnish on Earth that could possibly be an indication of the same being present in rock varnish on Mars.—M. C. NAGEL

Viewpoint:

No, rock varnish does not accurately record ancient desert wetness because it cannot be dated effectively and its mineral composition cannot exclusively be attributed to climate change.

The Miniature Universe of Rock Varnish and Its Relation to Ancient Climate Rock varnish is a microscopically thin coating of organic and mineral material that develops, over thousands of years, on exposed rock surfaces in many arid and semiarid environments. By examining samples of these ancient, microscopic universes collected from desert regions around the world, scientists hope to determine the age of rock surfaces and, in particular, to reconstruct climate changes in those deserts dating back to the last interglacial period. The ability to accurately determine ancient desert wetness would provide an important "missing link" in the theory of major climate changes on a global scale. However, by the beginning of the twenty-first century, scientific evidence remained inconclusive, at best, that rock varnish contains an accurate record of desert wetness.

Rock varnish is seldom more than 200 microns thick and samples are collected from tiny varnish-filled dimples, or microbasins, no more than .04 in (1 mm) in diameter in the surface of rocks. Thin slices or cores are cut from different areas of selected dimples for research purposes. Wallace S. Broecker and Tanzhuo Liu of the Lamont-Doherty Earth Observatory, Columbia University, in their article in *GSA Today*, explain that: "Rather than being measured in centimeters of accumulation per millennium, the sedimentary record in varnish is measured in micrometers of accumulation per millennium."

Exactly how rock varnish, often called desert varnish, develops remains uncertain; however, it is composed of layers of clay and deposits rich in minerals that appear to be formed of wind-blown detritus—loose materials such as rock particles and organic debris created by disintegration and destruction. Many researchers hypothesize that rock varnish also may contain a type of manganese-fixing bacteria living on the rock surface that digests organic debris. Other hypothesis relating to its formation also exist.

Dating Rock Varnish Regardless of mechanism of development, rock varnish forms in layers and, by studying these layers, researchers have attempted to determine the periods during which they developed. In one study, Steven L. Reneau and colleagues of the Earth and Environmental Sciences Division, Los Alamos National Laboratory, Los Alamos, New Mexico, collected rock varnish samples from the Cima volcanic field in the Mojave Desert, California. This location was chosen because studies have determined age estimates for a series of lava flows there, and this dating permitted varnish sampling from substrates of similar composition and under similar climatic conditions but of differing ages. In their article in the *American Journal of Science*, these researchers explained that varnish stratigraphy primarily is composed of concentrations of manganese (Mn), iron (Fe), silicon (Si), and aluminum (Al). In earlier work by other researchers, stratigraphy was assessed under the assumption that ancient environmental changes affected the Mn:Fe ratio. However, Reneau and colleagues found that the Mn:Fe ratios in upper layers of their samples appeared also to be affected by factors other than climate changes. Samples taken from the same location, and therefore presumably accreted (formed) under the same environmental conditions, varied widely in major elemental composition. The study therefore cautioned against correlating variations in stratigraphic composition to environmental and climatic changes.

The thickness of rock varnish was also assumed to indicate its relative age, as it tends to become thicker over time. In research published in *Quatenary Research* Reneau investigated samples collected from geomorphic surfaces of different ages in the Soda Mountains of the Mojave Desert. He discovered that varnish accretion on surfaces of similar ages and with similar characteristics could vary greatly over a distance as small as 1 km (approximately half a mile). This undermined the earlier assumption that rock varnish does not erode, but remains stable over the course of eons. He therefore recommended further research before varnish thickness can be correlated to ancient climatic conditions.

In their attempt to date rock varnish, Liu and Broecker studied its accumulation rate. They found a large variance between samples taken from the same site and concluded that "varnish thickness does not correlate with the age of the associated geomorphic feature." They also stated in their article in *GSA Today*: "Based on the examination of tens of thousands of varnished rocks in the field and thousands of thin sections in the lab, it appears to us that randomly sampled varnish surfaces rarely record reliably the full sequence of chemical events that transpired since the geomorphic surface was created."

It is not possible to date rock varnish radiometrically (radiocarbon dating, or ^{14}C). A collaborative investigation by W. Beck and seven other researchers from four different laboratories in the United States and Switzerland, published in *Science*, details the failure of this technique in attempts to date rock varnish. They refer to a study by Ronald Dorn, now at Arizona State University, who used this method. He found a mixture of type I (resembling coal) and type II (resembling pyrolyzed wood charcoal fragments) materials that made radiocarbon dating possible. However, type I materials had previously been dated at approximately 28,000 years old, while type II at only about 4,000 years. These huge age differences in the same minute specimens brought Dorn's research into question. A subsequent independent study by

Rock varnish covers the sandstone wall at Elephant Butte at Monument Valley in Arizona.
(© Tom Bean/Corbis. Reproduced by permission.)

Liu, followed by the collaborative effort of Beck and colleagues, tested samples from the same rock fragments studied by Dorn and found no such materials. Dorn's research and dating technique fell into serious disrepute, as has another of his dating techniques, the cation-ratio method proposed by him in 1983.

The inability to date rock varnish directly makes it virtually impossible to determine when its various layers formed. Therefore, although many researchers think that rock varnish record periods of desert wetness, determining when those periods occurred and correlating samples from different parts of the world to determine climatic events on a global scale is not possible. As Broecker and Liu concluded, rock varnish must be virtually precluded as a relative age indicator in geological and anthropological research.

Why Study Rock Varnish? Scientists believe—but cannot confirm—that certain areas that are now deserts were once much wetter, and some tropical areas much drier, during the glacial periods. Approximately 16,000 years BP (before the present), Earth's climate warmed and created climatic conditions much as we know them today. This warming trend, called the interglacial period, brought an end to the preceding glacial period (ice age) and was followed about 12,000 years BP by another glacial period. During the interglacial period, evidence suggests that Lake Victoria in Africa, dry during the last glacial period, became full; and that Lake Lahontan in the Great Basin of the United States suddenly shrank to ten times smaller than its original size. Scientists are attempting to find solid evidence that climate changes causing these phenomena occurred on a global scale and—because well-dated and detailed records of climate changes in deserts are difficult to find—many look toward the mini-universe of rock varnish for answers.

Flaws in Using Rock Varnish to Assess Ancient Desert Wetness Reneau believes that, although there are many reasons to suspect that stratigraphic variations in rock varnish chemistry should be, at least in part, related to climatic variations, evidence from his research into this area was inconclusive. Also, while Broecker and Liu firmly think rock varnish has the potential to confirm the theory of global climate change, they explain why available evidence is flawed:

First, scientists hypothesize that the higher the Mn content in varnish layers, the greater the wetness during the period of that particular layer's development. They admit, however, that this hypothesis remains unproven.

Second, precipitation may not be the only ingredient in the mix that causes rock-varnish development. While rainfall is the primary contributor, dust, dew, and aerosols also appear to be influencing factors.

Third, randomly sampled varnish seldom records reliably the full sequence of chemical events that have occurred from the time the surface on which it grows was created.

Fourth, as Reneau discussed, random samples are often adulterated by erosion, peeling, and "solution" events. During the nineteenth and twentieth centuries, a plethora of substances produced by humans have been released into the atmosphere that may change the chemical content and growth rate of rock varnish. For example, while aerosols are suspected to enhance varnish growth, acids threaten to dissolve accumulated varnish. Lack of careful selection of samples, therefore, may yield inaccurate or misleading results.

Fifth, methods of dating rock varnish have proven, at best, unreliable.

Concluding their article, Broecker and Liu ask the question: "Will rock varnish ever become a widely applied proxy for desert wetness?" They answer no—at least, they suspect, for the first decade of the twenty-first century, because too many unanswered questions remain pertaining to its formation and how its chemical composition relates to local environmental changes. —MARIE L. THOMPSON

Further Reading

Beck, W., et al. "Ambiguities in Direct Dating of Rock Surfaces Using Radiocarbon Measurements." *Science* 280 (June 1998): 2132–39.

Broeker, Wallace S., and Tanzhuo Liu. "Rock Varnish: Recorder of Desert Wetness?" *GSA Today* (August 2000): 4–10.

Dorn, R. I. "Uncertainties in the Radiocarbon Dating of Organics Associated with Rock Varnish: A Plea for Caution." *Physical Geography* 17 (1996): 585–91.

Liu, Tanzhuo, and Wallace S. Broeker. "How Fast Does Rock Varnish Grow?" *Geology* (2002): 183–86.

Lunar and Planetary Institute [cited July 19, 2002]. <http://www.lpi.usra.edu>.

National Oceanic and Atmospheric Administration Paleoclimatology Program [cited July 19, 2002]. <http://www.ngdc.noaa.gov/paleo/paleo.html>.

Reneau, Steven L. "Manganese Accumulation in Rock Varnish on a Desert Piedmont, Mojave Desert, California, and Application to Evaluating Varnish Development." *Quaternary Research* 40 (1993): 309–17.

Reneau, Steven L., et al. "Elemental Relationships in Rock Varnish Stratigraphic Layers, Cima Volcanic Field, California: Implications for Varnish Development and the Interpretation of Varnish Chemistry." *American Journal of Science* 292 (November 1992): 684–723.

"What Is Desert Varnish?" [cited July 19, 2002]. <http://minerals.gps.caltech.edu/FILES/VARNISH/Indes.html>.

Are the late Precambrian life forms (Ediacaran biota) related to modern animals?

Viewpoint: Yes, some species within the Ediacaran biota of the late Precambrian are the predecessors of modern animals.

Viewpoint: No, the late Precambrian life forms (Ediacaran biota) are not related to modern animals.

In 1946, Australian geologist Reginald Sprigg discovered fossilized remains of what turned out to be creatures from 544 to 650 million years ago. Their significance lay not merely in the fact that these were some of the oldest fossils ever discovered, but in their origins from what geologists call Precambrian times—long before the rise of most animal phyla found on Earth today. Scientists dubbed the creatures whose fossils Sprigg had discovered, which were apparently jellyfish-like forms in the shape of disks, Ediacara, after the Ediacara Hills where Sprigg had found them.

Other discoveries of Ediacara followed, most notably in Namibia, in southwestern Africa. This led to a great debate concerning the meaning of the Ediacara: were they completely different life forms, unrelated to animals that existed during the present geological era? Or were they precursors of modern creatures? Either answer, of course, is scientifically interesting, and strong arguments for each are made in the essays that follow. Before diving into those arguments, however, it is worthwhile to consider a few basics as to geological time, stratigraphy, relative and absolute dating, and taxonomy.

The expression *geological time* refers both to the extremely long span of Earth's existence, and to the ways that this is measured. Such measurement dispenses almost entirely with the terms to which most people are accustomed to using for the measurement of time. Years, centuries, even millennia are too minuscule to have any meaning within the context of the planet's 4.5 *billion*-year history. Therefore, geologists speak usually in terms of millions of years ago or billions of years ago, abbreviated Mya and Gya respectively.

Divisions of geological time, however, are not usually rendered in these absolute terms, but in relative terms, according to information gathered by studying rock layers, or strata. From such information, geologists have developed a time scale that uses six basic units, which are, in order of size (from longest to shortest): eon, era, period, epoch, age, and chron. None of these has a specific length, nor is there a specific quantity of smaller units that fits into a larger one. The only rule is that at least two of a smaller unit fit into a larger one; furthermore, the longer the unit, the greater the significance of the geological and/or paleontological (i.e., relating to prehistoric life forms) events that circumscribe that period. (For example, the end of the Mesozoic era before modern Cenozoic is one marked by the extinction of the dinosaurs, probably as the result of a meteorite impact on Earth.)

The term *Precambrian* encompasses about 4 billion years of Earth's history, including three of the four eons (Hadean or Priscoan, Archaean, and Proterozoic) of the planet's existence. Today we are living in the fourth eon, the Phanerozoic, which began about 545 Mya ago. Paleontologists know vastly more about the life forms of this eon than about those of any preceding one, though extremely simple life forms did exist prior to the Phanerozoic. However, to leave fossilized remains, a creature must have "hard parts" such as bones or teeth, and few life forms of Precambrian time met those qualifications.

Hence the significance of the Ediacaran biota, which left fossilized remains dating back to before the Phanerozoic eon. (The term *biota* refers to all the flora and fauna—plants or animals—in a given region. As we shall see in the essays that follow, this term is particularly advisable in the present context, since paleontologists are not certain whether Ediacaran life forms were plant, animal, or members of some other kingdom.) The Ediacaran biota dates from the latter part of the Proterozoic eon, in the Vendian or Ediacaran era. The latter preceded the Cambrian period, which began the Paleozoic era of the present Phanerozoic eon—thus explaining the significance of the term "Precambrian."

How do we know the age of the formations in which the Ediacaran biota were found? Through a combination of absolute and relative dating. Relative dating methods assign an age relative to that of other items, whereas absolute dating involves the determination of age in actual years or millions of years, usually through the study of radioactive isotope decay.

The principles of biostratigraphy, a subdiscipline concerned with the relative dating of life forms, were first laid out by English engineer and geologist William Smith (1769–1839). While excavating land for a set of canals near London, Smith discovered that any given stratum contains the same types of fossils, and strata in two different areas can thus be correlated or matched. This led him to the law of faunal succession, which states that all samples of any given fossil species were deposited on Earth, regardless of location, at more or less the same time. As a result, if a geologist finds a stratum in one area that contains a particular fossil, and another in a distant area containing the same fossil, it is possible to conclude that the strata are the same in terms of time period. This principle made it possible for paleontologists studying Ediacaran forms to relate the fossils found in Namibia, Australia, and other places.

In addition to the principles of geological and paleontological time classification at work in the study of Ediacara, there are the concepts of biological classification embodied in the discipline of taxonomy. As with the geological time-scale, there are a number of levels of taxonomic groups or taxa, from kingdom down to species, though we are concerned here only with the two largest significant groupings, kingdom and phylum. The number of smaller groupings within a larger group is, as with geological time periods, not fixed: for example, it is conceivable that a phylum could have just one species, if that species were so significantly different from all others as to constitute a distinct category. (This can easily be understood by reference to the taxonomy of political entities: though New York City is much larger than the Vatican City, the latter happens to be its own country, whereas New York is a city within a state within a country.)

Ediacara may or may not represent a distinct phylum, depending on the answer to the question of whether they were precursors to modern life forms. Even more fundamental, however, is the question of whether they were plants or animals. Although *plant* and *animal* are common terms, there are no universally accepted definitions. One of the most important characteristics of plants is the fact that they are capable of generating their own nutrition through the process of photosynthesis, or the conversion of electromagnetic energy from the Sun into chemical energy in the plant's body. Animals, by contrast, must obtain nourishment by consuming other organisms. This is a much more significant distinction than some of the most obvious-seeming ones, such as the ability or inability of the organism to move its body: thus sponges, which might seem like plants, are actually animals. —JUDSON KNIGHT

Viewpoint:

Yes, some species within the Ediacaran biota of the late Precambrian are the predecessors of modern animals.

Some species within the Ediacaran biota of the late Precambrian are the predecessors of modern animals. Although these soft-bodied organisms have occasionally been described as a "failed experiment" that died off before the Cambrian explosion, new fossil finds and scientific analyses indicate otherwise. Recent studies now show that the Ediacaran fauna not only persisted far longer in geologic time than originally thought, but also may well have given rise to many of the animals that currently live on Earth.

The notion that the Ediacaran animals are ancestors of modern life is not new. Discovered in Australia a century ago, the Ediacaran biota remained an enigma for about 50 years. Then paleontologist Martin Glaessner of the University of Adelaide asserted that they may be the missing link in the tree of life. Scientists had long pondered the evolutionary origins of the organisms associated with the Cambrian explo-

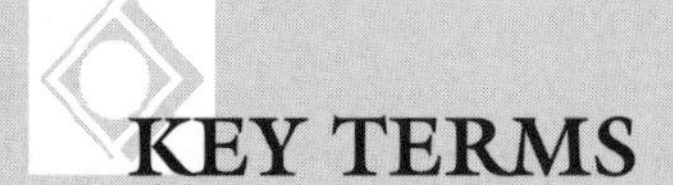

KEY TERMS

ARTHROPOD: The largest group of animals in the animal kingdom, characterized by a outer skeleton (exoskeleton) and jointed body parts (appendages). These include spiders, scorpions, horseshoe crabs, crustaceans, millipedes, centipedes, and insects.

CAMBRIAN: The earliest period of the Paleozoic era, between 544 and 505 Mya. It is named after Cambria, the Roman name for Wales, where rocks of this age were first studied.

CNIDARIA: Jellyfish, corals, and other stingers. The name comes from the Greek word *cnidos*, for stinging nettle. Cnidarian forms are very diverse, but all are armed with stinging cells called nematocysts. Thousands of species live in the oceans, from tropics to poles and from surface to bottom; a smaller number live in rivers and fresh-water lakes.

EDIACARAN FAUNA (ALSO VENDIAN BIOTA): Fossils first found in Australia's Ediacara Hills and later worldwide that represent the first true multicellular animals at the end of the Precambrian era. They have been variously considered algae, lichens, giant protozoans, and a separate kingdom of life unrelated to anything now living.

PHANEROZOIC: The time between the end of the Precambrian and today. The Phanerozoic (also called an eon) starts with the Cambrian period, 544 Mya, and encompasses the period of abundant, complex life on Earth.

PRECAMBRIAN: All geologic time before the beginning of the Paleozoic era. This includes 90% of all geologic time—from the formation of Earth, about 4.5 billion years ago, to 544 Mya. Its name means "before Cambrian."

PROTEROZOIC ("EARLY LIFE"): The last era of the Precambrian, between 2.5 billion and 544 Mya. Fossils of both primitive single-celled and more advanced multicellular organisms begin to appear in abundance in rocks from this era.

STRATIGRAPHY: The branch of geology concerned with the formation, composition, ordering in time, and arrangement in space of sedimentary rocks.

SYNAPOMORPHY: A shared (derived) trait, which suggests that two organisms are related evolutionarily.

TRACE FOSSILS: Record of the past behavior of organisms and include tracks, trails, burrows, and any other mark made by an animal walking, running, crawling, resting, or feeding on or within soft sediment such as sand or mud. Plants also leave traces.

VENDIAN: The latest period of the Proterozoic era, between 650 and 544 Mya, and sometimes called the Ediacaran period. The Vendian is distinguished by fossils representing a characteristic collection of complex soft-bodied organisms found at several places around the world.

sion, a period of spectacular diversification. Nearly all basic body plans appear during the Cambrian period which began about 540 million years ago (540 Mya), but where did they come from? Glaessner answered that the Ediacaran biota was the logical place to look. They represented the first definite multicellular animals and were prevalent in the late Precambrian era (also known as the Vendian Period). The scientific community embraced the idea and declared the Ediacaran biota as the forebear of modern animal life. A few vocal dissenters, including such outspoken scientists as Adolf Seilacher and Stephen Jay Gould, have challenged Glaessner's ideas, and his hypothesis has lost favor over the past two decades.

Now, however, paleontologists are finding more and more fossils that lived further and further into the Cambrian Period, and the view is shifting again toward the notion that at least some of the Ediacaran organisms were indeed the ancestors of modern life.

Overlap in Time One of the biggest arguments against the Ediacaran-ancestor hypothesis was that these odd-looking organisms became extinct several tens of millions of years before the Cambrian era, and therefore could not have given rise to modern phyla. Although many scientists suspected that at least some Ediacaran animals survived into the Cambrian, fossil proof was lacking. MIT geologist John Grotzinger is quoted in the *Discover* magazine as saying, "In sections around the world, you could walk up through a succession of layers that contained Ediacaran fossils, and then you wouldn't see any Ediacaran fossils for a long,

At Mistaken Point on the Avalon Peninsula, the Ediacaran animals were covered by volcanic ash.
(© James L. Amos/Corbis. Reproduced by permission.)

long time—I should say a great, great thickness—and then you would see the early Cambrian fossils." The gap suggested a lengthy separation in time between the occurrence of the two groups of organisms. The tide turned suddenly in 1994 when Grotzinger's research group went to a site in Namibia, which is located in southwestern Africa, and discovered a swath of Ediacaran fossils mingling with early Cambrian fossils, indicating that the two coexisted. Precise dating techniques verified the extension of Ediacaran animals into the next geologic period. Scientists now had solid evidence that the Ediacara lived during the time frame spanning 600–540 Mya, pushing their survival at least into the beginnings of the Cambrian explosion.

Faced with this new evidence, scientists realized that they might have been too hasty in their characterizations of the geological strata. Perhaps the Ediacaran animals coexisted with early skeletonized (shelled) animals in other areas, too, but simply left no trace in the rock. After all, the creatures were soft-bodied with no hard parts, and were therefore ill-suited to preservation. Most Ediacaran remains are merely impressions in sandstone beds. In addition, the Cambrian period appears to have ushered in the first predators, which would have greatly decreased the chances that an Ediacaran animal would have died intact. A digested Ediacaran creature leaves no trace in the fossil record. Simon Conway Morris of the University of Cambridge has gone so far as to say that skeletons arose in the Cambrian in response to pressure from predators.

Other scientists note that the disappearance of Ediacaran animals from the fossil record was likely caused by the disappearance of microbial mats. Explained James Gehling of the University of California—Los Angeles, "A form of 'death mask' resulted from bacterial precipitation of iron minerals in the sand that smothered decaying microbial mats and megascopic benthic organisms." He suggested that the mats provided a defense against erosion, while triggering mineral encrusting and sand cementation. As grazing organisms evolved and attacked the benthic mat communities, they began to disappear. Without the mats, the Ediacaran animals subsequently biodegraded without preservation.

Despite the poor odds, however, some Ediacaran animals were fortuitously preserved in the Cambrian strata. Preservation was typically a combination of the animals' habitat on the muddy sea bottom and the sudden deposition of sandy debris atop them. At Mistaken Point, Newfoundland, where a massive fossil bed is located, the Ediacaran animals were covered by volcanic ash. One scientist has even dubbed the Canadian site an "Ediacaran Pompeii" in reference to the volcanic ash from the eruption of Mount Vesuvius that buried and preserved the people of Pompeii back in A.D. 79.

Grotzinger's Namibia discovery was the first evidence of Ediacaran persistence past the Precambrian-Cambrian border, but it was not the last. Among a string of findings, James W. Hagadorn of the California Institute of Technology, and Ben Waggoner of the University of Central Arkansas, reported in 2000 that the Ediacaran fauna from the southwestern United States also survived to near the base of the Cambrian. They found numerous Ediacaran animals, including a variety of tubular Ediacaran organisms and a frond-like form known as *Swartpuntia*. Specifically, they noted that *Swartpuntia* persisted "through several hundred meters of section, spanning at least two trilobite zones." Scientists have also reported Cambrian Ediacaran fossils from numerous sites, including the famed Burgess Shale, a fossil bed located in the Canadian Rockies.

Odd-looking Ancestors The second prominent argument against the Ediacaran-ancestor hypothesis is simply that the Ediacaran animals are just too different from modern metazoan (multicellular animal) groups to be related. In the past decade, however, many research teams have pointed out enough similarities between specific pre-Cambrian and Cambrian forms to pronounce at least a few Ediacaran animals to have likely given rise to modern metazoans.

Classification of organisms is a tricky business, even among the extant species. When scientists are dealing with fossils, taxonomy is more difficult. When the fossils are as rare and incompletely preserved as the Ediacaran animals, it can be excruciating. However, the Ediacaran specialist Waggoner cautions against taking the easy route and placing unusual or incomplete organisms in new taxa that are unrelated to anything else. "[P]roblematic fossils must be incorporated into phylogenetic analyses. By definition, their synapomorphies (features that are shared) with other taxa will not be obvious; these synapomorphies must be searched for and documented," he recommended in a 1996 paper. For example, he cited the argument that Edicarans could not have given rise to modern phyla because many Ediacaran animals appear to exhibit glide reflectional symmetry and extant animals do not. In this type of symmetry, the organisms have mirror-image right and left sides, with one side slightly shifted so the two sides are off-kilter. Most modern animals have bilateral symmetry, with directly opposite right and left halves. Waggoner argues that the glide reflectional symmetry seen in some Ediacaran fossils may actually be just a result of shrinkage or distortion, and even if it isn't, that type of symmetry is not unknown among more evolutionary advanced animals, including Cambrian specimens from the Burgess Shale. In addition, he points out that while modern animals may not exhibit glide reflectional symmetry in their overall body plan, some do have such symmetry in their gross anatomy. In sum, the mere presence of glide reflectional symmetry does not preclude an organism from being a metazoan.

Other litmus tests have been proposed and later discounted. Adolf Seilacher, who challenged Glaessner's original Ediacaran-ancestor hypothesis, proposed that Ediacaran animals did not have mouths. Waggoner countered that at least two Ediacaran species, *Marywadea* and *Praecambridium*, clearly show digestive structures akin to those in arthropods. A third, *Metaspriggina*, also shows indications of a digestive tract.

Even Seilacher, who has disputed many Ediacaran-Cambrian species links, has proposed the existence of late-Precambrian organisms he calls *Psammocorallia*, which are sand-filled, cnidarian coelenterates. The fossil record also shows a diversification of worm-like Edicaran animals, such as *Helminthopsis*, *Cloudina*, and *Dickinsonia*, at the end of the Precambrian period. In a review of Vendian fossils found in the former Soviet Union and Mongolia, paleontologists Alexei Rozanov and Andrey Zhuravlev pointed out that organic "skeletons" appeared in Precambrian organisms, including *Redkinia* and *Sabellidites. Cloudina* is of particular interest since its fossil tubes show evidence of calcification. With the ability to precipitate calcium carbonate, this worm exhibited perhaps an early step toward the development of skeletal elements. Another species, the sedentary *Tribrachidium*, also shows signs of a mineralized skeleton.

Other late-Precambrian, or Vendian, organisms show marked similarities with animals of the Cambrian period. *Charnia*, a large and flat Ediacaran animal, had synapomorphies with the pennatulaceans, or sea pens, which are colonial coral-like organisms. *Eoporpita* is an animal with radially oriented tentacles that may demonstrate a phylogenetic relationship to Coelenterates, which also show radial symmetry. The disk-shaped *Arkarua adami* exhibits some primitive traits akin to echinoderms, which encompass the spiny, marine invertebrates. Several geologists and biologists believe the small, disk-shaped *Beltanelliformis* has ties to the sea anemones, because it is "strikingly similar to the base of some anemones and especially to the bases of Paleozoic burrows generally attributed to anemones," said geologist Guy Narbonne of Queen's University in Ontario, Canada. Likewise, other Ediacaran animals show some of the characteristics of trilobites or of sponges.

Several species show indications of relationships to arthropods. *Spriggina* had the head shield and segmented body that make it a likely arthropod ancestor. *Parvancorina* exhibited segmentation, a shield-like, ridged carapace, and at least 10 paired appendages as well as more than a dozen thinner appendages toward the rear. Noted Narbonne, "The similarity of *Parvancorina* to the Paleozoic arthropods of the *Marellomorpha* group may indicate that it is close to the ancestors of the Crustacea."

One of the most compelling Ediacaran fossils is *Kimberella*. Originally thought to be a box jellyfish, researchers now describe it as "a mollusc-like bilaterian organism" complete with a shell, a foot like the one that molluscs use for locomotion, and possibly a mouth. The authors of the research assert, "We conclude that *Kimberella* is a bilaterian metazoan ... plausibly bearing molluscan synapomorphies such as a shell and a foot.... This interpretation counters assertions that the Ediacaran biota represents an extinct grade of non-metazoan life."

Failed Experiment? Did the Ediacaran animals give rise to modern metazoans? The fossil evidence and scientific comparisons say that some did. Many species likely died off before or near the beginning of the Cambrian, some perhaps succumbing to predation or competition with the new species associated with the Cambrian explosion. Many others, however, survived and became the rootstock for modern animals. The research has clearly shown that the Ediacaran animals were in the right place at the right time to spark the Cambrian animal diversification, and exhibit striking similarities to some important features of Cambrian metazoans.

Although the geologic record is less than perfect and in some cases downright scant in the case of Ediacaran organisms, fossil finds and subsequent analyses have provided ample fodder against discounting the role of these soft-bodied Vendian animals in the Cambrian diversification. The link between the Ediacaran and Cambrian biota implies that the animal "explosion" so long described as a strictly Cambrian event may have started long before with the Ediacaran biota. Without a doubt, the debate over the role of the Ediacaran fauna in the evolutionary tree of life will persist for many years, and scientists will continue their quest for new fossils, additional insights through scientific investigations, and ultimately a better view into the past. —LESLIE MERTZ

Viewpoint: No, the late Precambrian life forms (Ediacaran biota) are not related to modern animals.

Before 1946, a huge gap existed in the fossil record. There seemed to be no complex animal fossils at all until the start of the Cambrian period, about 544 Mya, when nearly all animal phyla now on Earth appeared in a relatively short time span. Because of this gap, many people thought either that no animals evolved before the Cambrian, or that Precambrian animals couldn't leave fossils because they had no shells or hard skeletons.

Ancient Life Revised Then, in 1946, while exploring an abandoned mine in the Flinders Range of mountains near Adelaide, Australia, in the Ediacara (an Aboriginal expression meaning "vein-like spring of water") Hills, mining geologist Reginald Sprigg found fossilized imprints in quartzite and sandstone of what looked to have been a collection of soft-bodied organisms. Some were disk-shaped and looked like jellyfish, others resembled segmented worms, and some had odd, unrecognizable forms.

At first Sprigg thought the Ediacara fossils might be from the Cambrian (named for Cambria, the Roman name for Wales, where rocks of this age were first studied). But later work showed them to predate the Cambrian by about 20 million years or more. The fossils dated from the late Precambrian, between 650 to 544 Mya—a period variously known as the latest Proterozoic, the Vendian, or the Ediacaran.

The Proterozoic (meaning "early life") was the last era of the Precambrian, between 2.5 Bya and 544 Mya. Fossils of primitive single-celled and more advanced multicellular organisms appeared abundantly in rocks from this era. The Vendian (sometimes called the Ediacaran) was the latest period of the Proterozoic era, also between 650 and 544 Mya.

The first Ediacaran grouping was recorded, although not by that name, in Namibia (near Aus) in 1933, and reports of other Precambrian soft-bodied fossils had appeared in the scientific literature from time to time beginning in the mid-nineteenth century. But the fossils Sprigg found in the Ediacara Hills were the first diverse, well-preserved assemblage studied in detail. Since his discovery, paleontologists have found Ediacaran-age fossils at more than 30 places around the world, on every continent except (so far) Antarctica.

The best known Ediacara locations are Namibia, the Ediacara Hills in southern Australia, Mistaken Point in southeast Newfoundland, and Zimnie Gory on the White Sea coast of northern Russia. Fossils have also been found in Mexico, central England, Scandinavia, Ukraine, the Ural Mountains, Brazil, and western United States.

Some Ediacaran fossils are simple blobs that could be almost anything. Some are like jellyfish, worms, or perhaps soft-bodied relatives of arthropods. Others are harder to interpret and could belong to extinct phyla. Vendian rocks also hold trace fossils, possibly made by segmented worm-like animals slithering over sea mud. Trace fossils record past behavior and can include tracks, trails, burrows, and any other mark made by an animal or plant.

Analyzing trace fossils (a discipline called *ichnology*) has been called the Sherlock Holmes approach to paleontology. Even worms, whose bodies have no hard parts to become fossilized, can inscribe into sediments many kinds of trace, track, trail, and burrow patterns. Such patterns have been important in interpreting the Ediacaran fossils, because they tell about the nature of original surfaces and environments, and help paleontologists reconstruct the behavior and evolutionary transformations of soft-bodied life forms.

Today most paleontologists divide Ediacara fossils into four general groups, but there is no consensus on whether the fossils are animals, modern or otherwise. The groups are i) jellyfish or sea anemones, ii) frond-like organisms, iii) trace fossils of possible shallow burrows in mud, and iv) unusual forms.

After decades of research and investigation, paleontologists, paleobotanists, and paleogeologists still do not agree about what this group of fossils represents. Over time the fossils have been theorized to be ancestors of modern animals, algae, lichens, giant protozoans, and a separate kingdom of life unrelated to anything alive today.

Default Assumption From the first discovery of Ediacaran fossils in Australia, many investigators assumed that because the Ediacarans were preserved together and had certain similar characteristics, they must be the same sort of life form—that all the Ediacara fossils are members of the same high-level taxon and must fit the same plan of construction. But this default assumption is starting to crumble. According to Ben Waggoner, assistant professor of biology at Central Arkansas University in Conway, the long-standing debate about the nature of Ediacaran biota no longer has two sides—it has several.

Mistaken Point In 1967 geologist S.B. Misra discovered great numbers of unusual fossils of late Precambrian age on exposed rock surfaces

Michael Anderson shown with Precambrian fossils embedded in a rock formation along the Avalon Peninsula in Newfoundland.
(© James L. Amos/Corbis. Reproduced by permission.)

along the southern coast of Newfoundland's Avalon Peninsula. The most famous area is Mistaken Point, a crag at the southern tip of the Peninsula where more than 50 ships have wrecked. The rock slabs Misra saw held many imprints of soft-bodied organisms. Some resembled Ediacaran fossils found elsewhere, especially Charnwood Forest in central England, but most were unique to Mistaken Point and didn't look like any known life form.

The Mistaken Point fossils are also unique because they were preserved in large numbers in layers of fine volcanic ash, which created snapshots of the sea floor at the time they were preserved. And unlike most other Ediacaran biota, the Mistaken Point life forms seem to have lived in deep water, far below sunlight or surface waves.

Typical Mistaken Point biota were large frond-like leafy forms: some had stalks; some were bushy or cabbage-like; others had branching, tree-like or network-like shapes; and some looked like spindles and were long and pointed at both ends. There were also lots of large, lumpy disk-shaped fossils. Thanks to dating techniques for volcanic ash, the Mistaken Point fossils were shown to be 565 million years old, so far the oldest complex Ediacara fossils to be accurately dated.

There is at least one argument for the non-animal nature of the Mistaken Point fossils. Some Ediacara sites have trace fossils, which represent unambiguous evidence that animals were present at the time, although in most cases paleontologists can't link the trace fossils to specific animal body fossils. One exception is Newfoundland—not a single trace fossil has been found at the Ediacara sites there. This could mean that none of the fossil forms discovered there were living animals.

Quilted Organisms and Metacellularity By the 1980s paleontologists had described more than two dozen species from around the world, along with tracks on the sea floor, and more detailed studies and reconstructions of the life forms showed that some were not similar to jellyfish, corals, or worms.

At that time at least two researchers—German paleontologist Adolf Seilacher, a professor at Yale University and the Geological Institute at Tübingen University in Germany, and Mark McMenamin, a geology professor at Mount Holyoke College in Massachusetts—argued that the Ediacarans were unrelated to any living organisms and represented a new kingdom that fell victim to a mass extinction at the Vendian-Cambrian boundary.

Seilacher hypothesized that they were a failed experiment of evolution unrelated to modern animals. He called them vendobiota—a name some still use for the Ediacaran biota—and described them as large, thin, quilted air-mattress-like life forms.

Some experts, including Simon Conway Morris, professor of evolutionary paleobiology at the University of Cambridge, find the quilted Ediacaran structure to be unlike any animal form in existence today, and say it is "genuinely difficult to map the characters of Ediacaran fossils onto the body plans of living invertebrates; certainly there are similarities, but they are worryingly imprecise."

Because the fossil forms had no mouth, teeth, or jaws for biting or chewing, and no gut or way to eliminate waste, Seilacher suggested they took nutrients and handled wastes by diffusion from and into the sea water they lived in, or from internal organisms that lived by photosynthesis (drawing energy from sunlight) or chemosynthesis (drawing energy from chemical reactions). Mark McMenamin created the phrase "garden of Ediacara" to embrace this concept.

In a 1998 book of the same name, McMenamin theorized that the Ediacaran organisms were multicellular and related to animals, but not to those in existence today. According to his hypothesis, their body organization—a simple symmetry that divided their bodies into two, three, four, or five parts—was established early in development by individual cells that gave rise to separate populations of cells, which then grouped together to form the organisms.

Ancient Lichens But not everyone thinks the Ediacaran fossils necessarily belong to animals, extinct or otherwise. In 1994, University of Oregon paleobotanist Greg Retallack published an article in *Paleobiology* proposing that the Ediacarans were huge ancient lichens—symbiotic groups of algae and fungi—and that at the time there were "great sponges of vegetation all over, draping the landscape."

Retallack compared their thickness to that of much younger tree trunk fossils, and said the Ediacaran fossils seemed to have resisted compaction after they were buried, much like sturdy logs. Several features suggested that they were lichens—their size (up to 3.3 ft [1 m] across) and sturdiness; the lack of a mouth, digestive cavity, or muscle structure; and evidence from microscopic examination.

Rock lichens do live in the sea (shallow-water lichens) and on land, but some paleontologists who otherwise find Retallack's notion plausible for some of the Ediacaran life forms question whether lichens could have survived in the deep-sea Vendian deposits, like those found at Mistaken Point, where the deep sea environment was dark, rich in methane or sulfur, and low in oxygen.

Continuing Debate Paleontologists have debated the nature of the Ediacara fossils over the years since the fossils were discovered for two reasons: No one knows how or whether they are related to the Cambrian explosion of life forms, and no one knows how this group of apparently soft-bodied organisms came to be preserved in the first place. These questions have yet to be answered to everyone's satisfaction.

Even today, some paleontologists think at least some of the Ediacaran fossils could be related to Early Cambrian life forms and so to modern animals, and some agree with Seilacher and others that the fossils represent a failed experiment of evolution.

The question about preservation holds a potential argument for the failed-experiment view. If the Ediacaran life forms are unrelated to modern animals, their bodies may have been made of cells and tissues that differed substantially, and were more enduring, than those of soft-bodied creatures alive today.

Some taxa (named groups of organisms) are now known to have lived into the Cambrian period—since 1990 there has been a growing number of Cambrian-age discoveries of Ediacaran fossils from Cambrian deposits—and others may have evolved into different forms.

But most Ediacarans vanished from the fossil record near the beginning of the Cambrian. Some experts think this is evidence of mass extinction; others propose that conditions in the environment, especially the appearance of hard-shelled predatory animals with claws that could crawl and burrow, may have contributed over time to their decline. —CHERYL PELLERIN

Further Reading

Conway, Morris S. "Ediacaran-like Fossils in Cambrian Burgess Shale-type Faunas of North America." *Palaeontology* 36 (1993): 593–635.

"The Ediacara Biota: Ancestors of Modern Life?" Part of the online exhibit *The Dawn of Animal Life*, from the Miller Museum of Geology, Queen's University, Kingston, Ontario, Canada [cited July 19, 2002]. <http://geol.queensu.ca/museum/exhibits/ediac/ediac.html>.

Fedonkin, M.A., and B. M. Waggoner. "The Late Precambrian Fossil *Kimberella* is a Mollusc-like Bilaterian Organism." *Nature* 388 (August 28, 1997): 868–71.

Hagadorn, J. W., and B. M. Waggoner. "Ediacaran Fossils from the Southwestern Great Basin, United States." *Journal of Paleontology* 74, no. 2 (March 2000): 349–59.

Hagadorn, J.W., C. M. Fedo, and B. M. Waggoner. "Early Cambrian Ediacaran-type

Fossils from California." *Journal of Paleontology* 74, no. 4 (July 2000): 731–40.

"Introduction to the Vendian Period" [cited July 19, 2002]. <http://www.ucmp.berkeley.edu/vendian/vendian.html>.

Lipps, J. H., and P. W. Signor, eds. *Origin and Early Evolution of the Metazoa.* New York: Plenum Press, 1992.

MacNaughton, R. B., G. M. Narbonne, and R. W. Dalrymple. "Neoproterozoic Slope Deposits, Mackenzie Mountains, NW Canada: Implications for Passive-Margin Development and Ediacaran Faunal Ecology." *Canadian Journal of Earth Sciences* 37 (2000): 997–1020.

"The Mistaken Point Fossil Assemblage." Part of the online exhibit *The Dawn of Animal Life,* from the Miller Museum of Geology, Queen's University, Kingston, Ontario, Canada [cited July 19, 2002]. <http://geol.queensu.ca/museum/exhibits/ediac/mistaken_point/mistaken_pt.html>.

Monastersky, R. "The Ediacaran Enigma." *Science News* 148 (July 8, 1995): 28.

Narbonne, G. M. "The Ediacara Biota: A Terminal Neoproterozoic Experiment in the History of Life." *GSA Today* 8, no. 2 (1998): 1–6.

"The Oldest Known Animal Fossils." Part of the online exhibit *The Dawn of Animal Life,* from the Miller Museum of Geology, Queen's University, Kingston, Ontario, Canada [cited July 19, 2002]. <http://geol.queensu.ca/museum/exhibits/oldanim/oldanim.html>.

PaleoNet [cited July 19, 2002]. <http://www.ucmp.berkeley.edu/Paleonet/>.

Retallack, Greg. *Paleobiology* 20, no. 4 (1994): 523–44.

U.S. Geological Survey Paleontology Online Resources [cited July 19, 2002]. <http://geology.er.usgs.gov/paleo/paleonet.shtml>.

Wright, K. "When Life Was Odd." *Discover* (1997).

Was there photosynthetic life on Earth 3.5 billion years ago?

Viewpoint: Yes, morphological analyses and laser-Raman imaging have shown that photosynthetic life on Earth existed 3.5 billion years ago.

Viewpoint: No, it is more likely that life at the time was still using other, chemical, forms of energy to survive.

The two essays that follow debate claims concerning the possible early appearance of photosynthesis, which some scientists maintain originated on Earth 3.5 billion years ago (Gya). If this is true, it would be extraordinary, because Earth at that time was quite different from now, and indeed had only recently cooled after millions of years of battering by asteroids. For photosynthesis to have taken place at that time—and there is considerable evidence to suggest that it did—would be extremely remarkable, given the complexity of the photosynthetic process and the level of evolutionary advancement that it represents.

Photosynthesis is the biological conversion of light or electromagnetic energy from the Sun into chemical energy. It occurs in green plants, algae, and some types of bacteria, and requires a series of biochemical reactions. In photosynthesis, carbon dioxide and water react with one another in the presence of light and a chemical known as chlorophyll to produce a simple carbohydrate and oxygen. (A simple carbohydrate or simple sugar is one that, as its name suggests, cannot be broken down into any substance more basic. Examples include glucose or grape sugar, fructose or fruit sugar, and galactose.)

Though the thorough study of photosynthesis involves reference to exceedingly complex and (to the uninitiated) seemingly tedious biochemical information, in fact this process is one of the great miracles of life. In photosynthesis, plants take a waste product of human and animal respiration—carbon dioxide—and through a series of chemical reactions produce both food for the plant and oxygen for the atmosphere. Indeed, oxygen, which is clearly of utmost importance to animal life, is simply a waste by-product of photosynthesis.

Much of the debate in the essays that follow concerns the fossil record, which is extremely problematic when one is discussing life forms that existed during Precambrian time. The latter term refers to the first three of the four eons into which Earth's geological history is divided, with the Cambrian period marking the beginning of the present eon, the Phanerozoic, about 545 million years ago. The early Cambrian period saw an explosion of invertebrate (lacking a spinal column) marine forms, which dominated from about 545 to 417 million years ago. By about 420 to 410 million years ago, life had appeared on land, in the form of algae and primitive insects. This helps to put into perspective the time period under discussion in the essays that follow—about 3 billion years before the beginning of the reliable fossil record.

The term *fossil* refers to the remains of any prehistoric life form, especially those preserved in rock prior to the end of the last ice age about 11,000 years ago. The fossils discussed in the following essays are the oldest on

Earth, being the remains of single-celled organisms found in rock samples almost 80% as old as Earth itself. The process by which a once-living thing becomes a fossil is known as *fossilization*, wherein hard portions of the organism—such as bones, teeth, shells, and so on—are replaced by minerals. However, one of the controversial aspects of using the fossil record as a basis for forming judgments about the past (including the question of extremely early photosynthetic life) is that the fossil record is not exactly representative of all life forms that ever existed.

There is a famous story regarding a polling organization that produced disastrously inaccurate predictions regarding the outcome of the 1932 presidential elections. Their method of polling—using the telephone—would be perfectly appropriate today, but at the depths of the Depression, people who had phones were likely to be well-off. Those polled by this method reported overwhelmingly in favor of the incumbent, Herbert Hoover, but they were not a representative sample of the population. The vast majority of America—poor, uneducated, and hopeless—responded to the promises of Franklin D. Roosevelt, who won the election handily.

In the same way, the fossil record does not necessarily contain an accurate account of all life forms that existed at a particular point in the distant past. The majority of fossils come from invertebrates, such as mussels, that possess hard parts. Generally speaking, the older and smaller the organism, the more likely it is to have experienced fossilization, though other factors also play a part. One of the most important factors involves location: for the most part, the lower the altitude, the greater the likelihood that a region will contain fossils. Even so, all conditions must be right to ensure that a creature is preserved as a fossil. In fact, only about 30% of species are ever fossilized, a fact that scientists must take into account, because it could skew their reading of the paleontological record.

Also referenced in the arguments that follow is the study of DNA, or deoxyribonucleic acid, as a means of determining how life forms developed. All living cells contain DNA, which holds the genetic codes for inheritance—a blueprint for the organism. For organisms that reproduce asexually, through the splitting of cells, the DNA in ordinary cells also contains the blueprint for the offspring, whereas sexually reproducing organisms have special sex cells. The bacteria in question here would have reproduced asexually, as do bacteria today.

DNA is exceedingly complex, yet much of it varies little from organism to organism. For example, only about 0.2% of all human DNA differs between individuals, meaning that humans are 99.8% the same—and that all the variation that exists between people is a product of just 1/500th of the total DNA. Even between humans and chimpanzees, about 98% of the DNA is identical.

Photosynthesis, fossils, and DNA are just some of the concepts referenced in the lively discussion that follows. Compelling arguments are made for both sides, and while it might be easier, on the surface of it, to believe that photosynthesis could not have existed 3.5 Gya, the idea that it *could* have existed is too intriguing to be ignored. Furthermore, the fact that a number of prominent scientists have placed themselves on the affirmative side in this argument further recommends serious consideration of all the evidence. —JUDSON KNIGHT

Viewpoint:

Yes, morphological analyses and laser-Raman imaging have shown that photosynthetic life on Earth existed 3.5 billion years ago.

It may be hard to believe, but a six billionth of a meter-long microbe imbedded in a rock for 3.5 billion years is undergoing an identity crisis. Is it really a microbe that proves photosynthetic life was flourishing on Earth about one billion years earlier than scientists previously suspected? Or is it an impostor, a mere bubble-like structure or rock flaw that looks similar to a microbe but may be volcanic glass from a hydrothermal vent formed under water? The fierce debate surrounding the answers to these questions involves complex geochemical analyses. It also pits primarily two camps of scientists against each other, each claiming that what they see when they look through their microscopes is the truth.

The hotly debated topic is an important one. If photosynthetic life was thriving on Earth 3.5 billion years ago (Gya), this fact would ultimately require scientists to rewrite much of what they believe about Earth's early history, including the evolution of both the planet and life. For example, the 3.5-billion-year date indicates that life appeared soon (at least in geological terms) after asteroids stopped colliding en masse into Earth 3.9 Gya. The appearance of photosynthetic life at such an early point in Earth's history would also indicate that life evolved rapidly afterwards, probably in scalding pools as opposed to the warm springs that would have abounded a million years later after the Earth had a chance to cool down a bit.

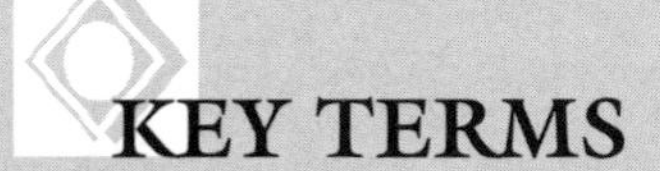

KEY TERMS

ABIOTIC: Not involving or derived from a living organism.
AEROBIC: Requiring oxygen.
ANAEROBIC: Not requiring oxygen.
ARCHAEAN: Relating to the earliest eon of geological history called the Precambrian, in which there was no life on Earth.
BACTERIA: Simple, single-celled life forms.
BIOTIC: Involving or derived from living organisms.
CATALYST: Something that helps a chemical reaction on its way but is not used up in the process.
CHERT: A hard, dark, opaque rock that looks like flint and consists primarily of a large amount of fibrous chalcedony (silica) with smaller amounts of cryptocrystalline quartz. Most often occurs as flint and sometimes in massive beds.
CYANOBACTERIA: A photosynthetic microorganism related to bacteria but capable of photosynthesis. They are prokaryotic and represent the earliest known form of life on Earth.
GENE: The "instructions" coded in a chemical called DNA that governs the growth and reproduction of life.
ISOTOPE: Different atoms of the same element that are chemically identical but have different weights.
KEROGEN: A complex fossilized organic material that is found in oil shale and other sedimentary rock. It is insoluble in common organic solvents and yields petroleum products on distillation.
OXIDIZE: To undergo a reaction in which electrons are lost, usually by the addition of oxygen.
PETROGRAPHY: The description and systematic classification of rocks and their composition.
PHOTOSYNTHETIC: Anything that uses sunlight to synthesize foods from carbon dioxide and water and generates oxygen as a by-product.
SPECTROSCOPY: Investigation and measurement of spectra produced when matter interacts with or emits electromagnetic radiation.

The Debate Begins The discovery of the oldest known fossil bacteria dates back to 1993 when William Schopf, director of the Center for the Study of Evolution and the Origin of Life at the University of California, Los Angeles (UCLA), and coworkers announced that they had discovered a diverse bacterial flora in the 3.465-billion-year-old Apex cherts (flint-like rock) in a greenstone belt near Marble Bar in Western Australia. They ultimately identified 11 different bacterial species, including cyanobacteria, which are known as the "architects of the atmosphere" because they are photosynthetic and produce oxygen.

Although Schopf's identification of the fossils as cyanobacteria was widely accepted by many scientists, others have disputed the findings because they are based largely on morphology, that is, on the bacteria's form and structure. Unfortunately, bacteria have little in the way of concrete morphological features that can be used for identification purposes, especially after being imbedded in rock for billions of years. As a result, it is extremely difficult to discern between real bacterial fossils and pseudofossil look-alikes.

However, Schopf's finding was later bolstered by analyses of the isotopic composition of carbon in Archaean sediments. These analyses have supported the general proposition that life could have begun 3.5 Gya. Cyanobacteria live in the water and can manufacture their own food. Because they are bacteria, they are extremely small and usually unicellular. However, they often grow in colonies large enough to see. These colonies are like sticky mats that become finely layer red mounds of carbonate sediment called stromatolites. The specimens Schopf analyzed were associated with isotopically light carbon (C), which indicates photosynthetic activity. Living photosynthetic organisms preferentially incorporate ^{12}C (six protons, six neutrons in the nucleus) as opposed to the rarer, heavier isotopes of ^{13}C and ^{14}C. Therefore, if a carbon-rich object has a greater amount of ^{12}C relative to ^{14}C than the surrounding rock, it is likely biologically produced carbon.

Although Schopf and his colleagues have been studying microscopic wisps of carbonaceous material, they carefully evaluated the evidence and came to the conclusion that this was indeed life. They based their reasoning on strict fundamental "rules" that Schopf and his colleagues had developed over years of studying and correctly identifying early small microorganisms. As Schopf stated in a presentation on microbe hunting and identification for a National Research Council workshop, it is not enough to deduce that an object must be biological because it does not look like a mineral or flaw. Positive evidence is needed; evidence strong enough to rule out plausible nonbiological sources. Schopf notes that "the best way to avoid being fooled by nonbiological structures is to accept as bona fide fossils only those of fairly complex form."

Although Schopf and his colleagues believed they had met these criteria and more, the morphological interpretation of data has kept the door open on a decade of debate. However, in the March 7, 2002, issue of *Nature*, Schopf and his colleagues presented evidence that strongly supports the view that life on Earth originated at least 3.5 Gya. The evidence is based on newly developed technology that enables scientists to look inside of rocks, determine what they are made of, and make a molecular map of any embedded structures.

Laser-Raman Imagery For paleontologists, laser-Raman spectroscopy is a tremendous breakthrough that gives them Superman-like x-ray vision. As Schopf explains: "Because Raman spectroscopy is non-intrusive, non-destructive and particularly sensitive to the distinctive carbon signal of organic matter of living systems, it is an ideal technique for studies of ancient microscopic fossils. Raman imagery can show a one-to-one correlation between cell shape and chemistry and prove whether fossils are biological."

In essence, Raman analysis provides both a two-dimensional image of the sample and its chemical composition. The spectrographic tool does this by bouncing laser beams off rock surfaces, creating a scattering pattern that is identical to the patterns created by organic molecules found in other fossils.

Schopf and his colleagues first tested the spectroscopic technique on four fossil specimens that had ages already established through other techniques and approaches. In their *Nature* article, Schopf and his colleagues state that the tests showed that laser-Raman spectroscopy can "extend available analytical data to a molecular level." Particularly, they note, "it can provide insight into the molecular make-up and the fidelity of preservation" of the kerogenous matter that makes up fossils. Kerogens are organic materials found in sedimentary rock.

Admittedly, the ancient samples under study are minute microscopic remnants, described by Schopf in the *Nature* paper as "graphitic, geochemically highly altered, dark brown to black carbonaceous filaments." But the laser-Raman analysis technique (developed by Schopf and his colleagues at UCLA and the University of Alabama) is highly sensitive to the distinct signal of carbonaceous matter so it can be used to characterize the molecular composition of fossils as small as 1 micron in diameter. The technique is so accurate that, as Schopf points out, it can achieve this analysis "in polished or unpolished petrographic thin sections, in the presence or absence of microscopy immersion oil, and whether the fossils analyzed are exposed at the upper surface of, or are embedded within, the section studied."

Schopf and his colleagues showed that the kerogen signal they identified did not result from contamination by immersion oil, which is used to enhance optical images. In fact, the laser-Raman technique could easily distinguish fossil kerogen signals from the possible spectral oil contaminants. Schopf and his colleagues also took great care to avoid contamination from graphite markers, which had been used to circle the fossils for identification. This source of contamination was further removed by the fact that some of the fossils were embedded as much as 65 microns in the quartz matrix, where they could not be affected by any surface contamination.

Schopf and his colleagues were also able to directly correlate molecular composition with filament morphology and establish that the filaments were composed of carbonaceous kerogen. Furthermore, they showed that the organic substance occurred in much greater concentrations in the fossils than in the "wispy, mucilage-like clouds of finely divided particulate kerogen in which the fossils . . . are preserved." These carbon molecules, which are the decay products of living bacterial cells, have firmly established the fossils' biogenicity.

Are Fossils in the Eyes of the Beholder? Despite Schopf's further evidence, another faction of paleontologists, led by Martin Brasier of Oxford University, have sounded off loudly against Schopf's findings, publishing a paper on their theory in the same March 7, 2002, issue of *Nature*. They believe that Schopf's fossils are nothing more than "marks" resulting from an ancient and unusual geological process involving hydrothermal vents for volcanic gases and the formation of rock from magma coming in contact with the cooler earth surface. Nevertheless, Brasier's group agrees that the marks or fossils (depending on your point of view) have a chemical composition that appears biological in origin.

Brasier and his colleagues try to explain away this biological confirmation by saying that these seemingly biological molecules are actually the result of interactions between carbon dioxide and carbon monoxide, which is released by hot, metal-rich hydrothermal vents. The group argues that these molecules were trapped in bacteria-like filaments as the hot rocks cooled. However Schopf and others point out that, if Brasier's hypothesis is correct, this type of material should be found abundantly but is not, or at least has not been so far.

In an *ABC Science Online* article, Professor Malcom Walter of Macquarie University's Australian Centre for Astrobiology notes that the chemical evidence combined with the fossil shapes was "strongly suggestive" that the marks are actually fossil bacteria. However, Brasier and his colleagues argue that the squiggly marks do not even look like other ancient microbes in that their shapes are too complicated. Brasier went on to argue that Schopf illustrated selective parts of the fossils that looked like bacteria while ignoring the bulk of other carbonaceous material. Walter responds that such variations are to be expected. "Long experience shows," says Walter, "that with preserved or fossilized bacteria, only a few cells maintain their shape. Brasier's argument overlooks this."

Schopf's group believes that Brasier's interpretation of the data is simply mistaken. As Schopf and his colleagues point out, Brasier's group is not as adept at looking at Precambrian microfossils as Schopf's group. Furthermore, the depth of focus using the laser-Raman technique can be confusing. For example, Brasier and his colleagues noted that often something that looked like a bacterium at one focal depth merely became weird shapes at other depths. But Schopf points out that Brasier is misinterpreting the readings and that the nonbacterial-like branching of these bacterial cell chains is actually the folding of chains.

The Facts Are Clear Schopf and his colleagues have used optical microscopy coupled with laser-Raman imagery and measurement of Raman point spectra to correlate chemical composition directly with observable morphology. As a result they have established biogenicity of the fossils studied through crucial indicators such as the fossils' kerogenous cell walls. The results will ultimately provide insights into the chemical changes that accompany organic metamorphosis.

A paleontologist, geologist, microbiologist, and organic geochemist, Schopf does not propose or follow an easy set of rules for verifying true microbial fossils as opposed to pseudofossils. When a recent claim for life arose involving fossils in a Martian meteorite, Schopf was one of the first skeptics to speak out and call for rigorous testing and analyses. He has carefully pointed out that techniques such as electron microscopy are not always reliable. However, he states unequivocally that the 3.5 billion-year-old fossils are real.

The fossils in the Apex cherts are structurally so similar to photosynthetic cyanobacteria that if one accepts they are indeed fossils, it is hard to argue that they are not cyanobacteria. "We have established that the ancient specimens are made of organic matter just like living microbes, and no non-biological organic matter is known from the geological record," says Schopf. "In science, facts always prevail, and the facts here are quite clear." —DAVID PETECHUK

Viewpoint:

No, it is more likely that life at the time was still using other, chemical, forms of energy to survive.

The main thrust of the argument that photosynthesis was already going strong 3.5 billion years ago (3.5 Gya) centers on several pieces of evidence indicating that the more complex form of photosynthesis using oxygen (aerobic photosynthesis) was already in place at that time. However, all the evidence in favor of this conclusion is uncertain. At best, the less complex form of photosynthesis (anaerobic photosynthesis) could have existed then, but it is more likely that life at the time was still using other, chemical, forms of energy to survive.

There are four broad categories of evidence that the earliest life on Earth left behind to document its existence. The first is the fossil record. The second are biomarkers—characteristic signatures life leaves behind in some of the molecules it comes into contact with. The third category is the geological record, as life can also bring about changes in the structure of the rocks that make up the planet. The final category comes from the record of life written in the very genes that control it. Unfortunately, the record of this evidence becomes increasingly sparse and difficult to interpret the farther back in time we go. By the time we are looking at what happened 3.5 Gya, interpreting the evidence is fraught with many uncertainties.

The Fossil Record The most obvious and supposedly strongest evidence for early aerobic photosynthesis comes from fossils. Fossils are any evidence of past life. Most people are familiar with fossils of bones or shells. However fossils of soft tissue are also possible. Fossils can be the unaltered original material of the organism, but more commonly the original material of the organism has been recrystallized or replaced by a different mineral.

The scanty fossil evidence that aerobic photosynthesis was thriving so early in Earth's history comes primarily from fossils found in 3.5 Gya rocks called the Warrawoona Group in Western Australia. J. William Schopf, Professor of Paleobiology at the University of California, Los Angeles, has matched the shapes and sizes of microscopic pieces of fossilized carbon with those of cyanobacteria, bacteria known to carry out aerobic photosynthesis. This assertion rests on the assumption that the external features of ancient bacteria indicate internal workings similar to modern bacteria, which is by no means certain. This is not the only problem, however. A recent paper published by Dr. Martin Brasier from the Faculty of Earth Sciences at the University of Oxford in the United Kingdom has shown that the shapes of the lumps of carbon supposed to be cells can all be formed by natural geological process acting on the carbon compounds that would have existed then.

Carbon Isotope Fractionation Just because cell-shaped carbon lumps can be formed by geological processes does not necessarily mean that they were. The second strand of evidence cited to indicate that the fossilized carbon derives from early life comes from biomarkers created through the process of isotopic fractionation.

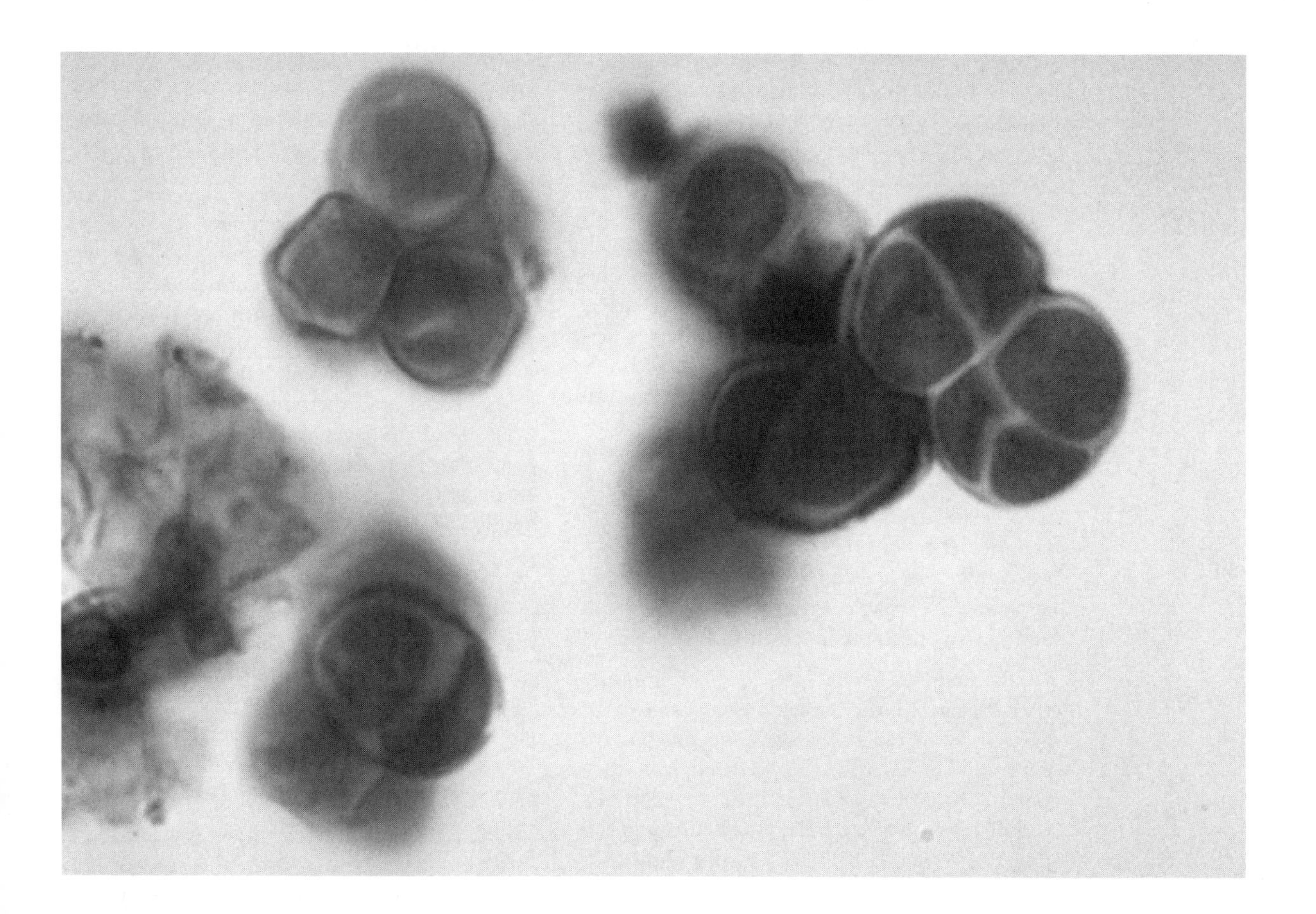

A microscopic view of cyanobacteria.
(© Lester V. Bergman/Corbis. Reproduced by permission.)

Isotopes are atoms of the same element that behave in an almost identical chemical fashion but have different weights. The carbon (C) from carbon dioxide in the air contains two isotopes: ^{12}C and ^{13}C. Photosynthesizing bacteria take in carbon dioxide from the air to make compounds containing mainly hydrogen and carbon (carbohydrates). These compounds provide the bacteria with food, as well as the building blocks for most of the structures that make up their cells. In photosynthesis, the lighter ^{12}C carbon reacts a little quicker than does the ^{13}C, so the bacteria contain slightly more of it. The carbon dioxide from the air is also used to make the rocks that surround fossilized bacteria, but the process that forms the rocks skews the ratio in favor of ^{13}C. Measuring the ratio of ^{12}C to ^{13}C between fossilized carbon and that of the surrounding rocks compared to a known ^{13}C standard provides an indication as to whether the carbon came from a living (biotic) or inorganic (abiotic) source.

Unfortunately, different photosynthesizers can create widely varying ^{13}C values that can overlap with abiotic values. There is also the added problem that over long periods of time, geological processes can alter the $^{12}C/^{13}C$ ratios, bringing yet more uncertainties into the equation. This has led some experts, including Dr. Roger Buick at the University of Washington, to question whether isotopic analysis qualifies as a valid means for testing for carbon from living organisms at all.

Were this not enough, several studies have also shown that the ^{13}C values for biotic carbon compounds can be mimicked by a type of chemical reaction called Fischer-Tropsch synthesis. This reaction uses an iron or nickel catalyst to help react hydrogen gas with either carbon monoxide or carbon dioxide to create hydrocarbon compounds. Fischer-Tropsch synthesis has been shown to occur under the conditions that would prevail around underwater vents. At these high temperatures, water is broken down to provide the hydrogen, and carbon oxides and the metal catalysts come from the rocks. This provides a plentiful supply of hydrocarbon compounds with ^{13}C values similar to those for biotic carbon compounds.

The Geological Record The final pieces of evidence for early aerobic photosynthesis are contained in the geological record in the form of two types of layered structures: banded-iron formations (BIFs) and stromatolites.

Stromatolites are finely layered rocks that can form through the action of different groups of microorganisms living together in huge colonies. Living stromatolites exist today, with aerobic photosynthesizers inhabiting the top layer, anaerobic photosynthesizers populating the layers below them, and other microorganisms that cannot survive in the presence of oxygen toward the bottom. The microorganisms secrete mucus, which binds rocky particles together to form the stacked rocks. Stromatolites are found in the fossil record extending all the way back to 3.5 Gya, and some of the more recent ones have been shown to contain the fossilized remnants of the microorganisms that lived within them. We know very little about how the more ancient stromatolites formed, however, and work by Professors John Grotzinger and Daniel Rothman, of the Department of Earth, Atmospheric and Planetary Sciences at the Massachusetts Institute of Technology (MIT), has shown that apparently identical structures can form easily through natural geological processes. Although this does not rule out microorganisms playing a part in early stromatolite formation, or at least inhabiting stromatolites, it cannot be relied upon without corroborating proof—from fossils, for example. However, no reliable fossil evidence has been found in any of the older stromatolites.

Banded-iron formations (BIFs) are brightly colored, layered rocks that are the source of most of the world's iron ore. Their distinctive red color comes from the oxidized iron they contain (the iron is described as oxidized because it has lost electrons, usually through the action of oxygen). BIFs first appear about 3.5 Gya and peter out around 2 Gya. Because of the particular way in which they form, BIFs are a good indicator of the level of oxygen in the atmosphere. The only mechanism previously thought able to explain the increase in BIFs was aerobic photosynthesis. A study by Professor Friedrich Widdel, Director of the Department of Microbiology at the Max-Plank-Institute for Marine Microbiology in Breman, Germany, has cast doubt on this by showing that anaerobic photosynthesis is also capable of oxidizing the iron and suggesting abiotic mechanisms that can do the same.

Evidence in Favor of Aerobic Photosynthesis All the evidence postulating aerobic photosynthesis at 3.5 Gya can therefore be explained away by either abiotic means or through anaerobic photosynthesis. This would not be so damning were it not for positive evidence indicating that aerobic photosynthesis only arose around 2 Gya—conveniently coinciding with the large increase in oxygen documented in the geological record. This evidence comes firstly from a biomarker called 2-methylhopanoids and secondly from genetic clues.

2-methylhopanoids are compounds that are found in the cell membranes (outer skin) of cyanobacteria. A paper published by Roger Summons, a Professor at the Department of Earth, Atmospheric and Planetary Sciences at MIT, has shown that if cyanobacteria undergo fossilization, the hydrocarbon derivatives of the 2-methlyhopanoids created are exceptionally stable, surviving to provide a record of the existence of aerobic photosynthesis. These derivatives have been shown to be abundant in carbon sediments created after 2.5 Gya but not before.

The final piece of the puzzle comes from the evolutionary history contained within genes. Genes are the instruction manuals for life. Passed on through reproduction and contained in every cell in every living creature, they control the main features and traits of living organisms. The instructions contained in genes are written in the sequence of the building blocks of a molecule called DNA. As life has evolved, new genes have arisen, but always by making copies of and changes to the structure of DNA in preexisting genes. Because evolution builds on what has gone before, once an effective and successful trait has arisen, it tends to be passed on relatively unchanged to successive generations. For this reason, the genes controlling the fundamental processes of life arose in the earliest and most simple life and still control the basics of what happens inside life today. That is why humans have many genes in common with organisms as diverse as geraniums, slugs, and chickens.

The DNA in different organisms can therefore be compared to provide a record of how different organisms are related, and even to give a rough guide to when and how their evolutionary paths diverged. Analysis of this type is not an exact science, however. DNA in different organisms changes at different rates, making the timing of specific events very difficult to pinpoint. Bearing these caveats in mind, though, most genetic analysis seems to point to a relatively early evolution for anaerobic photosynthesis and the much later appearance for aerobic photosynthesis at around 2 Gya.

A More Timely Appearance for Photosynthesis Taken together, this all suggests that aerobic photosynthesis did not exist 3.5 Gya. Proving that any form of photosynthesis did not exist at this time is much harder. Aside from fossils, the only evidence that could seemingly document its appearance is the carbon isotope record that has been shown to be so fallible in support of aerobic photosynthesis.

It is generally accepted that Earth formed about 4.6 Gya. However, until 3.9 Gya ago, evidence from the surface of the Moon shows us that the Earth's surface was continually bombarded by catastrophic impacts from meteors left over from the formation of the Solar System. These would almost certainly have wiped out any life that had managed to gain a foothold. If life could only have got started 3.9 Gya ago, the appearance of such a complicated trait as photosynthesis by 3.5 Gya implies a relatively rapid, early evolution for life on Earth. It seems far more reasonable to suggest that life was still using other chemical forms of energy production at 3.5 Gya. Anaerobic photosynthesis probably appeared some time after, culminating in the appearance of fully fledged aerobic photosynthesis sometime around 2 Gya. —JENS THOMAS

Further Reading

Brasier, Martin D., et al. "Questioning the Evidence of the Earth's Oldest Fossils." *Nature* 416 (March 7, 2002): 76–81.

Buick, R. "Microfossil Recognition in Archean Rocks: An Appraisal of Spheroids and Filaments from a 3500 M. Y. Old Chert-Barite Unit at North Pole, Western Australia." *Palaios* 5 (1990): 441–59.

Gee, Henry. "Biogeochemistry: That's Life?" *Nature* 416 (March 7, 2002): 73–76.

Grotzinger, John P., and Daniel H. Rothman. "An Abiotic Model for Stromatolite Morphogenesis." *Nature* 383 (1996): 423–25.

Gupta, Radhey S., Tariq Mukhtar, and Bhag Singh. "Evolutionary Relationships Amongst Photosynthetic Prokaryotes (*Heliobacterium chlorum, Chloroflexus aurantiacus, cyanobacteria, Chlorobium tepidum* and *proteobacteria*): Implications Regarding the Origin of Photosynthesis." *Molecular Microbiology* 32, no. 5 (1999): 893–906.

Hecht, Jeff. "Tiny Fossils May Be Earth's Oldest Life." *New Scientist* (March 6, 2002).

Hedges, S. Blair, et al. "A Genomic Timescale for the Origin of Eukaryotes." *BMC Evolutionary Biology* I (2001): 4.

Holm, Nils G., and Jean Luc Charlou. "Initial Indications of Abiotic Formation of Hydrocarbons in the Rainbow Ultramafic Hydrothermal System, Mid Atlantic Ridge." *Earth and Planetary Science Letters* 191 (2001): 1–8.

Kudryavtsev, Anatoliy B., and J. William Schopf, et al. "In Situ Laser-Raman Imagery of Precambrian Microscopic Fossils." *Proceedings of the National Academy of Sciences* 98, no. 3 (January 30, 2001): 823–26.

Robert, Francois. "Carbon and Oxygen Isotope Variations in Precambrian Cherts." *Geochimica et Cosmochimica Acta* 52 (1988): 1473–78.

Salleh, Anna. "Is This Life?" *ABC Science Online* [cited July 16, 2002]. <http://www.abc.net.au/science/news/space/SpaceRepublish_497964.htm>.

Schopf, J. William. "Microfossils of the Early Archean Apex Cherts: New Evidence of the Antiquity of Life." *Science* 260 (1993): 640–46.

———. *Cradle of Life; The Discovery of Earth's Earliest Fossils.* Princeton, NJ: Princeton University Press, 1999.

———. "Fossils and Pseudofossils: Lessons from the Hunt for Early Life on Earth." In *Size Limits of Very Small Microorganisms.* Washington, DC: National Academy Press, 1999.

———, et al. "Laser-Raman Imagery of Earth's Earliest Fossils." *Nature* 416 (March 7, 2002): 73–76.

Summons, Roger E., et al. "2-Methylhopanoids as Biomarkers for Cyanobacterial Oxygenic Photosynthesis." *Nature* 400 (1999): 554–57.

Widdel, Friedrich, et al. "Ferrous Iron Oxidation by Anoxygenic Phototrophic Bacteria." *Nature* 362 (1993): 834–36.

Is present global warming due more to human activity than to natural geologic trends?

Viewpoint: Yes, there is strong evidence that most of the global warming observed over the last 50 years is due to human activities.

Viewpoint: No, present global warming is not due more to human activity than to natural geologic trends.

Of all the issues relating to earth science, none is more politically volatile than that of global warming. Even the name is controversial, since it is not entirely clear, from the scientific data, that Earth really is warming. And even if it is, the causes are open to debate. Typically, when the term *global warming* is mentioned in the media, the implication is "global warming due to human activities," but this may only be part of the picture.

In the debate over whether humans are primarily to blame for global warming, the "yes" side has long been far more visible and vocal. This is due in large part to the prominence of the environmental movement, which, as it has gained strength in the period since the late 1980s, has exerted enormous influence through the media, entertainment, and popular culture in general. Overexposure of the "yes" viewpoint is thus a mixed blessing for scientists who hold to it purely on the basis of data, and not because it is the fashionable position. Yet as we shall see, there is a strong argument for the view that, first of all, global warming is proceeding in a more or less linear fashion, and secondly, that human activities are responsible for it.

The idea that humans are causing a radical change in the planet's temperature is not a new one; in fact, an early manifestation of this viewpoint occurred in the mid-1970s, at a time when the environmental movement was just starting to gain in political influence. At that time, however, winters were unseasonably cold—for example, the winter of 1976–1977 saw numerous cases of snowfall even in the southern United States—and the fear at the time was that human activities were causing Earth to cool down, not warm up. This illustrates just how difficult it is to analyze exactly what the climate is doing, and which direction it is headed. One thing is certain, however: Earth's climate is continually changing.

The planet has long been subject to ice ages, and to interglacial periods, which, as their name suggests, are relatively short segments between ice ages. All of recorded human history, which dates back only about 5,500 years, to approximately 3500 B.C. in Sumer, takes up just half of the current interglacial period, which has lasted for approximately 11,000 years. Regardless of whether Earth is warming up or cooling down now, or whether humans are responsible for it, it seems clear that within the next few thousand years, another ice age will befall the planet.

In the meantime, though, what is happening to the climate, and what role are humans playing? It is unquestionable that humans are pouring unprecedented amounts of pollutants into the atmosphere, including greenhouse gases such as carbon dioxide and carbon monoxide. We are also cutting down our forests at an alarming rate. Before humans began chopping down forests, Earth's combined vegetation stored some 990 billion tons (900 billion metric tons) of carbon, of which 90% appeared in forests. Today, only about 616 billion tons (560 billion metric tons) of carbon are stored in Earth's vegetation, and the amount is getting smaller as time passes. At the same time, the amount of carbon dioxide in the atmosphere has increased from about 270 parts per million (ppm) in 1850 to about 360 ppm in 2000, and again, the increase continues.

There is also the threat to the ozone layer that has resulted from the misuse of a particular type of catalyst, a substance that speeds up a chemical reaction without taking part in it. In the upper atmosphere of Earth are traces of ozone, a triatomic (three-atom) molecular form of oxygen that protects the planet from the Sun's ultraviolet rays. During the latter part of the twentieth century, it became apparent that a hole had developed in the ozone layer over Antarctica, and many chemists suspected a culprit in chlorofluorocarbons, or CFCs, which had long been used in refrigerants and air conditioners, and as propellants in aerosol sprays. Chlorine acts as a catalyst to transform the ozone to elemental oxygen, which is not nearly as effective for shielding Earth from ultraviolet light.

Due to concerns about the danger to the ozone layer, an international agreement called the Montreal Protocol on Substances That Deplete the Ozone Layer, signed in 1987, banned the production of CFCs and the coolant Freon that contains them. Even proponents of the global-warming theory (that is, those who support the position for scientific, and not political or emotional reasons) concede that since the signing of the protocol, the growth in the ozone-layer hole has decreased.

One of the biggest areas of disagreement over the global-warming issue is with regard to the greenhouse effect, which is also one of the least-understood natural phenomena under debate in the present context. Contrary to popular belief, the greenhouse effect is not something originally caused by human activity, nor is it necessarily detrimental; on the contrary, if there were no greenhouse effect, there could be no human life on this planet.

The Sun radiates vast amounts of energy, part of which is in the direction of Earth, but only a portion of which Earth actually absorbs. The planet then re-radiates a dissipated form of the same energy, known as long-wavelength radiated energy, to space. Water vapor, carbon dioxide, and carbon monoxide, as well as methane, nitrous oxide, and ozone, all absorb long-wavelength radiated energy as the latter makes its way up through the atmosphere, and when heated, these "greenhouse gases" serve to slow the planet's rate of cooling. Without the greenhouse effect, surface temperatures on Earth would be about 59° Fahrenheit cooler than they are, and much too cold for Earth's biological processes. Thus the greenhouse effect literally preserves life on the planet.

With the greenhouse effect placed in its proper context, it is still reasonable to question whether human activities are placing an excessive amount of carbon dioxide, carbon monoxide, and other greenhouse gases into the atmosphere. There is certainly a body of evidence to indicate that we are, and that this activity is having a deleterious effect on the planet's climate by causing undue heating. On the other hand, a single volcano can thrust more natural pollutants into the atmosphere than a thousand cities of the size of Los Angeles. The debate over global warming thus continues, with no obvious conclusion in sight. —JUDSON KNIGHT

Viewpoint:

Yes, there is strong evidence that most of the global warming observed over the last 50 years is due to human activities.

"We have met the enemy and he is us!" Walt Kelly's philosophical cartoon possum named Pogo was sadly remarking on the ways humans were trashing their environment on Earth Day 1971. The possum was staring at heaps of junk thrown on the forest floor. If Pogo reflected that comment today, he might not see as much trash on the ground, but he would feel the warmer weather that is also the unfortunate result of human activities. According to the Third Assessment Report of the Intergovernmental Panel on Climate Change (IPCC) released in 2001, "There is new and stronger evidence that most of the warming observed over the last 50 years is attributable to human activities."

In 1988 the United Nations set up the IPCC to figure out whether global warming was actually happening. The group has produced three reports, each more certain than the one before that global warming is in fact happening, and that the cause is due more to human activities than to natural events. Greenhouse gases accumulating from burning fossil fuels, and other gas-producing activities of industry, are blanketing the Earth and so causing the warming trend.

At the bidding of President George W. Bush, the National Research Council of the National Academy of Sciences studied the

A traffic jam in Los Angeles. Some researchers see the urbanization effect as a major cause for the warming over the past 100 years. *(© Andrew Brown/CORBIS. Reproduced by permission.)*

IPCC's findings and issued its own report in the summer of 2001. The news was grim. "Greenhouse gases are accumulating in Earth's atmosphere as a result of human activities, causing surface air temperatures and subsurface ocean temperatures to rise. Temperatures are, in fact, rising." In the United States, regional assessments were also made, and the results were the same. That is not to say global warming is uniform, however, because local variables, such as the proximity of the ocean or an ice field, play a role.

The Greenhouse Effect The greenhouse effect is not all bad. With the blanketing of greenhouse gases (water vapor, carbon dioxide, methane, nitrous oxide, and tropospheric ozone) the planet is kept about 59°F (15°C) warmer than it would be without the cover. Water vapor is by far the most abundant greenhouse gas. The greenhouse effect gets its name because a layer in the atmosphere acts like the glass in a greenhouse, letting the Sun's light energy in, but preventing reflected lower energy heat radiation from escaping.

All the greenhouse gases that occur naturally are also produced by human activities, thus increasing their levels in the atmosphere at an accelerating pace. In addition to the "natural" greenhouse gases are the fluorocarbons, which only appeared in the atmosphere when humans introduced them. Over the millennium before the Industrial Era, the atmospheric concentrations of greenhouse gases remained relatively constant, according to the IPCC report. The contribution of a particular gas to radiative forcing of climate change depends on the molecular radiative properties of the gas, the amount introduced into the atmosphere, and the residence time of the gas in the atmosphere. Gases that last a long time in the atmosphere can affect the atmosphere for decades, centuries, or millennia before they are naturally removed.

The concentration of carbon dioxide in the atmosphere has increased by more than one third since the start of the Industrial Era in the late eighteenth century. The atmospheric lifetime of carbon dioxide ranges from 5 to 200 years. Methane was at a pre-industrial concentration of about 700 parts per billion (ppb) and had risen to 1745 ppb by 1998. Methane has an atmospheric lifetime of 12 years. Nitrous oxide was about 270 ppb and rose to 314 ppb during the same period. It has a lifetime of 114 year. Three fluorocarbon compounds were not in the atmosphere until recent years, but one of them, chlorofluorocarbon, was at 268 ppb in 1998, with a lifetime of 45 years. Another, hydrofluoro-carbon, reached 14 ppb and has a lifetime of 260 years. The third, perfluoro-methane, reached 80 ppb by 1998. It has a lifetime of over 50,000 years, which, in human time, seems close to forever.

Chlorofluorocarbons are believed by most scientists to be responsible for depleting the ozone layer that exists at the top of the stratosphere, the second layer in the Earth's atmosphere. The upper stratospheric ozone layer shields the Earth from intense ultraviolet radiation that causes skin cancer. The troposphere, the region from the ground up to about 8.6 mi (14 km), is the first layer. All the weather takes place in the troposphere.

There was one very small bit of good news from the United States Environmental Protection Agency's (EPA's) Global Warming Site in January 2002: the growth rate of chlorofluorocarbons in the atmosphere has declined since the Montreal protocol was signed in 1987. This international treaty, signed by numerous countries across the world, was designed to phase-out the use of chlorofluorocarbons to protect the upper ozone layer.

KEY TERMS

AEROSOL: Extremely small particles of liquid or dust ejected into the atmosphere.

ANTHROPOGENIC: Caused by humans, as opposed to naturally occurring.

BIOMASS: Total amount of living organisms in a particular area.

CLIMATE CHANGE: A broader and more accurate term than either "global warming" or "global cooling" to describe shifts in average weather over long periods of time.

GREENHOUSE EFFECT: Natural phenomenon that traps heat near the Earth's surface by the action of greenhouse gases and that makes the Earth inhabitable by keeping its average surface temperature about 59°F (15°C) warmer than it would be without this action.

GREENHOUSE GAS: Any gas, such as carbon dioxide, methane, water vapor, or ozone, which has the property of absorbing infrared radiation within the Earth's atmosphere, thus retaining heat near the Earth's surface and creating the greenhouse effect.

INFRARED RADIATION: Energy emitted from all substances above absolute zero. We feel it as heat but do not see it.

OZONE: An unstable form of oxygen that has three atoms per molecule rather than the normal two.

RADIATIVE FORCING: The influence on the atmosphere produced by the greenhouse effect. Negative radiative forcing causes colder temperatures, positive causes warmer temperatures.

SOLAR IRRADIANCE: The amount of sunlight striking a particular area.

SULFATE AEROSOLS: Atmospheric particles created by the reaction of sulfur dioxide and other sulfide gases with a variety of substances. Sulfate aerosols reflect infrared radiation into space and thus contribute to global cooling.

Natural Factors That Affect Climate Changes

The climate has always been influenced by various factors. However, before the Industrial Era the changes were all due to natural causes. Solar irradiance is one example. The Sun has 11-year cycles of activity, with quiet periods and turbulent ones with active sunspots emitting storms of radiation. During periods of active sunspots, radio communications are often affected. Spectacular aurora displays in far northern and southern latitudes are also earthly evidence of solar storms.

Volcanoes play a part in temporary climate changes. Enormous ejections of toxic gases are produced as aerosols from explosive volcanoes such as the Mount Pinatubo eruption in the Philippines in 1991. That eruption produced some 20 million tons of sulfuric acid aerosols ejected into the stratosphere. The resulting upper atmosphere cloud covered the entire planet, reflecting sunlight and cooling the Earth by nearly 1°F (-17°C).

In 1815 Mount Tambora in Indonesia erupted, sending a cloud into the atmosphere that was four times the amount ejected by Mount Pinatubo. The year following the eruption has been described as "The Year without a Summer." Probably the greatest volcanic eruption to date is the 1883 Krakatoa eruption, which caused the Moon to appear blue for two years.

According to the IPCC, solar variation and volcanic aerosols are the two major natural factors that affect global warming. There are some anthropogenic aerosols, but they are short-lived. The major sources of anthropogenic aerosols are fossil fuel and biomass burning. These aerosols are associated with air pollution and acid rain near the Earth's surface.

Equatorial deforestation may have contributed to the rapid shrinking of Mount Kilimanjaro's ice cap, which lost over 80% of its area from 1912 to 2001.
(© Corbis. Reproduced by permission.)

There is one more natural factor to mention: the El Niño–Southern Oscillation (ENSO) phenomenon. A distinctive pattern of sea surface temperature changes in the Pacific Ocean is associated with large variations in weather around the world from year to year. The oceans act like a sink for carbon dioxide. The oceans absorb about 50% of the carbon dioxide released into the atmosphere. The exact amount depends on the ocean temperature. Therefore, the ENSO plays a role in modulating exchanges of carbon dioxide with the atmosphere.

Anthropogenic Factors That Affect Climate Changes Anthropogenic factors impacting the global climate are human population growth, urbanization, the energy supply mix and the efficiency with which energy is produced and consumed, deforestation rates, and emission rates of chlorofluorocarbons and other synthetic greenhouse gases.

According to Edward O. Wilson in *The Future of Life,* the world population reached 6 billion by October 1999. The rate of growth is a phenomenal 200,000 people each day. People born in 1950 will see the human population double in their lifetime, from 2.5 billion to over 6 billion. In 1800 there were only about 1 billion people, and as the Industrial Era went into full swing in 1900, the population was still only at 1.6 billion.

With the population expected to reach 8 billion by mid-twenty-first century, all the human activities that affect climate will increase proportionally. Already at the start of the twenty-first century, atmospheric carbon dioxide is at the highest level it has ever been in at least 200,000 years.

Likewise, urban influence cannot be ignored. As masses of people move into cities, the asphalt roads, tar roofs, and other urban features trap more heat than surrounding countryside. Some researchers see urbanization as a major cause for global warming in the past 100 years. Deforestation, the clearing of trees and vegetation to make cities, has further exacerbated the problem.

As populations increase, there is greater demand for fuel for heating and cooling, transportation, and for industry. Many countries' growing dependence on fossil fuels has produced not only political and economic problems, but environmental problems as well, increasing the amount of synthetic greenhouse gases in the atmosphere.

What Is in the Future? This is a difficult question to answer. Models developed by the IPCC and other research groups indicate that global warming will profoundly affect our quality of life. The factors that influence global climate are extremely complex and interrelated. There are some scientists who minimize our role

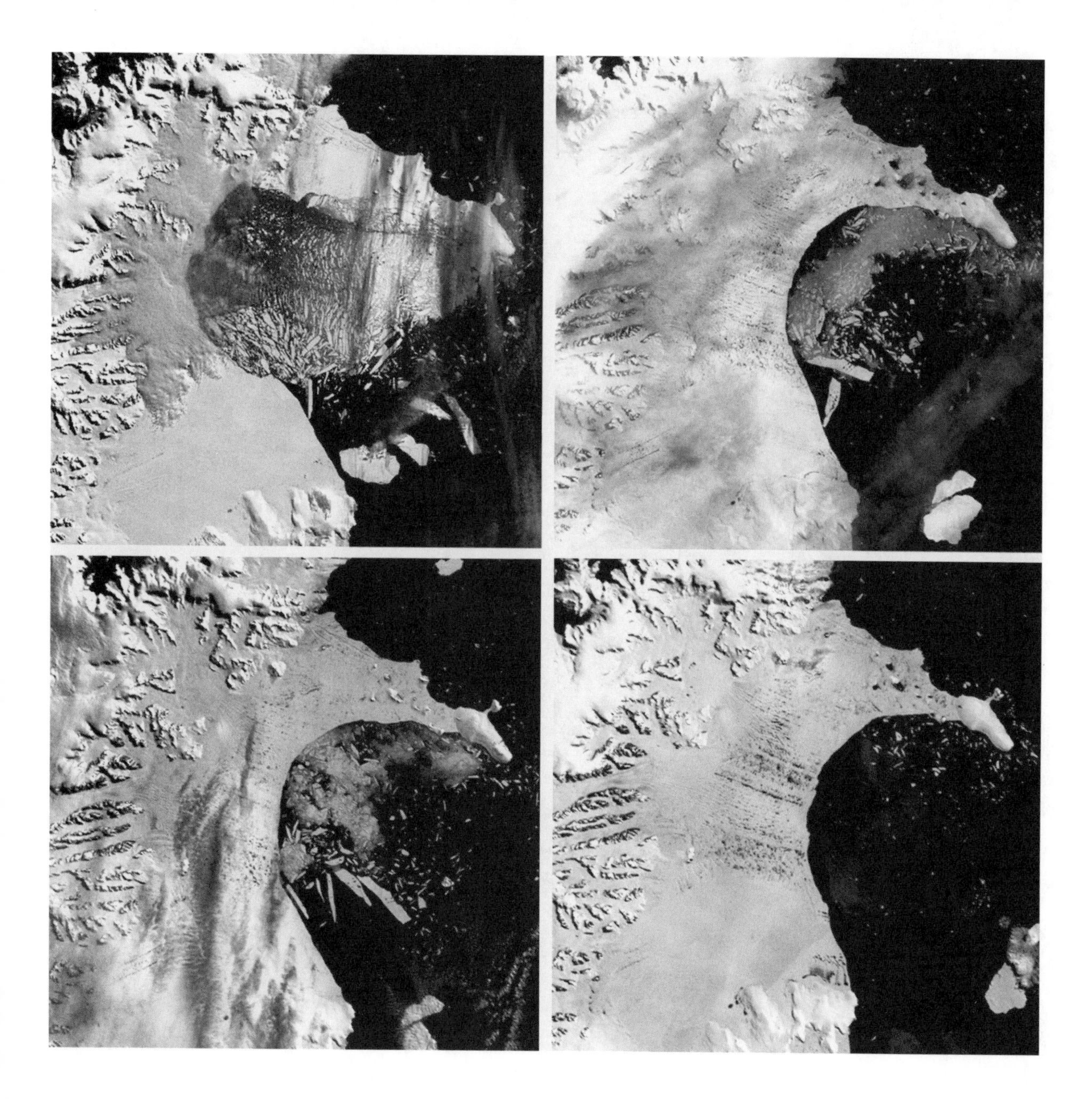

Satellite images taken in early 2002 show the progressive breaking apart of a section of the Larsen B ice shelf, on the eastern side of Antarctica.
(© AFP/Corbis. Reproduced by permission.)

in global warming. They contend that natural forces—such as solar energy output and the tilt of the Earth's axis—which are all beyond human control, dictate major climate changes.

In the near term, most scientists agree that we have been the cause for much of the change observed in the global climate since the start of the Industrial Era. If there is no moderation in population growth, fossil fuel use, and urban sprawl, global warming will continue to increase, weather events will be more extreme, and ocean levels will rise as glaciers melt. However, the alarm bells have been sounded and scientists from all over the world are working to address the problem. Citizens are also becoming more involved and knowledgeable about how they affect the environment. —M. C. NAGEL

Viewpoint:

No, present global warming is not due more to human activity than to natural geologic trends.

Human activities such as clear-cutting rain forests, burning fossil fuels, and driving cars contribute toward global warming. That is not in dispute. But human effects on climate are negligible compared with the effects naturally produced by the Earth, the atmosphere, and the Sun.

The term *global warming* can be misleading or even alarmist. It refers to only half the story of climate change. Gradual modifications of average worldwide surface temperature (AWST) over long periods of time are not only upward.

Volcanic eruptions are a natural cause of global temperature change. Shown here is Mount St. Helens, which erupted in 1980.
(© Corbis. Reproduced by permission.)

Equally important to a thorough understanding of the dynamics of the worldwide climate is the concept of "global cooling." A warming climate is a complex and erratic phenomenon, which may include or even pass over into long or short periods of global or regional cooling. Similarly, periods of global cooling can include or pass over into warming periods. Science clearly shows that AWST oscillates, sometimes in a warming trend, sometimes in a cooling trend. In nature, climate is never stable over long periods of time, and modifications of climate are seldom inexorably progressive.

Until fairly recently in geologic time, there was a long period when the Earth was too warm to allow the formation of polar ice caps. Since their appearance, the Earth has gone through continuously oscillating periods of warming and cooling. Ocean levels naturally fall and rise as the Earth alternately cools and warms, and as the size of the polar ice caps consequently increases and decreases. Sometimes the Earth is so cold that glaciers cover vast areas of otherwise inhabitable land. These periods are called ice ages. In the last 700,000 years, there have been at least seven ice ages, some lasting as long as 80,000 years. In between the ice ages are the interglacial periods, which are almost always shorter than the ice ages.

The current interglacial period began about 11,000 years ago. Since that time AWST has been about 59°F (15°C), plus or minus about two degrees. But 18,000 years ago, in the depth of the Great Ice Age, AWST was about 50°F (10°C).

In 1993 Danish geophysicist Willy Dansgaard reported that average Arctic surface temperature fluctuations of as much as 41° or 42°F (5° or 6°C) per hundred years were not uncommon during the Great Ice Age. The Arctic ice core drill sample shows that, about 10,700 years ago in a famous geologic incident called the Younger Dryas Event, just after the beginning of the current interglacial period, the average Arctic surface temperature dropped 45°F (7°C) in just 50 years.

Several periods of significant temperature fluctuation have occurred naturally during the current interglacial period. Two major warming trends were the Great Climatic Optimum, about 4000 B.C., when AWST reached about 63°F (17°C), and the Little Climatic Optimum, from about 800 to 1250, when AWST was about 61°F (16°C). AWST began dropping rapidly about 1300. By 1400 the world was in the Little Ice Age, from which it did not emerge until about 1850. AWST was about 57°F (14°C), except for a brief warming trend in the late fifteenth and early sixteenth centuries. The Thames froze regularly in winter, Greenland became uninhabitable, and in northeastern North America, 1816 was called "The Year without a Summer." Since the end of the Little Ice Age, the Earth has been generally warming, but not uniformly. AWST is difficult to measure accurately, but in 2000 it was between 59° and 60°F (15° and 15.5°C).

History of Iceland The recorded human history of Iceland is a clear illustration of how natural geological and climatological changes can affect people and their culture. The settlement of Iceland began with Norse landings around 874, during the Little Climatic Optimum, when the growing season was long enough to support several kinds of basic crops, and pasture was abundant for sheep and cattle. The Vikings named the place "Iceland" from sightings of drift ice in a fjord, not from the presence of glaciers. The Icelandic classic, *Njal's Saga*, written in the last half of the thirteenth century about events that happened in the tenth and eleventh centuries, speaks matter-of-factly of the existence of forests and lumbering in Iceland. Such statements would not be possible today. Trees still exist in Iceland, but they are maintained only with great difficulty and care. Because the average Icelandic surface temperature is lower now than in the Saga Age (930–1150), Iceland no longer has any natural stands of woods from which timber might be felled.

The Settlement Age (874–930) through the Saga Age was a golden era for Iceland in terms of both culture and climate. Civil strife in the Sturlung Age (1150–1264) resulted in Iceland's loss of independence, first to Norway in 1262, then to Denmark in 1397. The abrupt decline of Iceland was intensified by the onset of the Little Ice Age. The volcano Hekla erupted throughout the fourteenth century, most notably in 1389, darkening the skies, blocking the summer sun, and chilling the land. Pastures and farmland became unproductive, famine reigned, and many Icelanders starved or froze to death. The years from 1262 to 1874, which encompassed the Little Ice Age, held almost uninterrupted misery for Iceland. No Icelandic cultural, political, or economic renaissance was possible until the Little Ice Age began to abate in the mid-nineteenth century. Led by Jon Sigurdsson (1811–1879), Iceland recovered fairly quickly once global warming began, and the country regained limited independence from Denmark in 1874. Iceland remains vulnerable to its climate, and global cooling threatens it more than global warming.

AWST Fluctuations Statistical meteorological data did not become reliable until about 1880. Since then, AWST has increased about 33°F (0.5°C). This upward trend has not been consistent. From about 1880 to 1940 AWST rose, but from about 1940 to 1970 it fell, and since about 1970 it has been rising. There are also regional variations. The upper Northern Hemisphere cooled from 1958 to 1987, even while AWST increased.

No human activity began or ended the Great or the Little Ice Age or the Great or the Little Climatic Optimum. Anthropogenic, or "humanly caused," modifications of AWST did not begin until the Industrial Revolution during the late eighteenth century. These modifications have always been well within the range of natural fluctuations of AWST. Nothing that humans have ever done to AWST has approached even the maximum AWST of the Little Climatic Optimum. Measurements and predictions of anthropogenic warming of the Earth fall not only within the natural range of AWST fluctuations, but also within the relative error for such measurements and their subsequent predictions. In other words, the data are inconclusive that anthropogenic global warming is at all significant.

What causes the onsets and ends of ice ages and interglacial periods? No one knows for sure. Probably the greatest single factor in determining global cooling and warming cycles is the fluctuation of the Sun's energy output, over which humans obviously have no control. There may be a correlation between weather patterns on Earth and 11-year cycles of sunspots, which are indicative of changes in solar energy output.

Solar energy occurs in many forms, including visible light, ultraviolet radiation, and infrared radiation. Ultraviolet radiation is filtered by the layer of ozone in the stratosphere. The recent anthropogenic damage to the ozone layer above Antarctica by the emission of chlorofluorocarbons (CFCs) and other chlorine compounds is serious, but mostly irrelevant to the question of global warming. The portion of solar energy responsible for heating the Earth is infrared. Every physical body emits heat as infrared radiation.

Some of the earliest and most cogent investigations of variations in solar energy reaching Earth's surface were done by a Serbian engineer, Milutin Milankovitch (1879–1958). He theorized that cycles of global warming and cooling are due to a correlation of three factors: (1) the elliptical orbit of the Earth around the Sun, which varies from more elliptical to more circular on a period of 100,000 years; (2) precession of the axis on a 23,000-year cycle; and (3) variations of the angle of the Earth's axis between 21.6 degrees and 24.5 degrees every 41,000 years.

About half the solar energy directed at the Earth is reflected back into space by the atmosphere. Another 10% is absorbed by atmospheric gases. Of the 40% that reaches the Earth's surface, about two-thirds is absorbed and used as heat and about one-third is reflected back upward.

The so-called greenhouse effect is the absorption of infrared radiation, i.e., the trapping of heat near the Earth's surface, by atmospheric gases. The greenhouse effect is one important reason that the Earth is warm enough to support life. Contrary to misunderstandings popular since the 1980s, the greenhouse effect is natural and almost entirely beneficial to humans. It makes the planet inhabitable by keeping AWST about 59°F (15°C) warmer than it would be without this action.

Human Activity and the Greenhouse Effect

A major question in the global warming dispute concerns the extent to which human practices may exacerbate the greenhouse effect. The so-called greenhouses gases, such as water vapor, carbon dioxide, methane, ozone, and others, are mostly natural components of the atmosphere. The greenhouse effect works because of their natural properties and natural concentrations. Whatever humans may add to them hardly changes their relative concentrations. Humans have little effect on the concentration of the most abundant greenhouse gas, water vapor. The combustion of fossil fuels creates greenhouse gases, but it also creates by-products that contribute toward global cooling, such as sulfur dioxide and sulfate aerosols.

Anthropogenic emissions of carbon dioxide, one of the most significant greenhouse gases, increased dramatically in the twentieth century, chiefly as by-products of industry. Total annual industrial carbon dioxide emissions rose from about 0.5 GtC/yr (gigatons of carbon per year) to over 7 GtC/yr, i.e., about 1,300% from 1900 to 1992; but, in the same period, total atmospheric carbon dioxide concentration rose from about 295 ppmv (parts per million by volume) to about 360 ppmv, i.e., about 22%. Humans released twice as much carbon dioxide into the atmosphere in 1990 as in 1958, but very little of this carbon dioxide remained in the atmosphere to increase the overall concentration of greenhouse gases.

The main reason that atmospheric carbon dioxide concentration has not kept pace with industrial carbon dioxide emissions is that carbon dioxide is soluble in water. At 59°F (15°C), the solubility of carbon dioxide is 0.197 grams per 100 grams of water. Oxygen, by contrast, at the same temperature dissolves only 0.004802 grams in 100 grams of water. Most of the carbon dioxide emitted by human activities quickly dissolve in the upper few hundred feet of the world's oceans, and thus has no chance to accumulate in the atmosphere. The oceans are far from saturated with carbon dioxide. They serve as a "carbon dioxide sink," maintaining a nearly optimal level of carbon dioxide in the atmosphere.

A more troublesome way that humans increase carbon dioxide levels is by deforestation, which increases the amount of carbon dioxide in the atmosphere by reducing the number of trees and other green plants that use atmospheric carbon dioxide in photosynthesis. The forests serve as the second carbon dioxide sink. Equatorial deforestation may have contributed to the southward extension of the Sahara Desert and to the rapid shrinking of Mount Kilimanjaro's ice cap, which lost over 80% of its area from 1912 to 2001. But, however drastic, these anthropogenic modifications of tropical forests, deserts, and isolated mountain ice caps are just regional events from a climatological point of view, and have only insignificant effects on AWST.

Another climatological factor beyond human control is volcanic activity, a widespread, powerful, and completely natural phenomenon that generally has the effect of lowering atmospheric temperature after each eruption. When Mount Pinatubo in the Philippines erupted in 1991, it spewed such a quantity of radiation-blocking ash, gases, and sulfate aerosols into the air that it lowered AWST until 1994. The monstrous explosion of Krakatoa, between Java and Sumatra, in 1883 darkened skies across the world and is believed to have lowered global temperature for as long as 10 years. The eruption of another Indonesian volcano, Tambora, on the island of Sumbawa, in April 1815 contributed to "The Year without a Summer" in North America in 1816. We have already mentioned Hekla's continuing effect on Iceland.

None of the above arguments denies the danger facing such places as Venice, New Orleans, much of the Netherlands, and much of

Bangladesh, which may all disappear under the sea within the next few hundred years. That danger is quite real, but little or nothing can be done about it, as natural causes far outstrip anthropogenic causes of the current global warming trend. —ERIC V.D. LUFT

Further Reading

Adger, W. Neil, and Katrina Brown. *Land Use and the Causes of Global Warming*. Chichester, England: Wiley, 1994.

Budyko, Mikhail Ivanovich. *The Earth's Climate, Past and Future*. New York: Academic Press, 1982.

Burroughs, William James. *Does the Weather Really Matter? The Social Implications of Climate Change*. Cambridge: Cambridge University Press, 1997.

Christianson, Gale E. *Greenhouse: The 200-Year Story of Global Warming*. New York: Penguin USA, 2000.

Drake, Frances. *Global Warming: The Science of Climate Change*. London: Arnold; New York: Oxford University Press, 2000.

Horel, John, and Jack Geisler. *Global Environmental Change: An Atmospheric Perspective*. New York: Wiley, 1997.

Houghton, John. *Global Warming: The Complete Briefing*. Cambridge: Cambridge University Press, 1997.

Intergovernmental Panel on Climate Change [cited July 16, 2002]. <http://www.ipcc.ch>.

Johansen, Bruce E. *The Global Warming Desk Reference*. Westport, CT: Greenwood Press, 2001.

Kondratyev, Kirill Yakovlevich. *Climate Shocks: Natural and Anthropogenic*. New York: Wiley, 1988.

Lin, Charles Augustin. *The Atmosphere and Climate Change: An Introduction*. Dubuque, IA: Kendall/Hunt, 1994.

Mannion, Antoinette M. *Natural Environmental Change: The Last 3 Million Years*. London; New York: Routledge, 1999.

Newton, David E. *Global Warming: A Reference Handbook*. Santa Barbara, CA: ABC-CLIO, 1993.

Nilsson, Annika. *Greenhouse Earth*. Chichester, United Kingdom: Wiley, 1992.

Parsons, Michael L. *Global Warming: The Truth Behind the Myth*. New York: Plenum, 1995.

Philander, S. George. *Is the Temperature Rising? The Uncertain Science of Global Warming*. Princeton: Princeton University Press, 1998.

Ruddiman, William F. *Earth's Climate: Past and Future*. New York: W.H. Freeman, 2000.

Schneider, Stephen H. *Global Warming: Are We Entering the Greenhouse Century?* New York: Vintage Books, 1990.

Singer, S. Fred, ed. *Global Climate Change: Human and Natural Influences*. New York: Paragon House, 1989.

———. *Hot Talk, Cold Science: Global Warming's Unfinished Debate*. Oakland, CA: The Independent Institute, 1997.

Stevens, William K. *The Change in the Weather: People, Weather, and the Science of Climate*. New York: Delacorte, 1999.

United States Environmental Protection Agency Global Warming Site [cited July 16, 2002]. <www.epa.gov/globalwarming/>.

Wilson, Edward O. *The Future of Life*. New York: Random House, 2002.

Wuebbles, Donald J., and Jae Edmonds. *Primer on Greenhouse Gases*. Chelsea, MI: Lewis, 1991.

As the "next-generation" lithography (NGL) system, is extreme ultraviolet lithography (EUVL) more promising than electron beam projection lithography (EPL)?

Viewpoint: Yes, EUVL is more promising than EPL.

Viewpoint: No, EUVL is not necessarily more promising than EPL as the NGL system. Although a replacement for optical lithography must be developed, the nature of that replacement remains to be determined.

Long ago, the word *lithography* referred to a printing process, developed at the end of the eighteenth century, that made use of grease and water. Lithography is still used today for printing, but in the present context, the term refers to the application of lithography in etching the patterns for integrated circuits and transistors on computer microchips.

Essential to the operation of modern computers, chips replaced vacuum tubes during the 1960s, greatly expanding the power of computing machines. The typical computer chip—often called a *microchip*—is made of the element silicon, found in sand and glass, and contains integrated circuits. The latter either carry out instructions within a computer program, in which case the chip serves as a microprocessor, or, in the case of a memory chip, store data and the programs themselves. Transistors control the flow of electric current. The more transistors, and the more lines of electronic circuitry that can be etched onto the chip, the greater its computing power; hence the critical role of lithography in developing ever more powerful chips.

Near the end of the twentieth century, optical techniques, using ultraviolet light, represented the state of the art in computer-chip lithography. In the optical process, light in the ultraviolet spectrum—the wavelengths of which are shorter and higher in frequency than those of visible light—is focused through a mask, a transparent sheet with a pattern of opaque material corresponding to the finished circuit. The opaque material provided a template whereby the pattern could be etched in a silicon wafer to make a chip.

Optical lithography is capable or operating on a wavelength no smaller than 157 nanometers (nm), a nanometer being one-billionth of a meter or one-millionth of a millimeter. While this level of advancement fit the needs of the computer industry for much of the 1980s, by the 1990s it became apparent that optical lithography could not meet ever-increasing demands for chips with greater capacity.

The need to move beyond the bounds of optical lithography is driven in part by Moore's law. The latter is named after Gordon E. Moore (1929–), American engineer and cofounder of the Intel Corporation with Robert Noyce, coinventor of the integrated circuit. In a 1965 article for the journal *Electronics,* Moore stated that, based on developments to that point, the number of transistors on chips tended to double each year. This formulation became known as Moore's law, and though Moore later revised his figure to two years, subsequent developments in the computer industry led to the adoption of eighteenth months as the figure for the amount of time required for the number of transistors to double. In the last two decades of the twentieth century, the industry as a whole adopted the eighteen-month version of Moore's law as an informal and voluntary standard—a goal that its members have continued to meet through the beginning of the twenty-first century.

As a result of this standard, companies began to search for ways to move beyond optical lithography, and the search was on for a next-generation lithography (NVL). Two new methods presented themselves: extreme ultraviolet lithography, known as EUV or EUVL, and electron beam or electronic beam projection lithography, abbreviated as either EBL or EPL. Each offers its advantages and challenges.

Extreme ultraviolet radiation operates at approximately 13 nm, making it capable of printing transistor elements as narrow as the width of 40 atoms. Due to the extremely small wavelengths, use of EUVL requires technology far beyond the scope of that used with the old optical method. Because these ultra-fine wavelengths would be absorbed by optical lens materials, it is necessary to use reflective surfaces rather than lenses, and to coat these surfaces with some 80 alternating layers of silicon and molybdenum. By the beginning of the twenty-first century, the technology was in place to etch circuits as small as 70 nm using EUVL, and an experimental design was projected to create etchings less than half as narrow.

Impressive as EUVL is, it has its detractors, particularly among supporters of electron beam projection lithography. These supporters offered claims that their technology could match that of EUVL, producing resolutions below 35 nm. Furthermore, EPL had the advantage of using existing optical technology. Yet, as with EUVL, there are dangers of mask distortion due to heating, and EPL requires its own special coatings, in this case of tungsten.

By the end of the 1990s, the race was on for the development of NGL, and guiding the quest was International Sematech. The latter is an outgrowth of Sematech (*se*miconductor *ma*nufacturing *tech*nology), a consortium of 14 U.S. computer companies formed with the cooperation of the federal government in 1987 to reinvigorate an American semiconductor manufacturing industry then lagging behind foreign competitors. International Sematech has placed its support primarily behind EUVL, but some of its members, most notably Lucent Technologies and the Nikon Corporation, favor EPL.

As the twenty-first century dawned, corporate members of the computer industry could look ahead to a number of exciting new developments on the horizon, yet the capability of computers can ultimately be no greater than the power of the microprocessors that run them. Realization of this fact has spurred what some have called "the moon shot," the quest to develop NGL. In this effort, companies that normally compete with one another are working together, and though some entities have vested interests in EUVL or EPL, in general, the industry is concerned only with finding the technology that works best.

Indeed, there have been indications that the industry as a whole would be prepared to support both technologies, at least until one or the other emerges as the clear leader. Partisans in the EUVL–EPL debate—particularly supporters of EUVL—maintain that their side has already won; however, the pattern of the future is still not etched. —JUDSON KNIGHT

Viewpoint:
Yes, EUVL is more promising than EPL.

According to a consortium developing EUVL, microchips made with this technology could provide 10-gigahertz (GHz) processors. The speed of the fastest Pentium 4 processor on the market in 2002 is 1.5 GHz. To put it another way, EUVL NGL could lead to microprocessors that are 30 times faster than and have 1,000-fold greater computer memory than current processors. The EUVL technology is being backed by most of the "big guns" of the microelectronics industry, including the world's undisputed number one chip maker, Intel.

In December 1998, at a meeting attended by approximately 110 International Sematech representatives, EUVL was voted the "most likely to succeed" technology to produce the NGL for commercial microchip production. An electron-beam process, called SCALPEL and developed by Lucent, came in second. The Semiconductor Industry Association (SIA) also backs EUVL for NGL.

The 1998 vote of confidence was realized in April 2001 when Sandia National Laboratories/California hosted an EUVL milestone celebration for the launching of a prototype, called the engineering test stand (ETS). The ETS has six subsystems: a laser-produced plasma EUV source, condenser optics, projection optics, a mask, precision scanning stages, and a vacuum enclosure for the entire system.

Sandia is a multiprogram Department of Energy laboratory. It is operated as a subsidiary of Lockheed Martin corporation with main facilities in Albuquerque, New Mexico, and Livermore, California. The SIA is a trade association representing the U.S. microchip industry. More than 90% of United States–based semiconductor production is performed by SIA member companies.

International Sematech (*se*miconductor *ma*nufacturing *tech*nology) is a research consortium created from the globalization of Sematech, which was formed originally to reinvigorate the U.S. semiconductor industry. Member companies share expenses and risk in the precompetitive development of advanced manufacturing technologies, such as EUVL, for tomorrow's semiconductors.

Drivers of the Technology Gordon Moore, Ph.D., cofounder of Intel and author of Moore's law, and other Intel leaders believe the NGL system will be EUVL. They have backed their belief with major investment in an industry-government consortium that includes three U.S. Department of Energy national laboratories, Motorola, Advanced Micro Devices, Micron Technology, Infineon Technologies, and IBM.

Moore's law states that the amount of information stored on a microchip approximately doubles every year. Moore observed in 1965 that a plot of the number of transistors per square inch on an integrated circuit versus time showed the number of transistors doubling approximately every 18 to 24 months. The number of transistors that can be fabricated on a silicon circuit determines the computing speed of the circuit. The semiconductor industry took Moore's law as almost a mandate to maintain the pace. Moore explains that in the world of microchips, if things are made smaller, everything improves simultaneously. Performance and reliability increase, and the cost of an individual transistor decreases dramatically when more is packed in a given area on the chip. A problem is that optical lithography, the method used to produce the chips, has its limits.

Moore's law is the "engine" driving the microelectronics industry, and lithography is the technological driver of Moore's law. The capability of creating finer and finer features to fit more transistors on a chip depends on the capabilities of lithography. The present optical lithography is based on making an image by directing light on a mask, which is much like a stencil of the integrated circuit pattern. The image of the pattern is projected onto a semiconductor wafer covered with light-sensitive photoresist. Lenses are used to reduce and focus the image. Smaller features require shorter wavelengths. Semiconductor manufacturers can print line widths of approximately one-half of the smallest wavelength they can use. Optical lithography is expected to end at a practical limit of a wavelength of 157 nanometers (nm). Wavelengths in the EUV range, 5–25 nm, may some day be able to print a transistor element only 40 atoms wide. Wavelengths that short cannot be focused with lenses because they are absorbed by them, so all-new technology is needed.

Gordon Moore
(Photograph by Ben Margot. AP/Wide World Photos. Reproduced by permission.)

Intel has assumed the position of leader in driving new technology to adhere to Moore's law. In 1996, Intel began funding EUVL research and in 1997 established the consortium known as EUV LLC (Extreme Ultraviolet Limited Liability Company) with Motorola and Advanced Micro Devices. EUV LLC then joined with the Virtual National Laboratory, a partnership of Sandia, Lawrence Livermore, and Lawrence Berkeley, the three national Department of Energy laboratories. A quarter-billion-dollar pledge of private capital got the program under way. The investment return could be considerable. Worldwide semiconductor revenues may reach $1 trillion per year by 2012. U.S. industries are projected to have a 50% market share, or more.

The progress of EUV LLC attracted more companies to the consortium. Micron Technol-

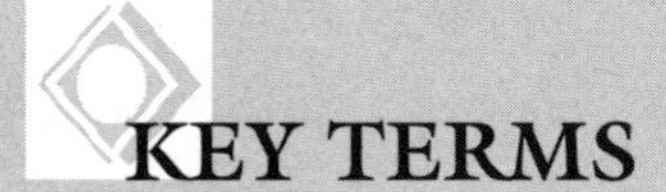

KEY TERMS

CHIP: Short for *microchip,* the very small silicon (semiconductor) modules that provide logic circuitry or memory for microprocessors. Transistors, the basic elements in integrated circuits, are interconnected with circuitry and etched on a microchip. Before etching, the module sometimes is called a *wafer.*

LITHOGRAPHY: A process of creating patterns on semiconductor crystals to be used as integrated circuits for the computer industry.

MASK: A transparent sheet containing a pattern of opaque material to be etched out with one of several lithography processes.

NANOMETER: One billionth of a meter. 50 nanometers is approximately 2,000 times narrower than the width of a human hair. Abbreviated nm.

OPTICAL LITHOGRAPHY: The method of lithography used at the end of the twenty-first century, to be replaced by a form of next-generation lithography (NGL). Optical lithography uses ultraviolet light and operates at a relatively wide wavelength of 157 nm.

PLASMA: An assembly of ions, atoms, or molecules in which the motion of the particles is dominated by electromagnetic interaction, but the plasma is electrically neutral. Plasma sometimes is described as the fourth state of matter.

PHOTORESIST: The photosensitive thin film applied to the surface of a wafer for lithography.

RESIST: A coating that protects against physical, electrical, or chemical action.

ogy and Infineon Technologies joined the team in early 2000. IBM joined in March 2001, when the company recognized EUVL as moving along rapidly in the race for the NGL. However, IBM had been doing and was continuing research on EPL.

Intel and IBM have different research and development strategies. Rather than relying on internal deep research, Intel invests in collaborations with universities and national laboratories to tap the best of the best in basic research and development resources. Intel has research laboratories, but it focuses on new processes that can be transferred into production. Intel's founders saw in earlier corporate experience too much money being spent on research that did not produce a corporate return. By backing the Virtual National Laboratory, the Intel consortium had a jump-start on EUVL research that has put it in front of the competition for the NGL.

Extreme Ultraviolet Lithography Technology

Basic research has been done at the three national laboratories. Each lab is making a unique contribution to the EUVL technology. Lawrence Livermore supplies expertise in optics design, precision engineering, multiple-layer coatings, and projection optics engineering. Sandia provides system engineering, development of the photoactive polymer thin film for photoresist, and light source development. Lawrence Berkeley contributes advanced light source capability for generation of EUV light, conducts defect inspection analysis, and performs EUV scattering experiments.

EUVL technology is built on experience gained from optical lithography, but it requires an entirely new technology. The short wavelengths are not transmitted through optical lens materials; rather they are absorbed. Therefore, instead of lenses, the reduction systems must use reflective surfaces, that is, mirrors, which have special coatings made up of 81 alternating layers of silicon and molybdenum. They were developed at Lawrence Livermore.

EUV radiation is strongly absorbed by most mask materials, so a highly reflective mask that contains the integrated circuit pattern is built into the system. The EUVL mask is produced by applying the same multiple-layer coating of molybdenum and silicon used on the mirrors to a flat, very-high-purity glass substrate. To that, a final EUV-absorbing metal layer is applied, and then the image of the circuit is etched away. The radiation reflected from this pattern mask enters an all-reflective camera containing four mirrors to reduce the image size and focus the image onto the wafer. With this setup, the ETS can print features as small as 70 nm. A six-mirror design for printing features down to 30 nm is in development. The entire lithographic process has to be performed in a vacuum because air absorbs radiation at this wavelength.

The Sandia laboratory developed the laser-plasma source for generating EUV radiation. The laser-plasma source is a commercial 6 kilohertz, 1,500-watt laser that focuses 5-nanosecond pulses of light onto a beam of xenon gas. As described in Sandia literature, the gas consists of weakly bound xenon clusters that contain many thousands of atoms each. The clusters are heated and vaporized. As the clusters absorb the laser energy, a jet of xenon gas is produced as a plasma that reradiates some of the energy at a 13-nm wavelength. The system includes a condenser optics box and a projection optics box. The main role of the condenser optic box is to bring as much light as possible to the mask and ultimately to the wafer. The more light delivered, the shorter is the exposure time, which translates to manufacturing more chips at a faster rate.

An issue during development of EUVL technology was the question of who owns the intellectual property. Inventions from a number of the labs were patented. When the work is entirely funded by the industrial consortium, the consortium usually owns the inventions. If the technology had to be licensed for use, EUVL would be prohibitively expensive.

The Challenges The challenges are many in the development of EUVL. As described in the September 2001 issue of the magazine *Research & Development,* the scope is so wide and the challenges are so great that this project has been called the semiconductor industry's "moon shot" of technology. The rewards are so high for developing the NGL that a number of countries are beginning to conduct research. In addition to the United States, Japan and France are among those developing EUVL. The Dutch tool supplier ASML dropped its involvement with EPL and may be the first non-U.S. company to join in the EUVL consortium.

Manufacturing the almost perfect mirrors needed for EUVL has been a challenge. In addition to being highly reflective, the mirrors must have uniform surface coatings. Any small lack of uniformity in the coatings would result in distorted patterns on the microchips. The Lawrence Livermore and Lawrence Berkeley labs have developed advanced molybdenum and silicon multiple-layer coatings that can reflect nearly 70% of EUV light at a wavelength of 13.4 nm. Each of the 81 layers is approximately 3.5 nm thick. These laboratories also developed an award-winning precision deposition system for applying the thin films to the curved mirrors needed for EUVL—the ultra clean ion beam sputter deposition system. The total thickness over the area coated has a surface deviation of less than one atom.

According to industry experts, the most complex challenges and greatest risks involve the mask for the NGL. Any defects would be replicated on the chips, damaging the complex circuitry. The national laboratories, using the ultra clean ion beam sputter deposition system, produce uniform, highly reflective masks with fewer defects than those produced with conventional physical deposition processes. The system consistently produces fewer than 0.1 defects per square centimeter. The goal is no more than 0.001 defects per square centimeter on the finished wafer blank.

The ETS represents a major milestone in that it has demonstrated the EUVL technology can work. Quality full-field images have been printed at approximately 100 nm, but at 80 nm the quality is not as good. There are still challenges in that the system is extremely complex and requires precision at the nanometer level. Another challenge is the power source. The program director of EUV LLC estimates the power source has to be 10 times more powerful than it is now to achieve the target throughput rate of 80 300-mm wafers per hour needed for cost-effectiveness.

The International Technology Roadmap for Semiconductors (ITRS) projects EUVL and/or EPL should be printing lines at approximately 50 nm by 2011 and 25 nm by 2014. Which technology will be the leader? Three national U.S. Department of Energy labs and the major chip makers are all betting on EUVL. Having Intel as driver of the EUVL technology helps. Peter J. Silverman, Intel's director of lithography capital equipment development, is quoted in the April 2001 *Scientific American*: "We fervently believe that there are not enough resources in the industry to develop both technologies."

ITRS is an assessment of the semiconductor technology requirements to ensure advancements in the performance of integrated circuits. The roadmap is produced by a cooperative effort of global manufacturers, government organizations, consortia, and universities to identify the technological challenges and needs facing the semiconductor industry over the next 15 years. It is sponsored by the SIA, the European Electronic Component Manufacturers Association, the Electronic Industries Association of Japan, and both the Korean and Taiwan Semiconductor Industry Associations. —M. C. NAGEL

Viewpoint:

No, EUVL is not necessarily more promising than EPL as the NGL system. Although a replacement for optical lithography must be developed, the nature of that replacement remains to be determined.

Today's optical lithography is outmoded and must go. That much, at least, is a point of agreement for all sides in the race to develop the NGL system. But which system? Among the candidates, EUVL has a number of supporters, including the SIA and International Sematech. But with test results that go back to the 1990s, many experts remain skeptical about the capabilities of EUVL. At the same time, a number of influential forces, including researchers at Bell Labs of Lucent Technologies and at Nikon Corporation, support a different form of NGL technology: EPL.

Whereas the term *lithography* might call to mind images of printers reproducing scenes of country life, this image belongs to another century. Twenty-first-century lithography is intimately tied to the computer industry, in which it is used to produce patterns on semiconductor crystals to be used as integrated circuits. Just as a book containing more lines and characters of text necessarily contains more information, the more patterns that can be written on the crystals, the more information they can process.

The present system of optical lithography uses light from deep within the ultraviolet spectrum, light with slightly shorter wavelength and higher frequency than ordinary visible light. These waves are focused through a mask, a transparent sheet containing a pattern of opaque material. The pattern of the opaque portion corresponds to that of the finished circuit, which is etched out of a silicon wafer.

As the semiconductor industry has progressed, demand has increased for ever more powerful semiconductors—that is, semiconductors containing more transistors—and this is where optical lithography has reached its limits. Because of the phenomenon of diffraction, or the bending of light waves as they pass around obstacles, patterns become blurred at very small sizes. As the sizes of patterns have shrunk to dimensions only slightly larger than the wavelengths of the light itself, the problem has become ever more serious.

Extreme Ultraviolet Lithography: High Hopes As the need for the NGL system became painfully apparent, manufacturers of semiconductors began the search for a system that would replace the outmoded optical variety. Beginning in the early 1990s, research and development laboratories began investigating the capabilities of EUVL, which uses a shorter wavelength than does optical lithography. Whereas optical lithography, in its most advanced forms, operates on a wavelength of 157 nm, the wavelength of EUV is less than a tenth of that, only 13 nm. (A nanometer is one billionth of a meter, or one millionth of a millimeter, meaning that there are 25,400,000 nanometers to an inch.)

Hopes for the future of EUVL have been high. These high hopes are reflected not only in the formation of EUV LLC, a consortium of companies supporting development of the technology, but also in the EUV LLC budget of $250 million. Experts in the semiconductor industry call the effort to develop EUVL the "moon shot," reflecting the challenges, expenses, and opportunities involved. Just as America's effort to reach the Moon in the 1960s was closely tied with defense interests, defense technology used in making precision mirrors has been adapted in the manufacture of circuits.

Whereas optical lithography is incapable of producing patterns smaller than 130 nm, supporters of EUVL in 2001 maintained that by 2005 it could operate in the 70-nm "node." EUV LLC program director Chuck Gwyn of Intel told *Research & Development,* "EUVL will be the production technology of the future." In the same article, Art Zafiropoulo, chief executive officer of Ultratech Stepper, says, "There is no question in my mind that this is the correct solution to producing devices that have geometries of 0.05 microns [50 nm] and lower. This is the only technology that will be successful." Zafiropoulo went on to observe, "Years ago, we didn't think we could put somebody on the Moon. It will take those kinds of resources and effort and brain power to do it. I would not rule out the ingenuity of American scientists or American companies. I think it's huge, but I love it."

Questions and Concerns Are all these high hopes warranted? The evidence suggests not, for three reasons. First, it is still too early to tell what technology will emerge as the dominant NGL. Second, the EPL method is at least as promising a candidate as EUVL for that dominant position. And third, EUVL has raised a number of concerns.

In the words of Skip Derra in *Research & Development,* "to many . . . the EUV 'alpha tool' is a big, complex system. Proving it can lay down one layer of a circuit pattern in a laboratory test is far removed from the intricacies and tradeoffs that rule real-world manufacturing environments." Henry Smith, professor of electrical engineering at the Massachusetts Institute of Technology, told Derra, "There are many people who are skeptical about EUV's viability in manufacturing."

Even Zafiropoulo, an effusive supporter of EUVL, indicated that the EUV LLC goal of producing ten beta or test machines by late 2003 or early 2004 was an ambitious one. He suggested this goal could not be met until 2007. There is also the matter of the expense: another EUV supporter, Juri Matisoo of the SIA, admitted to Derra that "EUV doesn't come free."

Of course, any technology costs money. That the EUV LLC goals may be a bit overly ambitious is not necessarily a sign of problems with EUVL itself. However, there are more serious concerns. For example, EUVL requires a power source with 10 times the capability of that available in 2001. In addition, detractors noted the complexity of EUVL as well as the fact that it requires extraordinary precision in operation. There is the matter of its high cost relative to other forms of technology. Dale Ibbotson of Agere Systems (a corporate spin-off of Lucent) told Derra he had found a much more cost-effective alternative to EUVL, that is, EPL.

Electron Beam Projection Lithography: A Strong Contender According to the 1998 *Guinness Book of World Records,* the world's smallest wire at that time was a 3-nm piece of nickel and chromium—so small that 8 million of them, laid side by side, would be an inch wide. The wire was made with EPL, a promising NGL

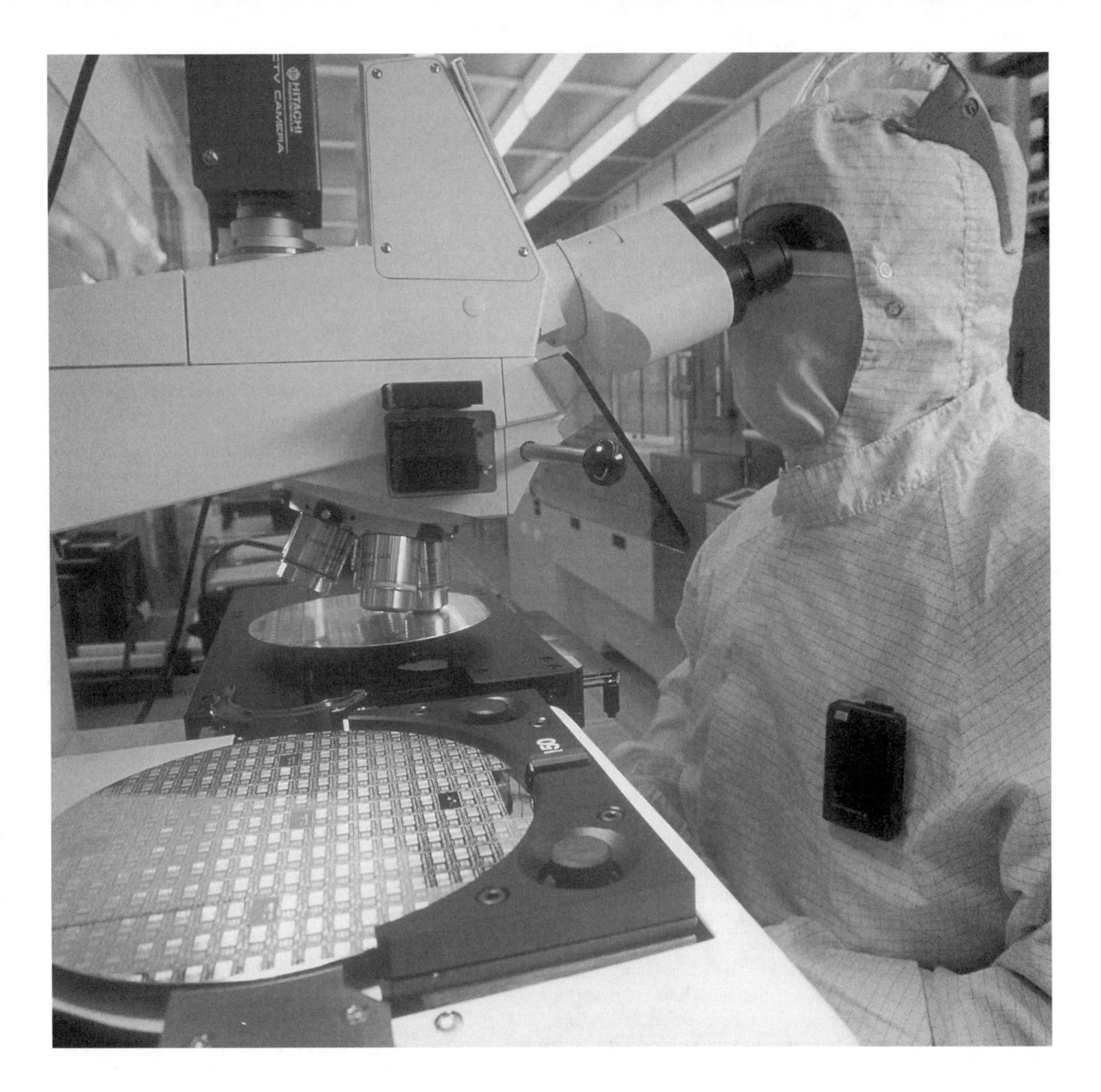

A worker tests silicon wafers using the optical lithography method. (© Kevin R. Morris/Corbis. Reproduced by permission.)

being promoted by a number of industry leaders. Among these is Lucent's Bell Labs, which in 1994 undertook efforts toward development of scattering with angular limitation projection electron beam lithography (SCALPEL).

Past attempts at using EPL had distorted the mask with heat, and the distortion prevented development of EPL as a practical means of chip production. SCALPEL, however, uses a mask pattern of tungsten, which because of its high atomic mass scatters electrons and causes little heat absorption. Errors are a serious problem in EUVL, which uses up to 80 coatings on a chip. Each coating must be perfect down to the atomic scale. By contrast, SCALPEL uses an electrostatic image deflector, which corrects errors in positioning of the projected image. SCALPEL also uses an annular aperture to protect against background exposure due to electron backscatter, or the rebounding of electrons from the resist. (A resist is a coating that protects against physical, electrical, or chemical action.)

SCALPEL offers the benefit of cost savings, because it uses much of the same equipment and many of the processes already used in optical lithography. As Lloyd R. Harriott, director of advanced lithography research at Bell Labs, told *Signal,* "When people have thought about moving to a new lithographic technology, they have usually thought about changing to a whole new infrastructure. But with SCALPEL, the resist mechanism is very similar to what is used in optical [lithography], and in fact uses the same materials. . . . Changeover will involve no huge disruption."

In February 2001, Nikon Research Corporation of America announced that its parent company, Nikon Corporation of Japan, had stepped up an EPL development program begun in conjunction with IBM in 1995. Said Nikon Corporation president Shoichiro Yoshida in a company press release, "We believe that EPL will provide our customers the capability to extend to sub–35-nm resolution." Ibbotson, quoted in the same document as a fellow supporter of EPL, stated ". . . we expect EPL to be especially suited to communications semiconductors, where the technology's speed, flexibility, and low-cost masks are key attributes."

The Future Is Unwritten Toward the end of the February 2001 Nikon press release is a statement that reveals something about the true nature of the EUVL-EPL debate. It is a quote from Mark Meilliar-Smith, president and chief executive officer of International Sematech, a consortium of 12 semiconductor manufacturers, among them both Agere and IBM, from seven countries. The ostensible purpose of the consortium is to advance semiconductor manufacturing technology with an eye toward what is best for the industry as a whole rather than one company or faction. Said Meilliar-Smith, "International Sematech has supported next-generation lithography for several years as an important part of our member companies' technology plans. *Our present programs cover both EUV and EPL*" [emphasis added].

In a similar vein, Aaron Hand of *Semiconductor International* reported after the fifth and final International Sematech NGL Workshop in August 2001, "The industry basically confirmed what had been supposed for some time—that it will proceed with both extreme ultraviolet lithography (EUVL) and electron-beam projection lithography (EPL)." Hand went on to note, "It was originally thought that the industry was going to have to settle on one NGL technology because it could not afford to support more than that. But the tune has changed along with the growing intensity of sticking with Moore's law," which calls for the doubling of the power in computer chips every 18 months. Paraphrasing John Canning of International Sematech, Hand noted, "Now the industry can't afford not to pursue two technologies."

In the annals of scientific discovery, then, the debate between EUVL and EPL is not likely to be remembered as a conflict between good and bad technology or between better and worse science. It is not like the distinction between the ancient atomic theory of matter, which was essentially correct, and the theory that matter is made up of four "elements," which was wholly incorrect. This is more like the conflict between supporters of the wave and particle theories of light in the period between the late-seventeenth and early-twentieth centuries, when it was discovered that light behaves both as a wave and as a particle. "All NGL methods have significant hurdles," Phil Ware of Canon USA, told *Semiconductor International*. "[Y]ou will find most of the requirements for the years beyond 2005 are listed in red boxes. This means that there is no known solution at the moment. Invention is required to meet these requirements." —JUDSON KNIGHT

Further Reading

"Cooperation Helps to Advance Scalpel." *Semiconductor International* 22, no. 3 (March 1999): 24–6.

Dance, Brian. "Putting EUV Optics to the Test." *Semiconductor International* 24, no. 8 (July 2001): 62.

Derra, Skip. "Can Lithography Go to the Extreme?" *Research & Development* 43, no. 7 (July 2001): 12. <http://www.rdmag.com/features/0107euvl.asp>

"Fight Is on for Top Process." *Electronic Times* (November 12, 2001): 3.

Hand, Aaron. "Commercializing NGL: The Push Forward." *Semiconductor International* 24, no. 14 (December 2001): 59–64.

Heller, Arnie. "Extreme Ultraviolet Lithography: Imaging the Future." *Science and Technology Review* (November 1999).

Jackson, Tim. *Inside Intel: Andy Grove and the Rise of the World's Most Powerful Chip Company.* New York: The Penguin Group (Dutton/Plume), 1998.

"Nikon Accelerates Electron Projection Lithography (EPL) Development to Meet Customer Demand at 70 nm." Nikon. February 28, 2001 [cited May 28, 2002]. <http://www.nikon.co.jp/main/eng/news/2001/epl01_e_01.htm>.

"Projection Electron Beam Lithography Breakthrough Achieved by Bell Laboratories Researchers." Lucent Technologies. November 5, 1997 [cited May 28, 2002]. <http://www.lucent.com/press/1197/971105.blb.html>.

Reed, Fred V. "Smaller, Faster Circuit Elements Bypass Previous Expectations." *Signal* 53, no. 1 (September 1998): 79–82.

Reid, T. R. *The Chip: How Two Americans Invented the Microchip and Launched a Revolution.* New York: Random House, 2001.

Stix, Gary. "Getting More from Moore's." *Scientific American,* April 2001.

Van Zant, Peter. *Microchip Fabrication: A Practical Guide to Semiconductor Processing.* New York: McGraw Hill, 2000.

Xiang, Hong. *Introduction to Semiconductor Manufacturing Technology.* Upper Saddle River, NJ: Prentice Hall, 2000.

Do finite element studies have limited usefulness, as they are not validated by experimental data?

Viewpoint: Yes, finite element studies have limited usefulness because they produce idealized data that may be useful for theoretical analysis but could produce serious errors in real-world machines or structures.

Viewpoint: No, most finite element studies are useful and are being validated in ever-increasing numbers of computer-aided design (CAD) applications.

In solving any problem, even those that arise in ordinary life, it is a common belief that the best way to attack the situation is to reduce it to smaller parts. This is the essence of the finite element study, a technique in engineering based on the idea of dividing structures into smaller parts or elements for the purposes of study and analysis. Finite element studies are based in finite element analysis (FEA), a set of mathematical principles underlying the entire process, and the finite element method (FEM), which is simply the application of these principles. The concepts of FEA and FEM are so closely linked that they are often used interchangeably in discussions of finite element studies.

Though the idea of reducing a problem to its constituent parts is an old one, FEA is a relatively new concept. It can be traced to the work of mathematicians and engineers working in the middle of the twentieth century, most notably German-born American mathematician Richard Courant (1888–1972) and American civil engineer Ray William Clough (1920–). Courant was the first to apply what would become known as the FEM when, in 1943, he suggested resolving a torsion problem (one involving the twisting of a body along a horizontal axis, as when turning a wrench) by resolving it into a series of smaller triangular elements. Clough, who coauthored a 1956 paper formulating the principles of the FEM, actually coined the term "finite element method" in 1960.

The occasion of the 1956 paper was a study of stiffness in wing designs for Boeing aircraft, and in the years that followed, the aircraft and aerospace industries would serve as incubators for the new technique known as FEM. The basics of FEM involve use of a number of techniques across a whole range of mathematical disciplines, particularly calculus. These include vector analysis, concerned with quantities that have both direction and magnitude; matrix theory, which involves the solution of multiple problems by applying specific arrangements of numbers in columns and rows; and differential equations, a means of finding the instantaneous rate at which one quantity changes with respect to another.

In performing finite element analysis, engineers build models using principles borrowed from graph theory and combinatorics, areas of mathematics pioneered by Swiss mathematician Leonhard Euler (1707–1783) in solving the "Königsberg bridge problem." Since Euler's method of resolving this brainteaser illustrates how engineers develop models based on FEA, it is worth considering very briefly how he did so.

The town of Königsberg, Prussia (now Kaliningrad, Russia), astride the Pregel River, included two islands. Four bridges connected the mainland with the first island, and an additional two bridges connected the second island with the shore, while a seventh bridge joined the two islands. The puzzle was this: could a person start from a particular point and walk along a route that would allow him or her to cross each bridge once and only once? In solving the problem, Euler drew a graph, though not a "graph" in the sense that the word is typically used. Euler's graph was more like a schematic drawing, which represented the four land masses as points or nodes, with segments between them representing paths. Using the nodes and segments, he was able to prove that it is impossible to cross each bridge without retracing one's steps at some point.

Models using FEA also apply nodes, which in this case refer to connections between elements or units of the model. A group of nodes is called a *mesh,* and the degree to which the nodes are packed into a given space is referred to as the *mesh density.* Engineers then take data derived from studies of the mesh and apply computer-aided analysis to study structural properties. This method is in contrast to what is called *analytical techniques,* or means that involve direct study of the structure in question.

FEA is only one of several techniques for mathematical model-building applied by engineers, but it is among the most popular. From its origins in the aircraft industry, its uses have expanded to include studies of shipbuilding, automobile manufacture (including crash tests), and various problems in civil and architectural engineering—for example, stress capacity of bridges. Thanks to the development of computer software that does the calculations for the user, application of FEA has also spread to other sciences—even, as noted below, paleontology.

Despite its wide application, FEA is not without controversy. Detractors maintain that FEA applies an undue level of abstraction to problems that are far from abstract—problems that, if miscalculated, could result in errors that would cost human lives. Furthermore, opponents of FEA hold that the underlying principle is invalid: whereas on paper it is possible to break a large entity into constituent parts, in the real world those parts are inseparable, and their properties impinge on one another in such a way as to affect—sometimes drastically—the operation of the whole.

Supporters of FEA, on the other hand, point out that there are many situations for which analytical techniques are simply inadequate. If a bridge can only be sufficiently tested when it is built, for instance, then that may well be too late, and the expenses and dangers involved far outweigh the possible risks that may result from the approximations necessary for FEA calculations. It is true that many engineering designs can be tested in model form by using a wind tunnel—another by-product of developments in the aerospace industry—but no method of pre-testing can be considered absolutely fail-safe.

The benefits of FEA to design are demonstrable, with a number of examples, cited below, of successful implementation in the development of aircraft, automobiles, and other products. Certainly there are risks, but these (according to supporters of FEA) reside more in the engineer's ability to apply the method than in the method itself. The mathematics involved are so complex that it is essential for the practitioner to understand what he or she is doing, regardless of whether that person possesses computer technology that can aid in calculations. —JUDSON KNIGHT

Viewpoint:

Yes, finite element studies have limited usefulness because they produce idealized data that may be useful for theoretical analysis but could produce serious errors in real-world machines or structures.

In solving engineering problems, a designer has at his or her disposal two basic choices: analytical techniques, which involve direct observation of the structure in question, and one or another form of mathematical model-building. Among the most widely used forms of the latter is the finite element method, also known as finite element analysis (FEA). The finite element method is a "divide and conquer" strategy: it breaks larger problems down into smaller ones, for instance by dividing a physical area to be studied into smaller subregions. At the same time, it calls for the development of approximations that further simplify the problem or problems in question.

If the finite element method sounds like a time-saver, that is because it is: it makes it possible for engineers to reduce insurmountable problems to a size that is conquerable. On paper, in fact, the finite element method is extremely workable, but "on paper" may be the method's natural habitat, since there are legitimate questions regarding its value in real-world situations. The phrase "close enough for government work," with its implication that minor errors will eventually become invisible if camouflaged in a large enough bureaucracy, might be applied to the finite element method. The latter, after all, is based in part on the idea that small irregularities in surfaces and measurements can and should be eliminated on paper, so as to make calculations easier.

This approach raises a very reasonable concern—namely, is a method that makes use of approximation good enough for design and analysis of real-world structures? On the surface of it, the answer is clearly *no.* It is clear enough intuitively that, given a certain number of irregularities that have been eliminated from calculations, the results may be in error by a degree significant enough to challenge the integrity of the larger structure. Assuming the structure in question is one with the potential to either protect or take human life—a bridge or an airplane, for instance—miscalculation can have serious results.

The Finite Element Method in Action The finite element method has its roots in engineering developments from around the turn of the nineteenth century. During that era, it first became common for bridge-builders and other engineers to apply the theory of structures, or the idea that a larger structure is created by fitting together a number of smaller structural elements. Today, of course, this concept seems self-evident, a concept as simple as assembling interlocking Lego blocks to form a model skyscraper, but at the time it was revolutionary. Of particular significance was the realization that, if the structural characteristics of each element could be known, it was theoretically possible—by algebraic combination of all these factors—to understand the underlying equations that governed the whole.

This matter of algebraic combination, however, proved to be a thing more easily said than done where extremely complex structures were involved. Engineering analysis of large structures hit a roadblock; then, in the mid-twentieth century, a new and promising method gained application in the aircraft industry. This new method was finite element analysis, itself a product of several developments in mathematics and computer science, including matrix algebra and the digital computer, which made possible the solution of extremely large and detailed problems.

The "finite elements" in the finite method are the simple regions into which a larger, more complex geometric region is divided. These regions meet at points called *nodes,* which are the unknown quantities in problems of structural analysis involving stress or displacement. Within the realm of finite element analysis, a number of methods exist for solving problems, depending on the nature of the factors involved: for example, in solid mechanics, functions of the various elements are used as representations for displacements within the element, a technique known as the displacement method.

An application of the displacement method could be a two-dimensional computer-aided design (CAD) model of an automobile before and after collision with another object. Finite element analysis would make it possible to evaluate the crumpling that would occur in each door panel, and in other discrete areas of the car body. Moreover, with the use of proper techniques, it is possible to plug in values for factors including the car's velocity, the velocity of the other object, the direction of impact, and so forth, so as to obtain an accurate model of how each part of the car responds to the crash. There are four essential stages to this process: generation of a model, verification of data, analysis of the data through generation of equations, and postprocessing of the output thus obtained so as to make it efficient for use.

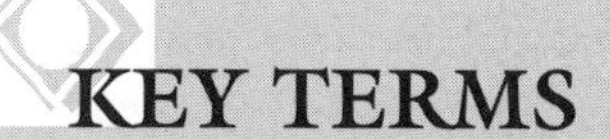

KEY TERMS

COMPUTER-AIDED DESIGN (CAD): The use of complex engineering or architectural software in design work, particularly for the purpose of providing two-dimensional models of the structure to be built.

DIFFERENTIAL EQUATION: A term used in calculus relating to an equation that expresses a relationship between functions and derivatives. Function (f) is defined as an association between variables such as would be expressed $f(x) = y$. Derivative means the limiting value of a rate of change of a function with respect to a variable, as could be expressed by dy/dx = a constant.

FINITE ELEMENT METHOD/FINITE ELEMENT ANALYSIS (FEA): A mathematical technique for solving complex engineering problems by dividing a physical area to be studied into smaller subregions, with variables represented by polynomial equations, and developing approximations of some values that further simplify the problems in question.

MATRIX THEORY: The algebraic study of matrices and their use in evaluating linear processes. A matrix is a rectangular array of numbers or scalars from a vector space.

NODE: In the context of finite element analysis, a node is the point at which subregions meet. In problems of structural analysis involving stress or displacement, nodes represent unknown quantities.

SCALAR: A quantity which has magnitude but no direction in space.

VECTOR: A quantity that has both magnitude and direction in space.

Though pioneered by the aircraft industry, the finite element method is applied today in shipbuilding, automobile manufacture, and both civil and architectural engineering. Engineers use it to predict levels of thermal stress, effects of vibration, buckling loads, and the amount of movement a building will experience due to wind and other factors. Paleontologist Emily Rayfield even used it to study the jaw of *Allosaurus fragilis,* a dinosaur of the Jurassic period. "I'm trying to see how this apparently carefully evolved bone structure translated into biting strength to see if it can further our understanding of how Allosaurus hunted and fed," she told *Mechanical Engineering*.

What's Wrong with This Picture? Like many users of the finite element method, Rayfield availed herself of software—in this case a program called *Cosmos,* produced by Structural Research and Analysis of Los Angeles, California—that performs the tedious and time-consuming computations involved. Such software, supporters of the method maintain, has made finite element analysis easier for the non-mathematician to apply. Wrote Ulises Gonzalez in *Machine Design,* "Admittedly, early FEA [software] versions were difficult to use. Engineers needed significant expertise to build and analyze models." However, "The required expertise level has since diminished so rapidly that most engineers can now produce accurate results. Expert analysts still have their place solving the most difficult problems and mentoring their juniors. But for the rest of the user audience, modern FEA packages automatically and intelligently deal with geometry, material, and control settings so users are free to be more creative."

As rosy as this picture may seem, the ease of operation associated with modern FEA software does not come without a price. In a *Design News* primer on such software for engineers, Bob Williams noted that "there's some tradeoff of speed for accuracy." Rayfield, in her calculations involving the Allosaurus skull, had to operate as though all the skull bones were fixed in place, when in fact she had reason to believe that some of the bones moved in relation to one another. This assumption was necessary in order to apply the finite element method. "It's faster than calculating stress by hand," observed Williams, discussing a finite element program applied for motion simulation, "but it uses a rigid-body motion program, so the engineer must accept certain assumptions. For instance, parts like gaskets are flexible, but a rigid-body program can't calculate that."

In using the finite element method, some accuracy is lost for the sake of speed and convenience. Assuming the design is for a machine or structure that has the potential to protect or take human life, the results can be grave indeed. According to Paul Kurowski in *Machine Design,* "Idealizing . . . a 3-D model [using the finite element method] eliminates small and unimportant details. Sometimes the process replaces thin walls with surfaces, or drops a dimension to work with a 2-D representation of the part. . . . The process eventually forms a mathematical description of reality that we call a mathematical model." But, as Kurowski noted, finite-element analysis "hides plenty of traps for uninitiated users. Errors that come from idealization . . . can be bad enough to render results either misleading or dangerous, depending on the importance of the analysis."

Kurowski explained the popularity of the finite method thus: "To analyze a structure, we solve its equations. Solving complex equations 'by hand' is usually out of the question because of complexity. So we resort to one of many approximate numerical methods," of which finite element analysis is the most widely used. Variables in a finite element model are represented by polynomial functions, which, if they described the entire model, would have to be extraordinarily complex. "To get around that difficult task," Kurowski went on, "the model (a domain) is split into simply shaped elements (subdomains). . . . Notice that a continuous mathematical model has an infinite number of degrees of freedom while the discretized (meshed) model has a finite number of degrees of freedom."

In other words, the finite element model, by breaking the problem into smaller pieces (a process known as *discretizing* or *meshing*), creates artificial boundaries between parts that are not separated in the real world. Two ballast compartments in a ship's hull, for instance, may be treated in a finite element model as though they were entirely separated from one another. However, if in reality there is a flow-through system such that overflow from one compartment enters the other compartment, then this can have a serious impact on issues such as the ship's center of gravity. Noting that "The allowed complexity of an element's shape depends on the built-in complexity of its polynomials," Kurowski observed that "Errors from restrictive assumptions imposed by meshing . . . can have serious consequences. For example, modeling a beam in bending [using just] one layer of first-order elements is a recipe for disaster."

Alternatives M. G. Rabbani and James W. Warner, writing in the *SIAM Journal on Applied Mathematics,* noted the shortcomings of finite element techniques for modeling the transport of contaminants through an underground water or sewer system. "The main disadvantage with the finite element method," they maintained, "is that its mathematical formulations are relatively complicated and follow set rules step by step. In a recent [early 1990s] investigation, it was found that such prescribed rules are not applicable everywhere." Similarly, Shyamal Roy, discussing what goes on in the initial stages of project planning, wrote in *Machine Design* that "when ideas are in flux and design goals are moving targets, CAD and FEA are not the most appropriate tools."

Roy went on to note that "Predesign work often prompts questions such as, when will we know how sensitive belt tension is to tolerances in pulley diameter and belt length? How much work should it take to optimize an extrusion profile so the material extrudes straight? How long will it take to determine section properties for a complex beam? Or, why is it necessary to wait until finishing a design to check for interferences in the mechanism? The common thread through all these questions is that design decisions are waiting on answers CAD programs"—and, by extension, finite element analysis—"cannot deliver." Roy went on to argue for modifications to premodeling design-analysis (PMDA) software, particularly by allowing opportunities for graphical solutions (i.e., drawings) that may not require the use of equations.

It should also be noted that the finite element method is not the only mathematical technique that can be used in engineering analysis. Alternatives include the boundary element, discrete, finite difference, finite volume, spectral, and volume element methods. With each mathematical method, however, there is still the potential for pitfalls that may result from relying too heavily on mathematical models. Not only are those models no substitute for real-world testing of materials, but when the model is based on approximations rather than exact figures, there is a danger that the resulting design may only be "close enough for government work"—which in many cases is unacceptable or even downright dangerous. —JUDSON KNIGHT

Viewpoint:
No, most finite element studies are useful and are being validated in ever-increasing numbers of computer-aided design (CAD) applications.

Finite element studies, which are identified more commonly as finite element analysis (FEA) and finite element method (FEM), have become a mainstay for computer-aided industrial engineering design and analysis. The difference between FEA and FEM is so subtle that the terms are frequently used interchangeably. FEM provides the mathematical principles used in finite element analysis. FEA software is robust and friendly. It is used by engineers to help determine how well structural designs survive in actual conditions, such as loads, stress, vibration, heat, electromagnetic fields, and reactions from other forces.

The majority of FEA are used by design engineers to confirm or choose between different systems or components. According to Steven J. Owen, Senior Member Technical Staff at Sandia National Laboratories, increasingly larger and more complex designs are being simulated using the finite element method. FEA was first applied in the aerospace and nuclear industries. The National Aeronautics and Space Administration (NASA) developed a powerful general purpose FEA program for use in computer-aided engineering. It would be accurate to say finite element studies have a *limitless* usefulness, for they are being validated in ever-increasing numbers of computer-aided design (CAD) applications.

Background Finite element studies are based on the idea that any structure, no matter how complex, can be divided into a collection of discrete parts, i.e., elements. This is not a new idea. Ancient mathematicians used it to estimate the symbol pi (π) from a polygon inscribed in a circle. What is relatively new is its application to structural analysis, fluid and thermal flows, electromagnetics, geomagnetics, and biomechanics.

Prussian mathematician Richard Courant is credited with having suggested in 1943 that a torsion problem could be solved by using triangular elements and applying approximations following the Rayleigh-Ritz method. Although he did not use the term finite element, Courant's work is often referred to as the beginning of FEM because he used the key features of the finite element method in his studies. The Rayleigh-Ritz method of approximation was developed earlier, and independently, by the two mathematicians, Lord Rayleigh (1842–1919) and Walter Ritz (1878–1909).

Some say FEM started with a classic paper titled "Stiffness and Deflection Analysis of Complex Structures," published in September 1956 by M. J. Turner, R. W. Clough, H. C. Martin, and L. J. Topp in the *Journal of Aerospace Science*. The paper was based on the combined research of Professor Ray Clough of the University of California at Berkeley, Professor Harold C. Martin of the University of Washington, and the other co-authors who were working with them at Boeing Airplane Company in 1953, analyzing the stiffness of a specific wing design by dividing the wing structure into triangular segments.

By 1960 Professor Clough had come up with the official name "finite element method," and by that time, there were computers powerful enough to meet the computational challenge. Also, by that time, NASA was working on the implementation of the finite element method using computers at the Goddard Flight Center. One of the NASA projects ultimately produced the powerful and proprietary program for FEM called *NASTRAN* (NASA STRuctural ANalysis).

How Does FEA Work? FEA is an analytical, mathematical-based computerized engineering tool. The mathematics of finite element analysis is based on vector analysis, matrix theory, and differential equations. Fundamental to FEA is the concept that a continuous function can be approximated using a discrete model. That is, FEA is based on the idea of building complicated objects from small manageable pieces. Element is the term for the basic subdivision unit of the model. A node is the term for the point of connection between elements. A complex system of points (nodes) make up a grid called a *mesh.* How closely packed the nodes are is described as the mesh density. Data on the mesh is what is used in computer-aided analysis of the material and its structural properties under study.

Using FEA, virtually any structure, no matter how complex, can be divided into small elements that simulate the structure's physical properties. The model created by FEA is then subjected to rigorous mathematical examination. FEA significantly reduces the time and costs of prototyping and physical testing. Computer-aided engineering has been extensively used in the transportation industry. It has contributed significantly to the building of stronger and lighter aircraft, spacecraft, ships, and automobiles.

Industry generally uses 2-D modeling and 3-D modeling, with 2-D modeling better suited to small computers. Today's desktop PC can perform many complex 3D analyses that required a mainframe computer a few years ago. A great deal of powerful software has been developed for finite element analysis applications. Among the new developments are mesh generation algorithms to tackle the challenges of smoothing surface domains. Where early FEA utilized only tens or hundreds of elements, users of FEA have pushed technology to mesh complex domains with thousands of elements with no more interaction than the push of a button.

Supercomputer vs. PCs for FEA In-depth studies of material synthesis, processing, and performance are essential to the automobile industry. Questions on how well lighter steels absorb energy compared to heavier materials during a crash are the types of problems that are being studied on supercomputers at Oak Ridge National Laboratory (ORNL). Researchers at ORNL have shown that complex models can be developed and modified to run rapidly on massively parallel computers.

While automotive design verification is provided dramatically from actual vehicle collision tests, a clear understanding of material behavior from vehicle impact simulation provides valuable information at a lot less expense. Computational simulations for the single- and two-car impacts have been run numerous times to identify deficiencies of existing vehicle models, and to provide data to improve their performance. The large, detailed finite element models are not feasible for single processors. According to an ORNL report on the analysis of material performance applications, for the best vehicle models to capture complex deformation during impact, it is not unusual to have 50,000 or more finite elements.

Finite element analysis of aerospace structures is among the research projects being carried out at the Pittsburgh Supercomputing Center, a joint effort of Carnegie Mellon University and the University of Pittsburgh together with Westinghouse Electric Company. The center was established in 1986 and is supported by several federal agencies, the Commonwealth of Pennsylvania, and private industry.

Charles Farhat, Professor and Director of the Center for Aerospace Structures at the University of Colorado at Boulder, has worked to develop sophisticated computational methods for predicting the "flutter envelope" of high performance aircraft and to encourage wider use of these methods in industry. Flutter is a vibration that amplifies. The potential of flutter to cause a crash challenges designers of high-performance aircraft. Wing flutter—partly the result of faulty structural maintenance—caused an F-117A Nighthawk, an operational "stealth" plane, to lose most of one wing and crash at an air show in September 1997.

Flutter envelope is a curve that plots speed versus altitude. It relates information on the speed not to exceed, even if the engines can do it, because any perturbation is going to be amplified, according to Farhat. Flutter involves the interaction between flow and structure. To predict it requires solving equations of motion for the structure simultaneously with those for the fluid flow. As computational technology improves, it is possible to use sophisticated modeling techniques, including detailed finite element structure modeling of aircraft. In 2001 Professor Farhat did flutter predictions for the F-16 using the Pittsburgh Supercomputing Center's new Terascale Computing System to validate the innovative simulation tool he developed.

The question of which is best—supercomputer or desktop PC—depends entirely on the application of FEA. If the job requires dismantling one model of a sport utility vehicle to do an FEA of each part and simulate crash tests, ORNL researchers with their massively parallel computers are needed to do the job. However, when the study involves flaw evaluation and stress analysis for materials and welds, a desktop PC can perform complex 3D analyses using available commercial software packages such as ABAQUS. Many jobs that would have required a mainframe or high-end workstation a few years ago can be done on a PC today. Even NASA's NASTRAN software has been adapted for use on a PC.

The NASA Connection The background for the development of NASTRAN, FEA software for computerized engineering, is one of NASA's Spinoff stories. The Spinoff stories describe developments that started as part of NASA's aeronautic and space research and have now become useful to other industries. In the 1960s the MacNeal-Schwendler Corporation (MSC) of Los Angeles, California, was commissioned to develop software for aerospace research projects, identified as the NASA Structural Analysis program, NASTRAN. In 1982, MSC procured the rights to market versions of NASTRAN. NASA says the transfer of technology to non-aerospace companies is viewed as important to developing U.S. competitiveness in a global market. Several types of licenses are available to use NASTRAN.

NASTRAN is described as a powerful general purpose finite element analysis program for computer-aided engineering. It has been regularly improved and has remained state-of-the-art for structural analysis. NASTRAN can be used for almost any kind of structure and construction. Applications include static response to concentrated and distributed loads, thermal expansion, and deformation response to transient and steady-state loads. Structural and modeling elements include the most common types of structural building blocks such as rods, beams, shear panels, plates, and shells of revolution, plus a "general" element capability. Different sections of a structure can be modeled separately and then modeled jointly. Users can expand the capabilities using the Direct Matrix Abstraction Program (DMAP) language with NASTRAN.

NASA continues to be involved. The Finite Element Modeling Continuous Improvement (FEMCI) group was formed at NASA to provide a resource for FEA analytical techniques necessary for a complete and accurate structural analysis of spacecraft and spacecraft components.

Some Applications of FEA PC software adapted from NASTRAN by Noran Engineering (NE) of Los Alamitos, California, is available for such applications as the analysis of the Next Generation Small Loader, which will replace an older aircraft loader used by the Royal Air Force (RAF) of the UK and the Royal Norwegian Air Force. As described by NE, the loader will have to be able to withstand both arctic and desert conditions as well as support large weights, withstand dock impacts, and various types of point loading. The FEA was completed in less than an hour with an Intel Pentium III 500 MHz computer.

FEA software identified as LS-DYNA was developed at Lawrence Livermore National Laboratory. Work on applications for stress analysis of structures subjected to a variety of impact loading began in 1976. The software was first used with supercomputers, which at that time were slower than today's PCs. The laboratory continued to improve the software. By the end of 1988 the Livermore Software Technology Corporation was formed to continue development of LS-DYNA. While research is still improving it, software is available for use in Bio-Medical applications. Serious head and neck injuries sustained by autocrash victims are simulated in LS-DYNA to aid in the design of airbags, head support, and body restraint systems. Other applications include simulations for airbag deployment, heat transfer, metal forming, and drop testing. In addition, the software is used in seismic studies, as well as for military and automotive applications.

As with any computer program, the results of using sophisticated applications is only as good as the input. FEA is described as a powerful engineering design tool. When using FEA software, it is exceedingly important for the user to have some understanding of the finite element method and of the physical phenomena under investigation. Finite element studies are becoming available as part of engineering programs in many universities, and research continues to make both FEM and FEA more useful. —M. C. NAGEL

Further Reading

Buchanan, George R. *Schaum's Outline of Finite Element Analysis.* New York: McGraw-Hill, 1994.

Gonzalez, Ulises. "Smarter FEA Software Unburdens Users." *Machine Design* 73, no. 5: 120–26.

Hughes, Thomas J. *The Finite Element Method: Linear, Static and Dynamic Finite Element Analysis.* Mineola, NY: Dover Publications, Inc., 2000.

Kurowski, Paul. "More Errors That Mar FEA Results." *Machine Design* 74, no. 6: 51–56.

Kwon, Young W., and Hyochoong Bang. *The Finite Element Method Using MATLAB.* Boca Raton, FL: CRC Press LLC, 2000.

Lander, Jeff. "Pump Up the Volume: 3D Objects That Don't Deflate." *Game Developer* 7, no. 12: 19.

Lee, Jennifer. "Behind the Leaky Faucet: Dissecting the Complex Drip." *New York Times* (6 March 2001): F-4.

Rabbani, M. G., and James W. Warner. "Shortcomings of Existing Finite Element Formulations for Subsurface Water Pollution Modeling and Its Rectification: One-Dimensional Case." *SIAM Journal on Applied Mathematics* 54, no. 3: 660–73.

Roy, Shyamal. "Don't Touch CAD Until You've Firmed Up the Concept." *Machine Design* 74, no. 4: 70–72.

Thilmany, Jean. "Digital Analysis of a Dinosaur's Head." *Mechanical Engineering* 123, no. 11 (November 2001): 18.

Williams, Bob. "Six Things All Engineers Should Know Before Using FEA." *Design News* 55, no. 24: 85–87.

Is the research emphasis on the wear of ultra-high molecular weight polyethylene bearing components in joint replacements warranted, given that the longevity of such replacements depends on a multitude of factors?

Viewpoint: Yes, the research emphasis on the wear of ultra-high molecular weight polyethylene bearing components in joint replacements is warranted: not only is physical wear the most important factor in the aging and malfunctioning of joint replacements, but it is closely tied with chemical and biological factors.

Viewpoint: No, the current research emphasis on the wear of polyethylene bearing components in joint replacements is unwarranted, given that alternative materials are available that can provide not only greater longevity but biological tolerance as well.

Prosthesis is the development of an artificial replacement for a missing part of the body, including arms and legs, and the area of medicine devoted to the study and development of prostheses is called *prosthetics.* Since ancient times, people have attempted to create substitutes for missing limbs, but the origins of prosthetics as a scientific discipline dates back only to the sixteenth century and the work of French surgeon Ambroise Paré (1510–1590).

Despite his efforts, Paré was ultimately faced with the limits of sixteenth-century knowledge, both in terms of antisepsis (keeping a body part free of germs) and engineering. Nevertheless, physicians of the early modern era were able to develop replacements for missing parts in the upper extremities—for example, metal hands fashioned in one piece, and later, hands with movable fingers. Still, the technology that would make it possible for a patient to actually move his or her own fingers the way a person with a whole limb would—in other words, through neuromuscular activity originating in the brain—still lay far in the future.

The two world wars of the twentieth century brought with them great improvements in prosthetics, including the development of more serviceable mechanical joints and the implementation of materials that were lighter in weight and therefore more usable. Today, prostheses are typically made of polyethylene, a dense form of plastic also used in manufacturing everything from electrical insulation to bottles. Other materials include metal alloys, such as cobalt-chromium, as well as ceramics, and often one of these is combined with polyethylene in creating a prosthetic device.

One of the biggest challenges in developing successful prostheses involves the interaction between the artificial material and the truncated body part to which it is attached. Below-knee prostheses, for instance, are held in place either by means of a strap that passes above the kneecap or by rigid metal hinges that attach to a leather corset wrapped round the thigh. Above-knee prostheses may be attached to a belt around the pelvis or slung from the shoulder, or they may be held in place by means of suction. In cases of amputation from the hip joint, it may be necessary to use a plastic socket. Arm prostheses may include a shoulder harness with a steel cable, which the patient can operate by shrugging the shoulder, thus tightening the cable and opening or closing the fingers of the artificial hand.

More sophisticated prostheses, developed in the latter half of the twentieth century, actually established an interface between the patient's muscles and the artificial limb. This is achieved through a surgical operation called *cineplasty.* It is also possible, in some cases, to use electric current to operate an artificial hand. In such devices, known as *myoelectrical control systems,* electrodes are built into the prosthesis itself and are activated by contractions of the patient's muscle.

In creating and using a device that attaches to a living body, there are a whole host of challenges, most of which are either physical or biochemical in nature. On the physical level, there is the problem of simply ensuring that the device will do what it is supposed to do, and at the same time withstand the ordinary wear that will be exerted on it in the course of daily use. Much more complicated, however, are the biochemical concerns, which can be extremely serious.

First of all, there is the matter of ensuring sterility, such that no foreign biological material is introduced to the patient's body. There is also a problem that crosses the lines between the physical and the biochemical, resulting from the friction between the body part and the prosthesis. This friction can and often does cause the displacement of microscopic particles from the prosthetic device—particles that may invade the patient's body or may serve as a medium for invasion by microbes.

Such are the problems that confront makers of prosthetic devices, who are engaged in an extremely complex and vital endeavor that brings together disciplines rooted both in the medical sciences and in engineering. Therein lies the basis for the controversy at hand, which revolves around the use of polyethylene components in artificial joints. In the essays that follow, several issues are examined. In particular, there are the questions of whether polyethylene is the best material for such components, and whether the present emphasis on physical wear in the testing of polyethylene prosthetic devices is warranted.

With regard to the second of these questions, regarding the matter of wear testing, it is worth considering whether physical concerns alone should weigh as heavily as they do in such studies, when the devices in question have an obvious and critical biological function. Unlike the primitive devices of Paré's time, which were as separate from the patient's body as an item of clothing, many modern prostheses are actually connected to living tissue. Therefore it is wise to question whether physical testing alone can protect against the possible biochemical effects that may result from the scattering of particles.

As to the first question, regarding the viability of polyethylene as a material, there is a body of opinion that maintains that metal alloys and/or ceramics might function more safely and efficiently in this capacity. Polyethylene has been in wide use for prosthetics since the early 1970s, but that period has also seen vast strides in the development of prosthetic technology. Improvements include the use of cobalt-chromium alloys in conjunction with polyethylene, but even this technology is subject to significant wear in a number of ways.

As the technology of prosthetic devices has changed, so has the market for them. Whereas in the past, patients equipped with prosthetic devices tended to be elderly, by the end of the twentieth century, prosthetic surgeons were outfitting artificial body parts for patients in middle age. Such patients, who in many cases lost their limbs in the course of vigorous physical activity, continue to be more active than the elderly prosthetic patient, even after being outfitted with the artificial limb. This change in the population makeup of prosthetic patients likewise poses new challenges to medicine and engineering in the quest to develop better and safer prosthetic devices. —JUDSON KNIGHT

Viewpoint:

Yes, the research emphasis on the wear of ultra-high molecular weight polyethylene bearing components in joint replacements is warranted: not only is physical wear the most important factor in the aging and malfunctioning of joint replacements, but it is closely tied with chemical and biological factors.

Just a few minutes into the classic 1967 film *The Graduate* occurs one of the most famous quotes in motion-picture history. A nervous guest of honor at a party thrown by his parents, Ben Braddock (Dustin Hoffman) is accosted by one of his parents' friends, a Mr. McGuire (Walter Brooke), who tells him, "I just want to say one word to you... just one word." That one word, which contains a large piece of the future for business, the economy, and even society, turns out to be—"Plastics." Though there is an unmistakably satirical edge to the interchange, the truth is that plastics have changed the quality of human life, and in few areas is this more apparent than that of prosthesis, or the development of artificial substitutes for missing limbs or other body parts.

The idea of developing prosthetic limbs goes back to Roman times, but the origins of medical prosthetics as a science dates only to the sixteenth century. These early prostheses were primitive affairs: metal hands that did not move, or even a hook in place of an appendage, an image that became attached to pirates and other exotic characters. Only in the wake of the twentieth-century world wars would scientists make the technological breakthroughs necessary to develop the type of workable prosthetics in use today. For that to become possible, it was necessary to make use of an entirely new type of material: plastics.

At the time of their initial appearance in the 1930s, plastics were among the first everyday materials that seemed to be completely artificial, in contrast to the wood, stone, cotton, wool, leather, and other natural materials that made up most people's worlds. Yet plastics are formed of hydrocarbons, or chemical compounds whose molecules are made up of nothing but carbon and hydrogen atoms, and since the presence of carbon and hydrogen identifies a substance chemically as organic, they are in fact organic materials.

Their unique physical characteristics result from the fact that plastics are made up of polymers, or long, stringy molecules that are in turn made up of smaller molecules called monomers. Plastics are by definition high-molecular weight substances, but by the latter part of the twentieth century, it became possible to create plastics of extraordinarily high molecular weight, which made them much more durable. Long chains of 10,000 or more monomers can be packed closely to form a hard, tough plastic known as high-density polyethylene, or HDPE, used, for instance, in making drink bottles—and, eventually, in constructing prostheses.

By the end of World War II, plastics had become the standard for prostheses, though usually the plastic had to be reinforced either by a metal frame or even by glass. By the time HDPE made its appearance, surgical prosthetics had begun to go far beyond the point of simply attaching fake limbs to real ones: in certain cases, physicians were able to achieve an interface between living and non-living material, for instance by attaching the biceps muscle to a prosthetic arm, thus giving the user greater control over movement. This interface, naturally, introduced an entirely new level of complexity to the nature of prosthetics, and raised concerns regarding the long-term usefulness of prosthetic body parts in light not only of possible physical wear, but of deterioration resulting from chemical and biological factors as well.

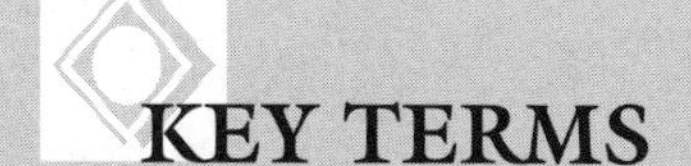

KEY TERMS

ACETABULAR: The cup-shaped socket in the hip bone.

CLINICAL STUDIES: Studies that involve or depend on direct observation of patients.

CYCLE: In the realm of prosthetics, a cycle is a completed movement of the body part—for instance, taking a step.

CYTOKINES: Immunoregulatory substances secreted by the immune system.

FREE RADICAL: An atom or group of atoms that has one or more unpaired electrons, such that it is highly reactive chemically.

HDPE: High-density polyethylene, a high-molecular weight variety of plastic, made up of long chains of 10,000 or more monomers packed closely to form a hard, tough substance.

HISTOLOGIC: Tissue structure and organization.

HYDROCARBONS: Chemical compounds whose molecules are made up of nothing but carbon and hydrogen atoms. Plastics are composed of hydrocarbons.

IRRADIATION: Bombardment with high-energy particles. In the case of prosthesis, this is for the purpose of sterilizing the prosthetic.

NEOPLASTIC: Having to do with the growth of non-functioning tissues such as tumors.

OSTEOLYSIS: The dissolution of bone; most often associated with resorption.

PHAGOCYTIZE: The ingestion of foreign material or bacteria in the body by phagocytes, a special kind of cell.

POLYMERS: Long, stringy molecules that are made up of smaller molecules called monomers.

PROSTHETICS: The branch of medical science devoted to the development and implementation of artificial substitutes for missing limbs or other body parts.

"Fighting Wear Has Become More Important Than Ever" It is now possible, with this background, to properly join the present debate over the research emphasis on wear—that is, on physical factors, as opposed to chemical or biological ones—in ultra-high molecular weight polyethylene joint replacements. Discussing the state of prosthetics at the end of the twentieth century, as well as this emphasis on wear testing, John

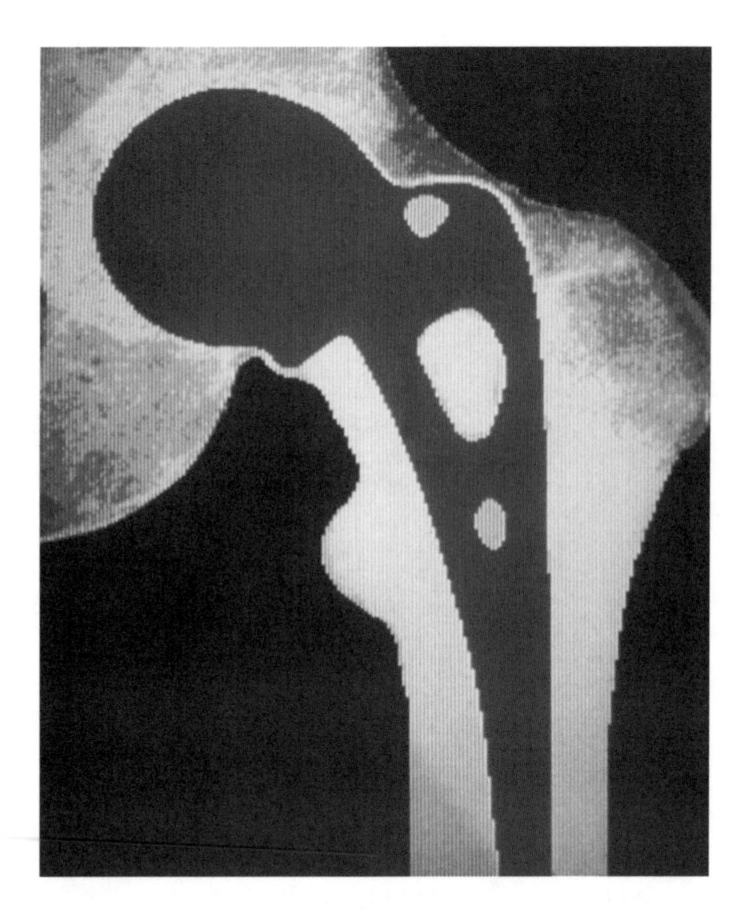

An illustration of a prosthetic hip joint.
(Custom Medical Stock Photo. Reproduced by permission.)

DeGaspari wrote in *Mechanical Engineering:* "Artificial hip joints must perform reliably for many years of use and millions of cycles," a cycle being a completed movement of the body part—for instance, taking a step. "A typical patient with a hip replacement joint may take as many as a million steps a year."

As DeGaspari went on to note, most modern hip replacement prostheses—and this is the case with numerous other prosthetic parts as well—are made from a combination of metal with ultra-high molecular weight polyethylene. However, not just any metal will do. According to William L. Healy and others in the *American Journal of Sports Medicine,* "First-generation, stainless steel hip and knee implants were associated with fracture rates that were not acceptable. Since the introduction of cobalt-chrome alloys and titanium alloys, implant breakage is not common."

Nor are plastics and metal the only materials necessary to developing a wear-resistant prosthetic. Reported Joseph Ogando in *Design News,* "Biomedical engineers once had feet of clay when it came to ceramic knees. Ceramics, with their low coefficients of friction and exceptional hardness, could certainly reduce the wear that shortens the life of artificial knees, but these brittle materials just could not handle the contact stresses found in even the best implant designs." Ogando went on to report that Smith & Nephew of Memphis, Tennessee, had created a new type of knee implant using a zirconium substrate with a zirconia surface—in other words, a metal interior with a ceramic exterior. "This hybrid material," he noted, "called 'oxidized zirconium,' pairs the mechanical properties of a metal with the wear-fighting capabilities of a ceramic."

Ogando also noted that "fighting wear has become more important than ever." This is a point made in many sources, each of which appears to take as a given the idea that wear should be the focus of prosthetic research. Thus, Timothy W. Wright began an article in the *American Academy of Orthopaedic Surgeons Bulletin* by declaring that "Wear-related failures of total joint arthroplasties [surgery to realign or reconstruct a joint] remain a serious problem." Likewise, Healy and his co-writers, noting that an international gathering of scientists from 35 nations and 44 states had declared the first 10 years of the twenty-first century the "Bone and Joint Decade," observed that "As a new millennium begins, new polyethylenes, ceramics, metals, and composite materials are being evaluated for improved wear characteristics at the bearing surface of joint replacements."

The Connection Between Physical, Chemical, and Biological Properties The writers in the *American Journal of Sports Medicine* also commented that "The wear factor and generation of particles may be the most important variable for patients and orthopaedic surgeons to consider when discussing athletic activity after joint replacement operation." This matter of "generation of particles" has causes that are tied to chemical factors, but the result appears (at least in part) in the form of physical wear.

In establishing the interface between mechanical prostheses and biological material, sterilization is essential. This sterilization has in the past been achieved through the use of gamma radiation, the bombardment of the polyethylene cup (which attaches to living tissue) with extremely high-energy particles that remove all possible contaminants. But this process of irradiation has the unintended consequence of removing hydrogen atoms from the hydrocarbon chains, resulting in the generation of free radicals. The latter term refers to electrons that have been released, bringing about a net electric charge and creating an unstable chemical situation. If oxygen, a highly reactive element that can bring about such corrosive effects as rust, comes into contact with the free radicals, the result is a chemical reaction that can be detrimental to the prosthesis. The polymer begins to lose molecular weight, and the surface of the polyethylene becomes less resistant to wear, such that infinitesimal particles of the plastic begin to break loose, creating inflammation in the body.

The above is an excellent example of how physical, biological, and chemical factors interact in prostheses, rendering these various factors ultimately inseparable. The final result, however, is physical, in the form of accelerated wear. This relationship between chemical and/or biological and physical factors is reinforced by findings published on the website of the Institute of Orthopaedics in London, on which it is stated (and corroborated with experimental data) that "The wear of polyethylene increases with the level of oxidation."

Wright, in the *American Academy of Orthopaedic Surgeons Bulletin,* likewise noted the relationship between physical and biological factors in causing wear. "Biologically," he observed, "the local reaction to wear debris"—the separation of particles from the prosthesis, as we have already noted—"results from the activation of phagocytosis by macrophages." In other words, the physical phenomenon of wear brings about biological damage through the action of microorganisms.

Protecting Against Biological and Chemical Complications From these findings and others, it is clear that the physical phenomenon of wear is inseparable, in a real-world sense, from chemical phenomena such as oxidization, or biological phenomena such as infection. Therefore, the direction of research regarding high-molecular weight polyethylene bearing components in joint replacement should remain focused on wear, since this is likely to be the source of biological or chemical complications.

What, then, is to be done? Or, to put it another way, how can prosthetic manufacturers develop materials that, by resisting wear, protect against oxidization, infection, and other chemical or biological factors? One promising area is in the crosslinking of polymers, a technique pioneered by polymer chemist Orhun Muratoglu at Massachusetts General Hospital in Boston. By heating polyethylene to 257°F (125°C)—slightly below its melting point—then irradiating it, the Massachusetts team has been able to recombine the free radicals, resulting in a product that shows no measurable signs of wear after 20 million cycles.

At the same time, research conducted by Thierry Blanchet, associate professor of mechanical engineering at Rensselaer Polytechnic Institute in Troy, New York, indicates that immersion of the prosthesis in an acetylene-ethylene-hydrogen atmosphere during heating produces a much more wear-resistant product. Other techniques are under investigation, and though many of them involve attempts to alter the chemical properties of the prosthetic, the emphasis is on wear. This is no doubt as it should be, since the prosthesis is ultimately a physical component. It must be biologically suited to the user, and must not induce chemical reactions that may render it less than fully effective, but its ultimate design is as a physical part of the body—an arm or hand that grasps, for instance, or a hip or leg that makes it possible for the user to walk like a person with two healthy legs. —JUDSON KNIGHT

Viewpoint:

No, the current research emphasis on the wear of polyethylene bearing components in joint replacements is unwarranted, given that alternative materials are available that can provide not only greater longevity but biological tolerance as well.

The Six Million Dollar Man once entertained television viewers with superhuman feats made possible by man-made replacement parts surgically implanted throughout his body. If the makers of these parts relied on the same materials primarily used in joint replacements today, the six-million-dollar man would be sitting in a rocking chair, unable to walk to the refrigerator without a cane. At best, he has been to see the doctors for numerous tune-ups and likely a few major repairs.

In the world of joint replacement prosthetics, especially those involving the hips and knees, polyethylene has been used for 30 years as a relatively successful bearing surface in combination with a metallic or ceramic counterface. The combination allows for total range of movement by forming a highly functioning articulation (the joint or juncture between bones or cartilages). Over the years, joint replacement manufacturers switched from using high-density polyethylene to ultra high molecular weight polyethylene (UHMWP) for improved mechanical properties and performance. UHMWP remains the most popular materials used in conjunction with a metal alloy like cobalt-chromium (Co-Cr) to form the articulation. UHMWP's higher molecular weight and fewer polymer chain branches result in a polymer with a higher abrasive resistance and good chemical resistance for durable wear properties. Nevertheless, UHMWP does have drawbacks, and emphasizing research to focus on the improvement of polyethylene is unwarranted, especially when other materials may be more durable and safer.

The Problem with the Current Emphasis The old saying, "don't put all your eggs in one basket," is most often used in terms of money and investments. But it applies to most aspects of life, including science. The use of polyethylene in joint replacements has served orthopedic patients well, but their long-term use has revealed several problems. For example, because joint prostheses are essentially bearings, friction and wear are major problems. High friction forces can produce shearing stresses and wear. The use of polyethylene and metal involves three types of wear. Abrasive wear results from the direct contact between metal and plastic components and can result in the metal abrading the plastic. Adhesive wear is similar to a welding and tearing of the contacting surfaces between the metal and polyethylene. Pitting wear is the most common type of wear and results from incongruent contact between the metal implant and plastic bearing. A major clinical concern in joint replacements has been the production of wear debris from contact between the articulating surfaces.

Although many efforts have been made to modify polyethylene to reduce wear, most have produced similar results as the conventional polyethylene. Furthermore, polyethylene components undergo oxidative degradation during manufacturing, which can lead to harmful changes in the material's physical and mechanical properties. Oxidative degradation of polyethylene can continue to cause components to be more and more brittle and less resistant to fracture and fracture-related wear damage.

Another approach used to improve polyethylene's wear resistance is to induce crosslinks into the polyethylene morphology. Crosslinking is a method involving either ionizing radiation or chemical methods to improve UHWMP's abrasion resistance. Essentially, it bonds molecules together to make a stronger material. However, many of the new crosslinked polyethylenes have reduced ductility and fracture resistance compared to the conventional polyethylenes. Despite all these attempts, no improved polyethylene has yet been proven in the clinical setting.

Durability has also become a major issue concerning the use of metal-polyethylene joint prosthetics. For many years, these prosthetics were used primarily in the elderly, that is, people in their 70s and older. However, joint replacement patients now include many who are in their 40s and 50s. These more active patients have placed extra stress on the replacement, and the metal-polyethylene combination has not stood up to the strain.

Studies have shown that polyethylene wear debris induces osteolysis as a long-term complication, especially in younger, more active patients. The metal and polyethylene combination can produce up to 40 billion particles per year. As this debris accumulates, it usually results in an aggregation of macrophages that attempt to phagocytize it. The resulting inflammatory response include the release of lytic enzymes, pro-inflammatory cytokines, and bone-resorbing mediators that result in osteolysis, or bone resorption. The osteolysis, in turn, causes aseptic (not involving infections or pathogens) loosening and the implant's clinical failure. One approach to this problem, for example, is to reduce the ball size of the polyethylene femoral in hip replacements. Although reducing the size of the polyethylene head components minimizes this wear, it has induced stability problems in some patients.

The Next Generation Orthopedic surgeons and scientists are looking at ways to increase prosthetic durability and reduce the volume of particulate debris released into the tissues surrounding the joints. This reduction, in turn, lowers the occurrence of osteolysis and aseptic loosening. Many researchers are focusing on bearing materials that wear significantly less than the standard metal-polyethylene combination. Their primary goal is to find a material or combinations of materials that have good wear properties and low coefficients of friction. The ultimate goal is to develop joint prosthetics that reliably relieve pain, allow unlimited function, and last throughout the patient's lifetime.

Although the use of metal-on-metal (M/M) bearings in joint prosthetics dates back to the late 1960s and ceramic-on-ceramic (C/C) bearings have been around for nearly as long, early primitive designs (primarily involving fixation of the stem or socket) led to inferior results compared to metal-on-polyethylene. However, recent advances in the use of metals and ceramics have led many researchers to consider replacing polyethylene with these materials. For example, scientific analyses of components made with these newly engineered parts show that the occurrence of wear can be at least 100 times less than polyethylene components.

Although poor design and manufacturing methods led to problems, M/M articulations have been undergoing a renaissance worldwide. First-generation metal-on-metal devices that have survived for more than two decades have shown low wear rates and very little change in surface finish and dimension. Studies of twenty- to thirty-year-old M/M hip prosthesis have shown a notable absence of osteolysis around the implant.

With today's modern technology, metal-bearing surfaces can be designed to be almost perfectly smooth and round, which further minimizes wear. In the case of hip replacements, M/M components can include larger head/cup sizes without large increases in wear volumes for improved range of motion and the use of the large implant sizes needed for surface replacement of the hip. Furthermore early findings have shown that M/M wear particles are in the 20–60 nm size range, which is an order of magnitude smaller than the polyethylene wear particles produced in the metal-polyethylene articulations. In fact the worst case estimate of combined femoral and acetabular (cup-shaped socket) linear wear in hip replacements was 4.2 microns per year, which is about 25 times less than the wear that typically occurs with polyethylene.

Clearly, close attention to design detail and manufacturing quality has led to revised interest in M/M joint prosthetics. These prosthetics have enormous potential for greatly reduced wear and fewer problems with osteolysis. Studies of second-generation M/M bearings have revealed very good results, including little evidence of metal wear and few wear particles found in histologic sections.

Ceramic-on-ceramic (C/C) bearings have been used in Europe since the early 1970s. Very early on, ceramic materials were associated with a large grain structure, which could lead to fractures. Overall, poor implant design, acetabular component loosening, and low-quality ceramic, resulting in fracture and debris generation, dampened enthusiasm for their use. However, like M/M bearings, early problems and failures were related less with the alumina material used than to inefficient design and, in some cases, the surgical procedure itself. Improved manufacturing processes have created a much stronger ceramic material with a small grain size. Overall, ceramic bearings made of alumina have shown to have the lowest wear rates of any of the bearing combinations, both in simulator cases and in clinical studies.

In a study conducted over 20 years beginning in 1977, C/C bearings showed excellent results in the younger population, with an 86% prosthetic survival rate at 15 years in patients less than 50 years of age. Osteolysis was found in less than 1% of patients. The linear wear rates were very low. Histological studies also confirmed that the little alumina ceramic debris that did occur was tolerated well biologically.

As a result of these and other findings, C/C is perhaps the most promising bearing material for use in young and active people. Some findings show that an exceptionally long-term survival rate without any activity limitations can be expected for C/C bearing prosthetics, with the fracture risk estimated to be at 1/2000 over a 10-year period. Furthermore, the design of the newer generation C/C bearings makes surgical implantation for correct positioning much easier.

The Biocompatibility Issue To be considered biocompatible, a material should be well tolerated in the body, causing little response from the person's (or host's) body, including metabolic, bacteriologic, immunologic, and neoplastic responses. Although all of the different materials used in joint replacements have been tested and approved in terms of biocompatibility, problems still occur. Extensive research into this area has shown that tissues and cells exposed to particulate polyethylene produce numerous inflammatory mediators associated with bone resorption (or loss). It is generally accepted that most cases of aseptic loosening and osteolysis are the result of macrophagic response to polyethylene wear particles.

In contrast, patients have shown a very low incidence of hypersensitivity to M/M implants made with a cobalt-chrome alloy. Even though studies have shown higher levels of cobalt and chromium in patient's hair, blood, and urine than in metal-polyethylene cases, researchers found that the cobalt ions are rapidly transported from the implant site and mostly eliminated in the urine. Nevertheless, chromium tends to be stored in the tissues for longer periods. Another concern with M/M implants was the question of carcinogenicity. Clinical studies to date have shown that CoCr implants do not induce cytotoxic effects. These findings are supported by epidemiological studies as well. Furthermore, ongoing studies of early M/M implants found that the clinical incidence of sarcomas in M/M patients is extremely low.

C/C bearings have the advantage of bulk alumina being chemically inert and, as a result, having excellent biocompatibility. Although a foreign body response is still induced by particulate alumina in tissues, the inflammatory mediators released by the immune system in response to particulates is generally less than those stimulated by polyethylene particles. As shown by cell-culture studies, particulate alumina is not toxic.

The Future of Joint Replacement Research

The adverse biological effects of wear debris from polyethylene bearing components coupled with high wear rates and shorter longevity paint a clear picture that research into joint replacement bearing materials should expand beyond polyethylene. Several alternatives to metal-polyethylene implants have been advocated, including the sole use of ceramics or metals. As a bearing surface, ceramic has great potential because of its hardness, which can tolerate rigorous polishing, resulting in greater scratch resistance. Furthermore, the bearing is wettable, which allows for less friction. Metal bearings have also shown superior durability compared to the use of polyethylene. The volume of wear debris in both metal and ceramic bearings is also expected to be minimal, resulting in less injury to the bone around the prosthesis.

Experience with metal-polyethylene bearings has shown a typical survival rate of about 15 years. Often, patients with these joint replacements return to a life of pain and restricted mobility that can be even worse than their presurgical conditions. Metal and ceramic bearings promise to prolong the durability of joint replacements, especially in younger patients. Of course, rigorous mechanical testing and hip simulator research are required, as are further controlled clinical trials for specific products. But by expanding research to focus more on metal-on-metal and ceramic-on-ceramic bearing components, joint prosthetics will one day last as long and be tolerated as well as the normal healthy joint. —DAVID PETECHUK

Further Reading

DeGaspari, John. "Standing Up to the Test." *Mechanical Engineering* 121, no. 8: 69–70.

"Durability of Biomaterials." Institute of Orthopaedics, London. May 30, 2002 [cited July 22, 2002]. <http://www.fractures.com/institute/bme/durabil.htm>.

Healy, William L., Richard Iorio, and Mark J. Lemos. "Athletic Activity After Joint Replacement." *American Journal of Sports Medicine* 29, no. 3: 377–88.

Jazrawi, Laith M., and William L. Jaffe. "Ceramic Bearings in Total Hip Arthroplast: Early Clinical Results and Review of Long Term Experiences." *Arthroplasty Arthroscopic Surgery* 10, no. 1 (1999): 1–7.

Lerouge, S., et al. "Ceramic-Ceramic vs. Metal Polyethylene: A Comparison of Periprosthetic Tissue from Loosened Total Hip Arthroplasties." *Journal of Bone and Joint Surgery* 79 (1997): 135–39.

McNeill, Bridgette Rose. "Getting Hip to Joint Replacement." Southwestern Medicine (The University of Texas Southwestern Medical Center at Dallas). May 30, 2002 [cited July 22, 2002]. <http://www.swmed.edu/home_pages/publish/magazine/hip/joint.html>.

Morawski, David R., et al. "Polyethylene Debris in Lymph Nodes after a Total Hip Arthroplasty." *Journal of Bone and Joint Surgery* 77-A, no. 5 (May 1995).

Ogando, Joseph. "A Knee for the Long Haul." *Design News* 56, no. 12: 76–77.

Rimnac, Clare M. "Research Focuses on Polyethylene Wear." *The American Academy of Orthopaedic Surgeons Bulletin* 49, no. 1 (February 2001).

Waddell, James P. "Improving the Durability of Total Joint Replacement." *Canadian Medical Association Journal* 161 (1999): 1141.

Wagner, M., and H. Wagner. "Medium-Term Results of a Modern Metal-on-Metal System in Total Hip Replacement." *Clinical Orthopaedics* 379 (October 2000): 123–33.

Wright, Timothy. "Investigators Focus on Wear in Total Joints." American Academy of Orthopaedic Surgeons (AAOS). May 30, 2002 [cited July 22, 2002]. <http://www.aaos.org/wordhtml/bulletin/dec00/fline1.htm>.

Now that the Human Genome Project is essentially complete, should governmental and private agencies commit themselves to the Human Proteome Project, which would study the output of all human genes?

Viewpoint: Yes, governmental and private agencies should now commit themselves to a Human Proteome Project because of the many practical benefits such work could bring, from improvements in rational drug design to the discovery of new disease markers and therapeutic targets.

Viewpoint: No, governmental and private agencies should not commit themselves to a Human Proteome Project; the intended endpoint of the project is unclear, and the battles over access to the data might be even more intense than those that marked the Human Genome Project.

The term *genetics* was coined at the beginning of the twentieth century to signify a new approach to studies of patterns of inheritance. Within about 50 years, the integration of several classical lines of investigation led to an understanding of the way in which the gene was transmitted and the chemical nature of the gene. Subsequent work revealed how genes work, the nature of the genetic code, and the way in which genetic information determines the synthesis of proteins. During the course of these investigations, classical genetics was largely transformed into the science of molecular biology.

For most of the first half of the twentieth century, scientists assumed that the genetic material must be protein, because no other species of biological molecules seemed to possess sufficient chemical complexity. Nucleic acids, which had been discovered by Friedrich Miescher in the 1860s, were thought to be simple, monotonous polymers. The work of Erwin Chargaff in the 1940s challenged prevailing ideas about DNA and suggested that nucleic acids might be as complicated, highly polymerized, and chemically diverse as proteins. However, because the structure of DNA was still obscure, experiments that indicated that DNA might be the genetic material could not explain the biological activity of the gene. This dilemma was resolved in 1953 when James D. Watson and Francis Crick described an elegant double helical model of DNA that immediately suggested how the gene could function as the material basis of inheritance. Indeed, molecular biologists called the elucidation of the three-dimensional structure of DNA by Watson and Crick one of the greatest achievements of twentieth-century biology.

Based on the Watson-Crick DNA model, researchers were able to determine how genes work, that is, how information stored in DNA is replicated and passed on to daughter molecules and how information in DNA determines the metabolic activities of the cell. In attempting to predict the steps involved in genetic activity, Watson stated that: DNA → RNA → protein. The arrows represent the transfer of genetic information from the base sequences of the nucleic acids to the amino acid sequences in proteins. The flow of information from DNA to RNA to protein has been called the Central Dogma of molecular biology.

Further research demonstrated that specific mutations in DNA are associated with changes in the sequence of amino acids in specific proteins that result in a wide variety of inherited diseases. By the 1980s scientists had established the existence of some 3,000 human genetic diseases, which suggested that developments in genetics would make it possible to manage many diseases at the gene level. In theory, the ability to produce genes and gene products in the laboratory should allow "replacement therapy" for the hundreds of genetic diseases for which the defective gene and gene product are known.

In the 1970s, scientists learned how to cut and splice DNA to form recombinant DNA molecules and place these molecules into host organisms, and then clone them, that is, induce them to reproduce within their hosts. The commercial potential of recombinant DNA and genetic engineering was almost immediately understood by scientists, university development officials, venture capitalists, and entrepreneurs. Genentech (Genetic Engineering Technology) was established in 1976 for the commercial exploitation of recombinant DNA techniques. In 1980, the U.S. Patent Office granted a patent for the recombinant DNA process. This was the first major patent of the new era of biotechnology. Scientists confidently predicted that the techniques of molecular biology would make it possible to treat genetic diseases, manipulate the genetic materials of plants, animals, and microorganisms, synthesize new drugs, and produce human gene products such as insulin, growth hormone, clotting factors, interferons, and interleukins. However, such predictions also raised questions about the potential uses and abuses of recombinant DNA, cloning, genetic engineering, gene therapy, and genetically modified foods.

Refinements in the techniques of molecular biology made it possible to plan and execute the Human Genome Project (HGP), an international effort dedicated to mapping and sequencing all of the bases on the 23 pairs of human chromosomes. In 1988 an international council known as the Human Genome Organization (HUGO) was established to coordinate research on the human genome, and the Human Genome Project was officially launched in 1990. Some scientists called the project the "Holy Grail" of human genetics, but others warned against diverting the limited resources available to biological science to what was essentially routine and unimaginative work. As the project proceeded, however, new techniques for rapid and automated sequencing and data analysis demonstrated the feasibility of the project. Another aspect of the HGP was the growing influence of research methodologies known as "discovery science" and "data mining," which involve generating and utilizing enormous databases.

In 1998 the consortium of scientists involved in the HGP found itself in competition with Celera Genomics, a corporation founded and directed by J. Craig Venter, a former NIH researcher. On June 26, 2000, leaders of the public genome project and Celera held a joint press conference to announce the completion of working drafts of the whole human genome. Both maps were published in February 2001. By January 2002, Celera announced that it would shift its focus to drug development, because its original plan to sell access to its genome databases was not as profitable as anticipated.

With the successful completion of the first major phase of the Human Genome Project, scientists could direct their energies to the daunting task of analyzing the tens of thousands of human genes and their relationship to the hundreds of thousands of human proteins. Based on experience gained during the quest for the complete sequence of the human genome, scientists suggested creating a complete inventory of human proteins and calling this effort the Human Proteome Project (HUPO).

Australian scientists Marc Wilkins and Keith Williams coined the term *proteome* in 1995. Proteome stands for the "set of PROTEins encoded by the genOME." Although nucleic acids have received so much attention since Watson and Crick introduced the DNA double helix, proteins are actually the hard-working macromolecules that do most of the tasks needed for life. Proteins are the essential structural elements of the cell, and they serve as enzymes, hormones, antibodies, and so forth. Because the relationship between genes and proteins is dynamic and complex, the proteome can be thought of as the total set of proteins expressed in a given organelle, cell, tissue, or organism at a given time with respect to properties such as expression levels, posttranslational modifications, and interactions with other molecules. The term proteomics is routinely used as the name of the science and the process of analyzing and cataloging all the proteins encoded by a genome. Some scientists think of proteomics as "protein-based genomics."

Scientists argue that instead of establishing a mere catalog of proteins, proteomics would create a complete description of cells, tissues, or whole organisms in terms of proteins, and the way that they change in response to developmental and environmental signals. Many practical benefits are anticipated from the analysis of such patterns of proteins; for instance they could lead to major improvements in rational drug design and the discovery of new disease markers and therapeutic targets.

The first major conferences devoted to the possibility of establishing a major proteome project was held in April 2001 in McLean, Virginia. Comparing the challenges raised by proteomics to those involved in the Human Genome Project, the organizers called the conference "Human Proteome

Project: Genes Were Easy." The founders of HUPO hoped to bring academic, commercial, national, and regional proteome organizations into a worldwide effort to study the output of all human genes. However, skeptics warned that the intended endpoint of the proteome project was more nebulous than that of the genome project. Cynics warned that battles about access to the data generated by public and private groups might be even more intense than those that marked the rivalry between the public genome project and Celera.

Publication of the first draft of the human gene map in 2001 was accompanied by claims that the Human Genome Project would provide the complete script from which scientists could read and decode the "nature" of the human race. More sober evaluations suggested that the HGP had produced a parts list, rather than a blueprint or a script. Although advances in proteomics may indeed lead to new drugs and therapeutic interventions, the feasibility and potential of the Human Proteome Project are still points of debate, as indicated by the following essays. —LOIS N. MAGNER

Viewpoint:

Yes, governmental and private agencies should now commit themselves to a Human Proteome Project because of the many practical benefits such work could bring, from improvements in rational drug design to the discovery of new disease markers and therapeutic targets.

Human Proteome A gene is a piece of DNA. The 35,000 or more genes that each person carries are the working units of heredity that pass from parents to child. A genome can be defined as all the DNA in a given organism.

The genome's entire job is to direct cells to make, or express, proteins. The genes in each cell in a human body use DNA-encoded instructions to direct the expression of one or many proteins. So 35,000 genes may generate millions of proteins, each of which is then modified in many ways by the cellular machinery.

In the words of Stanley Fields, professor of genetics and medicine at the University of Washington-Seattle, "... Proteins get phosphorylated, glycosylated, acetylated, ubiquitinated, farnesylated, sulphated, linked to ... anchors ... change location in the cell, get cleaved into pieces, adjust their stability, and ... bind to other proteins, nucleic acids, lipids, small molecules, and more."

All these modifications make it possible for proteins to become hair and nails, enzymes and connective tissue, bone and cartilage, tendons and ligaments, the functional machinery of cells, and much more. So the genome supplies these building blocks of life, but proteins do all the work, and, according to David Eisenberg, director of the UCLA-DOE Laboratory of Structural Biology and Molecular Medicine, they do it "as part of an extended web of interacting molecules."

Compared with the genes in the human genome, proteins are so complex and dynamic that some experts say there can be no such thing as an identifiable human proteome—defined as all the proteins encoded in the human genome—and so no basis for a human proteome project.

Others hold that a true understanding of the genetic blueprint represented by the human genome can't begin until all the genes are not only mapped but annotated (explained), along with their products—proteins. And even though the proteome does exist in near-infinite dimensions, it is possible to start planning and identifying short-term milestones and measures of success, and prioritize specific stages of a human proteome project.

The History of Proteomics According to Dr. Norman Anderson, chief scientist at the Large Scale Biology Corp., what is now proteomics started in the 1960s as the first large-scale engineering project—the original Molecular Anatomy Program at Oak Ridge National Laboratory in Tennessee. It was supported by the National Institutes of Health (NIH) and the Atomic Energy Commission, and aimed to produce as complete an inventory of cells at the molecular level as technology would allow.

The sponsors were especially interested in discovering and purifying new pathogenic viruses, and technology developed in the program included a systematic approach to fractionating (separating into different components) proteins, large-scale vaccine purification methods, high-pressure liquid chromatography, and high-speed computerized clinical analyzers.

In 1975, after a new technology called high-resolution two-dimensional electrophoresis was introduced, the program moved to Argonne National Laboratory to take advantage of image-analysis techniques and emerging automation.

Beginning in 1980, an attempt was made to launch a Human Protein Index (HPI) project as

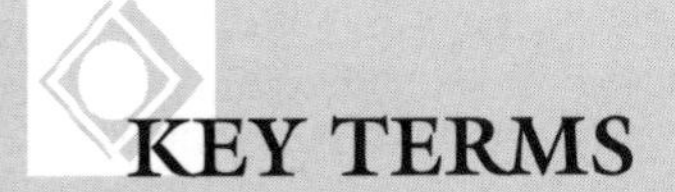

KEY TERMS

BIOINFORMATICS: The use of computers in biology-related sciences. Even though the three terms bioinformatics, computational biology, and bioinformation infrastructure are often used interchangeably, bioinformatics typically refers to database-like activities. Creating and maintaining sets of data that are in a consistent state over essentially indefinite periods of time.

DEOXYRIBONUCLEIC ACID (DNA): The chemical inside the nucleus of a cell that carries the genetic instructions for making living organisms.

DNA SEQUENCING: Determining the exact order of the base pairs in a segment of DNA.

GENE: The functional and physical unit of heredity passed from parent to offspring. Genes are pieces of DNA, and most genes contain the information for making a specific protein.

GENE EXPRESSION: The process by which proteins are made from the instructions encoded in DNA.

GENE MAPPING: Determining the relative positions of genes on a chromosome and the distance between them.

GENOME: All the DNA in an organism or cell, including chromosomes in the nucleus and DNA in mitochondria.

HIGH-PRESSURE (PERFORMANCE) LIQUID CHROMATOGRAPHY: Protein molecules are separated according to their physical properties such as their size, shape, charge, hydrophobicity, and affinity for other molecules. The term high-performance liquid chromatography was coined to describe the separation of molecules under high pressure in a stainless-steel column filled with a matrix.

HUMAN GENOME PROJECT: An international research project to map each human gene and to completely sequence human DNA.

MASS SPECTROGRAPH: An instrument that separates beams of ions according to their mass-to-charge ratio, and records beam deflection and intensity directly on a photographic plate or film.

NUCLEOTIDE: A structural component, or building block, of DNA and RNA. A nucleotide is a base (one of four chemicals: adenine, thymine, guanine, and cytosine) plus a molecule of sugar and one of phosphoric acid.

PROTEIN: The "Central Dogma" of molecular biology states that DNA, a gene, is transcribed into RNA, which is translated into protein. The RNA nucleotides are repeated in various patterns of three, called codons, which indicate a specific amino acid. The order and type of codons in a RNA sequence determine the amino acids and resulting protein.

PROTEOMICS: The study of protein function, regulation, and expression in relation to the normal function of the cell and in disease.

TWO-DIMENSIONAL GEL ELECTROPHORESIS: Allows protein biochemists to cut isolated spots out of a gel and sequence their amino acids.

2D POLY-ACRYLAMIDE GEL ELECTROPHORESIS (PAGE): Conventional two-dimensional electrophoresis method of resolving proteins. Proteins are first separated using isoelectric point (electric charge of molecule) and then by molecular weight.

SEQUENCE: A series of biological molecules that are connected. A two-dimensional sequence of amino acids can form a protein. A two-dimensional sequence of nucleotide bases forms DNA.

a serious national objective. In 1983 some HPI proponents proposed launching a dual effort—sequencing the human genome and indexing proteins as parallel projects. The Human Genome Project was first to succeed, partly because the basic gene-sequencing technology was widely available.

Now—thanks to advances in mass spectrometry and automation, and the expansion of many protein-related technologies and systems—a coordinated program in proteomics could finish what the Human Genome Project started in 1983.

With this in mind, in February 2001 the Human Proteome Organization (HUPO) was launched, with the official formation of a global advisory council of academic and industry leaders in proteomics. HUPO is not necessarily meant to become the official Human Proteome Project, although it could become such a framework.

It was formed in part to help increase awareness of proteomics across society, and cre-

Eric Lander, Robert Waterston, James Watson, and Francis Collins at a press conference announcing the sequencing of the human genome.

(© Corbis. Reproduced by permission.)

ate a broader understanding of proteomics and the opportunities it offers for diagnosing and treating disease. It was also formed as a venue for discussing the pros and cons of creating a formal project, and for shaping such a project if it goes forward.

As a global body, HUPO wants to foster international cooperation across the proteomics community and promote related scientific research around the world. Participants include those representing government and financial interests, to make sure the benefits of proteomics are well distributed. Advisory council members are from Europe, North America, the Far East, and Japan. Special task forces in Europe and Japan represent HUPO at a regional level.

In April 2001, HUPO held its first three-day meeting in McLean, Virginia, and more than 500 members of the global proteomics community attended. The international conference was called *Human Proteome Project: Genes Were Easy.*

Genes Were Easy Topics at the April meeting included such basics as funding and fostering, scope and scale, financial implications, current proteomic efforts and lessons, patent and insurance issues, and potential solutions. Throughout the meeting a series of round table discussions by the big guns in international proteomics tackled thornier issues and posed tough questions.

Questions from *Round Table 1: Lessons from the Human Genome Project* included: What agreements are needed on data reporting and data quality? Who will own the information? Would a formal project spur technological innovation? What happens if there is no formal project? What kind of funding would be required and where should it come from? What should be the benefits from the project?

Questions from *Round Table 2: Lessons from Current Efforts-Defining the End Result of the Human Proteome Project* included: Is it a complete list of all human proteins? Is it a full database of protein-expression levels? Would it include healthy and disease-related protein expression? Would it include protein expression in response to drug treatment? Is it a full database of human protein-protein interactions? Is it a complete understanding of the function of each protein?

Discussions in Round Table 3 focused on the major technical challenges posed by a human proteome project. These included: capability to handle the full range of protein expression levels; reproducibility of protein expression studies; reducing false positives in yeast 2-hybrid studies; and significantly increasing high-throughput capacity and automation.

Six months later, in October 2001, HUPO convened a planning meeting to review the state of the art in proteomics and consider how to further knowledge of the human proteome. The HUPO meeting was held in Leesburg, Virginia, and sponsored by the National Cancer Institute (NCI) and the Food and Drug Administration (FDA).

Meeting attendees agreed that the constantly changing human proteome has quasi-infinite dimensions, and that initiating a large international human proteome project would be much harder than the human genome project because there can be no single goal, like sequencing the human genome. Other complicating factors include the more complex and diverse technologies and resulting data, and a need for more sophisticated data-integration tools for the proteome than for the genome.

Still, the group ambitiously defined the goal of proteomics as identifying all proteins encoded in the human genome; determining their range of expression across cell types, their modifications, their interactions with other proteins, and their structure-function relationships; and understanding protein-expression patterns in different states of health and disease.

Just because a proteome project is harder to conceptualize and carry out than a genome project doesn't mean it shouldn't be done. So the attendees recommended short-term milestones and measures of success. They agreed that private-public partnerships should be formed to work on specific, affordable, and compelling areas of scientific opportunity as individual projects with defined time frames.

They agreed that a proteome project should combine elements of expression proteomics, functional proteomics, and a proteome knowledge base.

Expression proteomics involves identifying and quantitatively analyzing all proteins encoded in the human genome and assessing their cellular localization and modifications.

Functional proteomics involves defining protein interactions, understanding the role of individual proteins in specific pathways and cellular structures, and determining their structure and function.

A proteome knowledge base involves organizing proteome-related data into a database that integrates new knowledge with currently scattered proteome-related data, leading to an annotated human proteome.

They also agreed that technology development should be an integral component of the project, that all users should have open access to data that comes from the project, that strong consideration should be given to integrating proteome knowledge with data at the genome level, that the project should have definable milestones and deliverables, and that a planning and piloting phase should precede the project's production phase.

The Alternative to Cooperation The debate continues about whether a Human Proteome Project is possible, even while HUPO and the proteomics community hammer together an increasingly solid foundation for such a global project. According to the HUPO literature, this careful and consensual planning is helping "harness the naturally occurring forces of politics and markets, technology and pure research" that will make an international proteomics effort possible.

Now that the human genome is sequenced, the next logical step is to understand proteins—the products of genes—in the context of disease. With the tools of proteomics, specific proteins can be identified as early, accurate markers for disease. Proteins are important in planning and monitoring therapeutic treatments. Understanding protein-expression patterns can help predict potential toxic side effects during drug screening. Proteins identified as relevant in specific disease conditions could have an important role in developing new therapeutic treatments.

Genomics spawned a multibillon-dollar research effort and many commercial successes. But, according to HUPO founding member Prof. Ian Humphery-Smith, "proteins are central to understanding cellular function and disease processes. Without a concerted effort in proteomics, the fruits of genomics will go unrealized."

At the National Institute of General Medical Sciences (NIGMS), John Norvell, director of the $150-million Protein Structure Initiative, found that when researchers had total discretion to pick their own projects, they studied proteins with interesting properties and ignored proteins whose structures could illuminate the structures of many similar proteins. "We decided an organized effort was thus needed," he said.

"Proteome analysis will only contribute substantially to our understanding of complex human diseases," says proteomics researcher Prof. Joachim Klose of Humboldt University in Germany, "if a worldwide endeavor is initiated aiming at a systematic characterization of all human proteins."

For the fledgling Human Proteome Project and the resulting understanding of how proteins work together to carry out cellular processes—what Stanley Fields calls the real enterprise at hand—the alternative to international cooperation through a systematic proteomics effort, says Ian Humphery-Smith, "is sabotage and self-interested competition." —CHERYL PELLERIN

Viewpoint:

No, governmental and private agencies should not commit themselves to a Human Proteome Project; the intended endpoint of the project is unclear, and the battles over access to the data might be even more intense than those that marked the Human Genome Project.

Less than 10 years ago, while scientists were in the throes of sequencing the human genome, Marc Wilkins of Proteome Systems in Australia invented the word "proteome" to describe the qualitative and quantitative study of proteins produced in an organism, which includes protein structure, protein interactions, and biochemical roles, such as in health and disease. Now at the start of the twenty-first century, as the human genome sequencing project is near completion, the proteome research project is being discussed, but with caution and debate surrounding the cost, feasibility, and usefulness of the project.

Trying to translate and comprehend the chemical alphabet sequence of the 3 billion nucleotide base pairs of the human genome DNA in the Human Proteome Project follows the "Central Dogma" of molecular biology—DNA is transcribed into RNA, which is translated into protein. (The nucleotide base pairs of DNA code for RNA, which, in turn, codes for protein). In addition to gaining basic information about proteins—their structure, cellular interactions, and biochemistry—biotechnology and pharmaceutical companies are particularly interested in investigating a vast number of proteins the human genome encodes as potential drug target candidates.

So what are the perils of the Proteome Project? Critics cite the complexity of the human proteome, technology, and money.

If sequencing the 3 billion base pairs of the human genome seemed like a huge project, the Proteome Project is much more complex and involves far more data. In addition, the Human Proteome Project investigates many aspects of proteins and is not just a single task, like sequencing in the Human Genome Project. The projected 34,000 or so genes encode 500,000 to 1 x 106 (million) proteins. The DNA in the 250 cell types in the human body, such as skin, liver, or brain neuron, is the same, but the proteins in the different cell types are not. The amount of protein can vary according to cell type, cell health, stress, or age. At different times and conditions, different subproteomes will be expressed. Scientists expect that in the next 10 years several subproteome projects will take place, rather than the "entire" human proteome project being done. The subproteome projects, for example, would investigate proteins found in one tissue, body fluid, or cell type at a separate moment of life. Will there ever be a complete catalogue of human proteins? According to Denis Hochstrasser at the Swiss Institute of Bioinformatics, we would never know.

The scientific research community agrees that a variety of technologies and technological developments are necessary to investigate the proteome. For example, it is slow and tedious to excise protein spots from commonly run two-dimensional polyacrylamide gel electrophoresis (PAGE) experiments by hand. In addition, big hydrophobic proteins do not dissolve in the solvents currently used in 2D PAGE experiments. It is also hard to distinguish very large or very small proteins in 2D PAGE experiments. Automating the running and analysis of 2D gel technology would also allow a 2D gel experiment to be run in days rather than months or years. Hoping to make the process of separating and identifying proteins easier, Hochstrasser and colleagues are developing a molecular scanner that would automate the separation and identification of thousands of protein types in a cell.

Developments in mass spectrometry and bioinformatic technologies will also be necessary to further the proteome project. Although mass spectrometry is useful for molecularly characterizing proteins, the technique can fail to detect rare proteins. Also the cost of a mass spectrometer is huge. In 2002, the cost of a mass spectrometer is estimated to be $500,000. Nuclear magnetic resonance (NMR) tools that enable scientists to take dynamic, "movie" data of proteins instead of static snapshots will also be necessary. In order to store and analyze all the information on proteins, researchers will also need to develop bioinformatic computer databases.

More words of caution about the proteome project are uttered when the amount of money necessary to do various parts of the project are estimated. Each of the genome sequencing projects, which leads to the discovery of more proteins and proteome projects, is a multi-million-dollar endeavor. Celera raised 1 billion dollars for a new proteomics center, and the University of Michigan has received $15 million in grants from the U.S. National Institutes of Health (NIH). Each of the components of a Japanese proteome project has million-dollar budgets, which is typical for a proteome project anywhere in the world. 3,000 protein structures in five years will cost $160 million. Technology development will cost $88 million. The synchrotron alone will cost $300 million. Critics of the proteome project feel that the era of big public science initiatives, such as sending a man to the Moon, especially for a program that is not well-defined, is or should be over. —LAURA RUTH

Further Reading

Abbott, Alison. "Publication of Human Genomes Sparks Fresh Sequence Debate." *Nature* 409, no. 6822 (2001): 747.

Ashraf, Haroon. "Caution Marks Prospects for Exploiting the Genome." *Lancet* 351, no. 9255 (2001): 536.

Begley, Sharon. "Solving the Next Genome Puzzle." *Newsweek* 137, no. 8 (February 19, 2001): 52.

Bradbury, Jane. "Proteomics: The Next Step after Genomics?" *Lancet* 356, no. 9223 (July 1, 2000): 50.

"Celera in Talks to Launch Private Sector Human Proteome Project." *Nature* 403, no. 6772 (2000): 815–16.

Eisenberg, David. "Protein Function in the Post-genomic Era." *Nature* 405 (June 15, 2000).

Ezzell, Carol. "Move Over, Human Genome." *Scientific American* 286, no. 4 (2002): 40–47.

———. "Proteomics: Biotech's Next Big Challenge." *Scientific American* (April 2002): 40–47.

Fields, Stanley. "Proteomics in Genomeland" *Science* 291, no. 5507 (2001): 1221.

Gavaghan, Helen. "Companies of All Sizes Are Prospecting for Proteins." *Nature* 404, no. 6778 (2000): 684–86.

Human Genome News. Sponsored by the U.S. Department of Energy Human Genome Program [cited July 20, 2002]. <http://www.ornl.gov/hgmis/publicat/hgn/hgnarch.html#fg>.

HUPO Human Proteome Organization [cited July 20, 2002]. <http://www.hupo.org/>.

HUPO Workshop, Meeting Report, October 7, 2001 [cited July 20, 2002]. <http://www.hupo.org/new/document_report.html>.

Pennisi, E. "So Many Choices, So Little Money." *Science* 294, no. 5 (October 5, 2001): 82–85.

Petricoin, Emanuel. "The Need for a Human Proteome Project—All Aboard?" *Proteomics* 5 (May 1, 2001): 637–40.

"The Proteome Isn't Genome II." *Nature* 410, no. 6830 (April 12, 2001): 725.

Schrof Fischer, Joannie. "We've Only Just Begun: Gene Map in Hand, the Hunt for Proteins Is On." *U.S. News & World Report* 129, no. 1 (July 3, 2000): 47.

Vidal, Marc. "A Biological Atlas of Functional Maps." *Cell* 104 (February 9, 2001): 333–39.

Are genetically modified foods and crops dangerous to human health and to the environment?

Viewpoint: Yes, genetically modified foods and crops, which result from techniques that may have profound, unanticipated, and dangerous consequences, are dangerous to human health and the environment.

Viewpoint: No, genetically modified foods are not dangerous; they are essential for the world's future.

In the 1970s, scientists learned how to cut and splice DNA to form recombinant DNA molecules and began developing techniques that made it possible to isolate, clone, and transfer genes from one organism to another. Scientists hoped that the new techniques of molecular biology, often referred to as genetic engineering, would be used to treat genetic diseases. But it was soon clear that genetic engineering could also be used to manipulate the genetic materials of plants, animals, and microorganisms for commercial purposes.

Responding to widespread fears about the safety of recombinant DNA technology and genetic engineering, the City Council of Cambridge, Massachusetts, proposed a ban on the use of gene splicing experiments in area laboratories. Other cities and states debated similar bans. However, the development of guidelines for recombinant DNA research and the successful use of the technique to produce valuable drugs, like human insulin and erythropoietin, diminished public anxieties. Recombinant DNA techniques also led to the development of transgenic organisms and genetically modified (GM) crops.

By the 1990s, GM crops were being field-tested or commercialized in the United States, Canada, Argentina, and Australia. Critics of the new technologies called the products of GM crops "Frankenfoods" and insisted that GM crops were dangerous to human health and the environment. The Cartagena Protocol on Biosafety, sponsored by the United Nations Convention of Biological Diversity, was adopted in 2000. The goal of the Protocol was to regulate the trade and sale of genetically modified organisms. However, in Europe and the United States concerns about the safety of genetically modified foods led to protests, boycotts, and demands for bans on GM foods.

Genetic modification of plants and animals usually involves the addition of a gene that provides a desirable trait not ordinarily present in the target variety. Those who support the use of genetically modified plants, animals, and foods believe that GM products will become essential components of our future food supply. Particularly in areas of the world threatened by malnutrition, famine, and starvation, GM crops seem to promise benefits in terms of costs, safety, availability, and nutritional value of staple foods. Predicting a global population of 7 billion by 2013, demographers warn that agricultural biotechnology and productivity will become increasingly critical in the not-too-distant future. Genetically modified plants and animals could be used as a source of drugs, vaccines, hormones, and other valuable substances. The addition of appropriate genes may also help accelerate the growth of plants, trees, and animals, or make it possible for them to grow and flourish in harsh climates.

Based on studies of GM foods and comparisons with conventional food sources, most scientists consider GM products to be safe for the environment and for human consumption. Indeed, many common food sources, such as potatoes and manioc, contain chemicals that can be toxic or dangerous to some or all consumers. Conventional foods, such as peanuts, are extremely dangerous to people with allergies, while others contain chemicals that may cause cancer or birth defects. Part of a plant may be toxic, such as rhubarb leaves, whereas other parts are safe to eat. Moreover, supporters contend that GM crops are subjected to higher level of scrutiny than plant varieties developed by traditional breeding techniques.

Molecular biologists note that critics of GM foods seem to ignore the difference between genes and gene products. Therefore, opponents of GM foods claim that the product of an exotic gene will be present in every cell of the GM organism, so that consumers will be subjected to dangerous levels of both the exotic gene and the protein it encodes. Such arguments fail to take into account the fact that although every cell in the organism contains the same genes, not all parts of the organism contain the same gene products. For example, all cells contain the gene for hemoglobin, but this protein is only found in red blood cells.

Genetic engineering has been used to create plants that synthesize their own insecticides, plants that are resistant to herbicides, viral diseases, spoilage, and cold, and plants that are enriched with vitamins or other nutrients. For example, a strain of rice known as "golden rice" was modified to contain higher levels of a vitamin A precursor. Although half of the world's population depends on rice as a staple food, rice is a poor source of many essential vitamins and micronutrients. The United Nations Children's Fund predicts that eliminating vitamin A deficiency could prevent one to two million deaths each year among children aged one to four years. Advocates hope that GM foods will be looked at positively in the light of "nutritional genomics" fighting malnutrition rather than the source of "Frankenfoods" designed to increase the profits of commercial enterprises. They note that research on golden rice was funded by grants from the Rockefeller Foundation, the Swiss Federal Institute of Technology, and the European Community Biotech Program.

Critics of genetic engineering believe that the potential economic advantages of GM plants, animals, and foods are encouraging the development of organisms and foods that may be dangerous to humans and to the environment. For many millennia, farmers have used classical selection and hybridization techniques to produce plants and animals with desirable characteristics, but they were unable to deliberately transfer genes from one species to another. Although scientists may hope to effect the very specific transfer of one highly desirable gene, critics argue that the impact of the exotic gene might have profound, unanticipated, and dangerous consequences. Indeed, although the genomes of many species have now been sequenced, much remains to be discovered about genetic regulation. The addition of exotic genes into widely used crops raises the possibility that GM foods could provoke allergic reactions in susceptible individuals. Another concern raised by critics is that once GM crops are common, pollen from these plants might be accidentally transferred to other plants, with unpredictable results.

At the very least, opponents of GM foods want to have such foods clearly labeled so that people will be able to choose whether or not they want to consume them. Survey of public attitudes towards biotechnology in the 1990s suggested that GM foods were much more controversial in Europe than in the United States. Although there was more press coverage of the issue in Europe, much of it more positive than press reports in the United States, about 30% of Europeans expressed opposition to GM foods.

The history of biotechnology regulation in Europe and the United States has taken different paths. In the United States, most of the major regulatory issues were settled by 1990. Generally, biotechnology products were regulated within existing laws and procedures. In Europe, however, regulators attempted, rather unsuccessfully, to establish new regulatory procedures to deal with biotechnology issues. This approach created more widespread public debate and uncertainty. Surveys indicated that Americans had higher levels of trust in their regulatory agencies than was the case in Europe.

The dream of mixing characteristics from different sources to produce the best offspring is very ancient, but the outcome of classical or modern genetic techniques is not always predictable. According to playwright George Bernard Shaw, a beautiful actress once proposed that they should have children together. She was sure that the children would have her beauty and his brains. Shaw, however, warned her that the children might suffer by inheriting her brains and his face. Scientific literacy remains a problem, as demonstrated by the response to a survey on attitudes towards biotechnology carried out in 1997 in the United States and Great Britain. Many of the respondents believed that GM tomatoes contained genes, but natural tomatoes did not. Since the introduction of recombinant DNA technology in the 1970s, it has been very difficult to conduct an informed debate about biotechnology issues, but it is difficult to imagine a scientific debate more polarizing than the one being waged over the safety of GM foods. —LOIS N. MAGNER

Viewpoint:

Yes, genetically modified foods and crops, which result from techniques that may have profound, unanticipated, and dangerous consequences, are dangerous to human health and the environment.

Proponents of genetically modified (GM) organisms, particularly crops and foodstuffs, maintain that the genetic modifications are intended to make the crops more hardy and healthful, and to make more food available to more people in need. Though such altruism may, in fact, apply in some cases, it is far closer to the truth to say that corporate profit is the real impetus behind the GM movement. Yet there are very real dangers inherent in GM crops and foods. These dangers, described below, threaten human and animal health and the health of ecosystems. As you will see, proliferation of GM organisms poses a dire threat to the future of agriculture and, perhaps, to the survival of natural ecosystems.

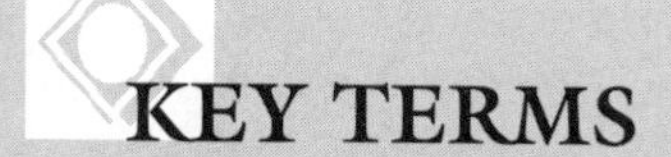

KEY TERMS

ARABLE LAND: Terrain that is fit to be used as farm land, that can be easily cultivated.

DNA (DEOXYRIBONUCLEIC ACID): A nucleic acid found primarily in the chromosomes that contains an organism's genetic information. DNA is the hereditary material of all organisms, which pass on their traits through their DNA.

GENETICALLY MODIFIED (GM) FOODS: Foods that contain some plant or animal material that has been genetically altered. Usually, this means that a gene, a piece of DNA, has been added that provides the organism with a trait it would not ordinarily have.

HERBICIDE: A chemical that can destroy plants or at least inhibit their growth; the targets of herbicides are usually weeds.

KINGDOM: The highest level of organism classification. The kingdoms are Monera, Protists, Fungi, Plants, and Animals.

PESTICIDE: A chemical that kills pests, most often insects.

PHYLA (SINGULAR, PHYLUM): The major groups into which organisms are scientifically classified; a high-level category just below kingdom and above class.

TRANSGENE: An alien gene inserted into the DNA of an organism (a trans [ferred] gene) through genetic engineering.

What Is Genetic Modification?

Classical Agricultural Manipulation. For the last 10,000 years or so—since the Agricultural Revolution—humans have been growing and raising their own food. From the very beginning, farmers have tried to improve their crops and livestock by selective breeding. For example, farmers would save seeds from those tomato plants that bore the most and the sweetest fruit. They would plant only these seeds and, through ongoing selection of seeds from plants with the most desirable traits, they would change and improve the yield, the hardiness, and the quality of the crops they grew. This kind of classical manipulation of agricultural products has been going on for centuries.

Cross-breeding is another kind of classical manipulation of crops. A farmer might notice that one tomato plant yielded many fruits and that a different tomato plant was highly tolerant of cold. The farmer might cross-breed these two plants—transferring the pollen of one to the pistil of the other—to yield a tomato plant that both yielded many fruits and that was tolerant of cold. This kind of crop manipulation involves the mingling of genes from two closely related plants—all the genes involved in this cross come from tomatoes.

What farmers could never do was cross-breed or transfer genes from two totally unrelated organisms. That's the realm of genetic modification.

Biotechnology Defined. Genetic engineering—biotechnology—is the process in which a gene from one organism is inserted into the genome of a completely dissimilar and unrelated organism. The inserted gene causes all or some of the cells of the host organism to produce a particular substance (a protein) or to manifest a particular trait.

For example, in classical selective breeding, a farmer trying to breed a variety of tomato that is tolerant of extreme frost has rather limited options. That's because there is a limit to the degree of cold tolerance inherent in the genes of any tomato variety. In the genetically modified world, however, a scientist can insert into the tomato plant a gene from a bottom-dwelling fish that codes for the production of anti-freeze. Thus, the tomato might conceivably grow in the Arctic because the fish gene causes the plant to produce a kind of anti-freeze.

Proponents of genetic engineering claim that it permits them to transcend the natural genetic limits of a particular class of organisms; for example, the tomato plant's limited genetic ability to tolerate freezing temperatures. Increasingly, however, scientists and the public are questioning the wisdom of such manipulations.

YOUR GM CUPBOARD

The following are just a few of the everyday foods that contain genetic modifications, primarily for the production of pesticides or for herbicide resistance.

Soy products. Soy flour, soy oil, lecithin, vitamin E, tofu, veggie burgers and soy-based meat substitutes, soy sauce; products that contain soy or soy oil (comprising 60% of all processed foods): ice cream, yogurt, infant formula, sauces, margarine, baked goods, chocolates, candies, fried foods, pastas; also shampoo, bubble bath, cosmetics.

Corn products. Corn flour, corn oil, corn starch, corn sweeteners, corn syrup; products that contain corn or corn derivative: vitamin C, tofu, chips, candy, ice cream, infant formula, salad dressing, tomato sauce, baked goods, baking powder, alcohol, vanilla, margarine, powdered sugar, enriched flour, pastas.

Canola products. Canola oil; products containing canola oil: chips, peanut butter, crackers, cookies.

Potatoes. Fresh potatoes, processed or restaurant potatoes and potato products: chips, vegetable pies, soups.

Tomatoes. Fresh tomatoes; products containing tomatoes: sauces, purees, pizza, lasagna, and most Italian and Mexican foods.

Dairy products. Cows are treated with hormones (BGH—bovine growth hormone and are fed GM feed, for example, corn products): cheeses, butter, sour cream, yogurt, ice cream, cream cheese, cottage cheese.

[source: Friends of the Earth: <www.foe.org/safefood/foodingredients.html>.]

—Natalie Goldstein

Genetic engineering's ability to transfer genetic material—not within or even between species, but across entire phyla or kingdoms; in fact, from anywhere in nature—is extremely worrisome. Many scientists believe that this compromises the essential identity and integrity of individual species. Further, because genetic modification is concentrated among crops, which are grown openly on the land, there is a grave risk that these GM organisms cannot be contained, but will spread their altered and, literally unnatural, genes throughout natural ecosystems—with unforeseen, but likely dire effects.

There are dangerous misconceptions and numerous risks inherent in genetic engineering. These include: (1) misunderstanding genetic interactions; (2) risks to human health; (3) gene escape; and (4) testing, regulation, and labeling. Each is discussed below.

Genetic Misconceptions

The One-to-One Myth. Genetic engineers operate on the assumption that genes and traits have a direct, one-to-one relationship. They genetically manipulate organisms based on the idea that single genes are distinct, determinate, and stable, and that they function in isolation from other genes. For example, they believe that a fish gene for cold tolerance codes only for that one trait, and if it is inserted into another organism it will confer on it only that one trait. This idea is considered extremely simplistic and patently false.

Genetic material works in conjunction with other genetic material. A gene inserted into DNA is affected by the other genes that make up the DNA: it interacts with these other genes, and it is influenced by these genes and the proteins they produce. The organism itself affects all its genes, and, as the environment affects the organism, so too does the environment affect the genes. Thus, simply inserting a gene for a trait does not mean that that trait will be expressed purely or in a way that the scientists intended. And it certainly does not mean that that gene will reside in the organism's DNA without affecting the genes surrounding it. There are no data to indicate the nature of the changes the inserted gene has on the organism's natural genetic makeup. Yet scientists know for a fact that genes interact with and affect each other. And everyone knows that genes mutate from time to time. As yet, we have no way of knowing how an alien, inserted gene will affect an organism's DNA over time. If the inserted gene is later shown to have disastrous effects on the organism, there is no way to undo the genetic damage.

The "Junk" Myth. Genetic engineering also operates under the outdated assumption that it is safe to insert alien genes into the 95% of genetic material that was once called "junk genes." Scientists noted that only about 5% of an organism's genes seem to actively code for a known protein and purpose. That's why the remainder of the genes, whose purpose they could not figure out, were called "junk."

Recent research has shown that these "junk" genes are anything but. The research strongly indicates that "junk" genes are very likely a vital "backup" for the more active genes. Further, in the 1990s, studies show that interference with these random genes can have dire effects. If these genes are altered, they tend to mutate. If a gene (whether alien or from another region of the organism's own DNA) is inserted into a region of random genes, the rate of genetic mutation increases sharply. In some experiments, the genetic mutations were associated with disease.

Protesters demonstrate against genetically modified organisms in England.
(© Adrian Arbib/Corbis. Reproduced by permission.)

Health Effects

There are many potential, or at least unresearched, health risks involved in GM crops and foods.

Eating Alien Gene Products. There may be a risk attendant upon consuming foods that contain genes from what are generally non-food organisms.

There is a grave risk involved when genes known to be harmful to living things are inserted into consumables. This is the case for GM potatoes and other crops (particularly soy and corn). These GM crops have had inserted into their DNA a gene that produces a naturally occurring pesticide. The pesticide, Bt (*Bacillus thuringiensis*), is toxic to certain insect pests. When the gene is inserted into the crop seed, every cell of the growing plant contains this gene. When people eat these GM foods, they ingest the pesticide.

The GM plant is naturally resistant to the insect pests that would otherwise devour it, so the genetic modification helps farmers get a good crop and allows crops to be grown with fewer pesticide applications. Yet there is little or no research on the human health effects of a lifetime of eating pesticide-laden foods. A pesticide is a chemical that is toxic to at least some living things. What is the effect of a lifetime of ingesting GM foods?

Allergies. In the 1980s, biotech scientists genetically engineered a new strain of celery, which grew better and lasted longer in the supermarket produce section. Alas, it had to be withdrawn after people who handled it or ate it broke out in a severe rash. A potato produced in the 1990s, genetically modified to enhance its nutri-

Rapeseed plants in flower. In the late 1990s a genetically modified strain escaped from an experimental field.
(© Gunter Marx/Corbis. Reproduced by permission.)

tional value—a worthy endeavor—ended up making consumers violently ill. It turns out that the genetic modification that boosted the potato's nutrients also caused the plant to produce toxic levels of glycoalkaloids. The genetic engineers had no idea that the simple insertion of a single gene would have this unforeseen and dangerous outcome.

Also in the 1990s, genes from Brazil nuts were inserted into soy beans to enhance their protein content—again, a worthwhile goal. But the beans were not labeled to show they were genetically engineered. Trusting consumers who ate the extra-nutritious beans had no idea that they contained the nutty genes. People allergic to Brazil nuts fell ill, some seriously. Tests done on non-human animals did not reveal this side effect.

In 1998, researchers tested the health effects of a crop plant into which a virus gene had been inserted. (The virus gene gave the plant resistance to a bacterial plant disease.) Mice were fed the GM plant. The scientists found that in the process of digestion, some of the virus genes escaped into the animals' bloodstream. The more GM plants the mice ate, the greater the blood level of the virus. This research shows that alien genes in GM foods can escape into the body. What scientists don't yet know is the health effects caused by alien virus genes that remain in and circulate through the body.

Finally, there are the health effects that arise directly from the corporate profit motive. For example, Monsanto is a world leader in the creation and production of genetically modified crop seeds. The company also manufactures agricultural chemicals; Roundup®, for example, is an herbicide made by Monsanto. Its GM division has developed "Roundup-ready" plant seeds, which yield crops that are impervious to the plant-killing effects of Roundup®. Thus, a farmer can douse his fields with far more herbicide than he would otherwise, because he knows that his cash crop will be unaffected. Yet when that cash crop is harvested and sold as food, it contains far more toxic herbicide than it would have had if it been naturally herbicide sensitive. The plant food is sold to consumers as being identical to and as safe as any non-GM food, yet it contains far greater quantities of herbicides, which are known toxins. Again, the question must be raised about the health effects of lifelong consumption of herbicide-laden foods.

Gene Escape At the heart of genetic engineering lies an unavoidable contradiction. On one hand, genetic engineers operate under the assumption that genes are universally transferable, even across entire kingdoms and phyla of unrelated organisms. On the other hand, they reassure a worried public that once alien genes are transferred into an organism they cannot and will not spread further—into non-target, wild organisms, for instance.

Research has shown that alien genes inserted into an organism—transgenes—have escaped and become incorporated into wild organisms. The effects of gene escape are unknown, but most scientists believe they will likely be catastrophic.

In the late 1990s, British researchers found that transgenes from GM rapeseed plants in a 123-acre experimental field escaped and were incorporated into the genetic material of weeds growing more than one mile away. Further, when the crop was harvested, some seeds naturally fell on the ground and were eaten by birds. Evidence was found that the birds had transferred the transgenes to plants many miles away.

In another study, researchers discovered that plants genetically engineered to withstand the effects of heavy applications of herbicides did well in an herbicide-doused field. Yet in a "natural environment" in which herbicide was not used, the plants did very poorly. The scientists concluded that if this transgene escaped into the wild (where, of course, herbicide is not used) and was incorporated into wild plants, these wild plants would also do poorly (due to lack of herbicide in the natural environment). With transgene-infected wild plants growing poorly, wildlife and the entire ecosystem will suffer and may even face collapse.

A research report published in the September 1999 issue of *Nature* revealed that, contrary to the assertions of the biotech industry, genetic material from GM plants was far more likely to escape into the wild than were genes from non-GM plants. The scientists state that an organism's natural genes are more securely "attached" to the chromosomes because they have evolved in response to other genes and the environment over millions of years. The research underlines the inevitability that alien transgenes will escape and become incorporated into the genetic material of wild plant species and varieties. Any mutations or negative effects the transgenes have on wild plants will be irreversible. Once genes escape and spread through the wild, there is no way to recall or stop them. Since we don't know what effects transgenes will have on wild plants, it is surely wise and prudent to study their potential effects thoroughly before permitting their widespread use.

The "Super" Syndrome Some scientists insist that the inevitable escape of transgenes into the wild will have disastrous effects not only on ecosystems, but on agriculture as well. Consider—

Super-weeds. Some GM crop plants are genetically engineered to resist the effects of herbicides. What happens when, inevitably, the genes for herbicide tolerance are transferred from crops to pesky weed species? Once weeds become resistant to herbicides, there'll be no stopping them. In a way, the herbicide-resistant genes inserted into crops contain the "seeds" of their own destruction. Sooner or later, their herbicide-resistant genes will escape and be incorporated into the very weeds the herbicides are intended to kill. Then what?

Super-bugs. Among the most popular and widespread applications of genetic engineering is the transfer of genes to plants that confer resistance to pests and plant diseases. Inevitably, these genes escape to wild plants, which incorporate the gene and themselves become resistant to insect pests and plant diseases. As is by now well known, insect pests and the bacteria and viruses that cause disease reproduce so rapidly that sooner or later—usually sooner—they become immune, or resistant, to whatever is trying to do them in. It is therefore only a matter of time before the GM crops succumb to new, improved Super-pests or Super-diseases—insects, bacteria, and viruses that have developed resistance—immunity—to the genetic modification. What do we do then to combat infestation and disease? What happens when these Super-bugs descend on the natural environment?

(It's interesting to note that in the entire history of humans' fight against crop pests, and despite an enormous arsenal of increasingly lethal chemical pesticides, not one insect pest species has gone extinct. They just adapt a resistance to every chemical thrown at them!)

Testing, Regulation, and Labeling Scientists and a large segment of the public are legitimately worried about the widespread use of GM crops and foods. The general public may be less aware of the ecosystem-wide ramifications of transgene escape, but people are increasingly concerned about the human health effects of GM foods. For this reason, they are asking that GM foods and products that contain GM foods be labeled as such.

In the mid-1990s, senior scientists with the U.S. Department of Agriculture's (USDA) biotech division reported to agency heads that GM foods are so different from non-GM foods that they pose a serious health risk. The senior biotech scientists recommended that the USDA establish a protocol for testing this novel type of food. The scientists criticized the biotech industry's own safety data, which were accepted without question by the USDA, as "critically flawed" and "grossly inadequate." However, the USDA ignored these recommendations, publicly insisting that GM foods are identical to non-GM foods and therefore need no special safety testing or labeling. Memos that subsequently came to light (via lawsuit) showed that the federal government had directed the USDA to "foster" promotion and acceptance of GM foods. Thus, the USDA ignored the recommendations of its own senior scientists and compromised the health of the American people.

To date in the United States (unlike Europe), the reasonable request for product labeling has been either ignored or simply shot down. Despite the very real dangers of allergic reactions to foods containing genes from known allergens—not to mention the risks inherent in pesticide-laden foods—neither corporations nor the government agencies charged with assuring us a safe food supply will agree to label GM foods. In what is widely regarded as inaction in support of corporate profit, the government refuses to respect its citizens' right to make an informed choice. (A wonderfully told, and very telling, tale of the shenanigans that surround GM foods and the U.S. agencies that, supposedly, regulate their safety is Michael Pollan's article "Playing God in the Garden.")

Conclusion Genetically modified foods and crops pose unforeseen and potentially catastrophic threats to wild plants and to ecosystems because of the inevitability of transgene escape and the evolution of resistant weeds, pests, and disease organisms. Misconceptions about how genes work within organisms pose the additional threat of mutations, which may threaten human health. Human health is threatened more directly by GM plant foods that contain known toxins. In sum, genetic manipulation of organisms for profit without adequate study of the consequences for human life or the environment is a very dangerous and potentially catastrophic undertaking. As Pollan described it: "Biological pollution will be the environmental nightmare of the twenty-first century . . . [it] is not like chemical pollution . . . that eventually disperses . . . [it is] more like a disease." —NATALIE GOLDSTEIN

Viewpoint:

No, genetically modified foods are not dangerous; they are essential for the world's future.

Genetically modified (GM) plants and animals have only been available for a few years, but already they are becoming key elements of the food supply. The reason for the rapid acceptance of genetically modified foods by farmers and food producers is that they provide several advantages, including cost, safety, and availability in areas of the world where people are starving. These advantages will only become more and more apparent as their presence in the food supply becomes more dominant. Such dominance is not a matter of "if," but of "when." A survey of the different types of GM plants and animals being developed indicates why this is the case, and also indicates why prospects are good that these organisms are not only useful but safe to both humans and the environment.

It should be noted at the start that "safe" is always a relative term. No food, no matter how "natural" it may be, is without risks. Eating too much of anything can cause problems—a man who for months ate tomatoes at almost every meal ended up turning orange from the pigments in this vegetable. A number of natural foods contain toxic substances, though usually in very small quantities. For example, carrots and potatoes both carry toxins, but these aren't a problem in the quantities they are usually eaten. Such plants have never been subjected to the scrutiny that GM foods have received. In the United States, the Environmental Protection Agency (EPA) demands information on the safety of GM crops—both to humans and to the environment—before these crops are planted in open fields. This is a greater level of review than is given to plants that are created by more traditional breeding programs, such as crossing different strains of a plant to produce seeds carrying novel sets of characteristics. This type of breeding has been going on for centuries, with little concern about the safety of the plants created in this way. Yet there can be problems with this approach. For example, a carrot strain produced by such the traditional breeding program was found to be high in toxins, and the same was true of a new potato strain.

Protecting Against Pests One approach to the genetic engineering of plants is to insert a

African scientists have created a virus-resistant form of cassava, a yam-like plant that's a staple crop in many parts of Africa.

(© Wolfgang Kaehler/Corbis. Reproduced by permission.)

gene into a plant that makes the plant more resistant to pest damage. Many plants produce chemicals that taste unpleasant or are toxic to insects or slow insects' growth. These genes could be transferred to crop species that don't ordinarily possess them. Farmers could then grow plants that carry their own insecticides with them. This can reduce the spraying of insecticides that are sometimes toxic to humans as well as to other insects. In China alone, 400–500 cotton farmers are killed each year in accidents involving pesticides. Also, using such GM plants could make growing plants cheaper, since farmers wouldn't have to buy as much insecticide, and there would be less damage to non-target insects.

These benefits are more than just speculation or wishful thinking; there is solid evidence for each of them, and it comes from one of the earliest efforts to genetically engineer crop plants. A bacterium called *Bacillus thuringiensis* makes a substance that kills many common insect pests. Lures containing this substance have been used for years in gardens to keep down insect populations. Biologists took the gene for this chemical and inserted it into corn cells. The resulting corn plants, called Bt corn, make the substance and thus carry their own natural insecticide with them. Bt corn is now planted widely in the United States, as is Bt cotton. The fact that these plants have only been available for a few years, and that they have already been so widely accepted by farmers, indicates the value of this genetic modification. In addition, Bt corn has been found to be comparable to regular corn and doesn't cause any ill effects in humans who eat it.

Many of the problems that plague crop plants are not as obvious as insects. Invisible viruses can also destroy plants. Thanks to genetic engineering, biologists are now able to create plants that are resistant to viral diseases. African scientists have created a virus-resistant form of cassava, a yam-like plant that's a staple crop in many parts of Africa. Now they are working on a GM form of the sweet potato that is virus-resistant. These projects indicate that GM crops are not just for wealthy nations and may even prove to be more important in underdeveloped countries. Many of these nations have rapidly growing populations, some of whom are already living on subsistence diets, often low in important nutrients. And food crises are likely to worsen in the future as populations continue to increase. In such situations, GM crops may be the only way to provide sufficient, highly nutritious food to stave off malnutrition. Rather than being a threat to health, such foods may instead be essential to the well-being of large numbers of people. Still another way to improve yields is to prevent spoilage, a reason much food is lost before it gets to those who need it. Specific genes have been identified that slow ripening, so that vegetables and fruits with such genes can be transported to market and sold before they spoil.

Fighting Weeds In addition to inserting genes that provide protection against pests and thus cut down on the use of pesticides, biologists have also found a way to make the use of herbicides more effective and less costly. Herbicides are chemicals that kill plants, and therefore can be employed as weed killers. The problem with their use in farming is that they also kill a lot of crop plants. However, biologists have identified genes that make plants resistant to herbicides, so the plants grow well even in their presence; these genes have now been transferred to many crop plants. This means that farmers can use a weed killer, and still have a good crop. This is economically important because overall, crop yields are reduced by 10–15% due to weeds, which compete with the crop plants for water, soil nutrients, and sunlight. Also, besides increased crop yields due to less competition from weeds, there is less need for tilling, so erosion, a constant problem for farmers, is reduced.

Health Benefits In some cases, GM crops may provide real benefits to human health. A strain of rice has been developed that contains a gene to make beta-carotene, a nutrient the human body can convert into vitamin A. Vitamin A deficiencies are painfully common in many Third World countries, including countries where rice is the staple crop. This new type of rice, called golden rice because the beta-carotene gives grains a yellow color, could provide distinct health benefits. A vitamin A deficiency can cause birth defects, and in children and adults it can also cause skin problems, and perhaps more importantly, night blindness. Public health researchers calculate that curing vitamin A deficiencies could significantly reduce deaths due to night-time accidents in Third World countries. Many of these deaths can be linked to vision difficulties related to vitamin A deficiency. Another strain of GM rice, one that increases the iron content of the grain, could also be a boon to health, since iron deficiency is extremely common throughout the world, particularly among women and children. In the future, genes for many other nutrients could be added to a wide variety of crop plants, making many foods much more nutritious. This is particularly crucial in countries where most people are poor and the variety of foods available to them is limited.

GM food crops could also aid human health more indirectly. Plants can be genetically engineered to make molecules that could be used as drugs. For example, carrots have been engineered to produce a vaccine against hepatitis B. This vaccine can be produced in more traditional ways, but it is very expensive. A plant source for the vaccine would significantly decrease its cost. Research is also being done on developing GM plants that would produce a wide variety of vaccines, hormones, and other drugs that are currently expensive and difficult to make. Plants could produce these substances in large quantities, and purification of these substances from plant materials would be relatively easy and therefore economical. This would mean that many drugs now almost totally unavailable in Third World countries because of their cost would be obtainable by much larger numbers of people who desperately need them.

What should also be mentioned here are the benefits of genetically engineered animals as well as plants. Animals with added genes for growth will grow more quickly on less feed, making the health benefits of the high-quality protein in meat available to more people. The same argument can be made for adding genes that increase milk production in cows. Efforts are also underway to create GM animals that could serve as sources for vaccines and other drugs.

Protecting against Harsh Environments Most of the world's available high-quality arable land, that is, land well-suited for growing plants, is already being used as farm land. With a rapidly growing world population that already includes over 1.5 billion people suffering from malnutrition, finding land on which to grow enough food to feed everyone well is going to become an increasingly difficult problem. Yet converting the world's rain forests to farmland is not considered an environmentally responsible answer to the problem. The poor soil of rain forests doesn't support agriculture for more than a few years, so farmers are forced to then destroy more rain forests, leading to spiraling environmental deterioration. Also, the loss of rain forests means the loss of many species of plants, animals, and microscopic organisms, thus depleting the Earth's biodiversity and worsening the present mass extinction many biologists consider the most severe in Earth's history. That's why the benefits of GM crops in increasing yields per acre by reducing crop loss due to weeds and pests are important, and that's why it's also important to extend agriculture into areas where the weather has been considered too harsh for large-scale agriculture in the past.

In such unpromising environments, GM crops can be invaluable. Plants can be engineered to be resistant to frost, thus extending the growing season in colder regions. Genes have also been identified that provide drought and heat resistance, making hot, dry lands more attractive for agriculture. Such areas may also benefit from GM plants with genes that allow them to grow in highly saline environments. This also means that salt water could be used for irrigation, something that has been impossible up to now because crop plants can't ordinarily tolerate high salinity.

It is undeniable that some GM organisms may cause environmental problems and may have some health consequences. But the monitoring of these plants by the Environmental Protection Agency means that problems are dealt with before the plants are used widely. Balanced against these difficulties, the dire consequences to the environment and human health of not using GM organisms are much greater. Malnutrition will become more widespread, more rain forests will be destroyed, and life-saving drugs will continue to be priced outside the means of a majority of the world's population. No innovation is without consequences, but in the case of GM foods, the benefits so outweigh the dangers that to hesitate is to court environmental and human disaster. —MAURA C. FLANNERY

Further Reading

Anderson, Luke. *Genetic Engineering, Food, and Our Environment.* White River Junction, VA: Chelsea Green, 1999.

Butler, Declan, and Tony Reichhardt. "Long-Term Effect of GM Crops Serves Up Food for Thought." *Nature* 398 (April 22, 1999): 651.

Campaign for GM Product Labeling [cited July 16, 2002]. <www.thecampaign.org>.

Charles, Douglas. *Lords of the Harvest: Biotech, Big Money, and the Future of Food.* Cambridge, MA: Perseus Publishing, 2001.

GM Food Safety [cited July 16, 2002]. <www.prast.org/indexeng.htm>.

Gura, Trisha. "The Battlefields of Britain." *Nature* 412 (August 23, 2001): 760.

Ho, Mae-Won. *Genetic Engineering: Dream or Nightmare? Turning the Tide in the Brave New World of Bad Science and Big Business.* New York: Continuum Press, 2000.

Lappe, Marc. *Against the Grain: Biotechnology and the Corporate Takeover of Your Food.* Milford, CT: LPC Pub., 1998.

"Many Links for GM Food Safety" [cited July 16, 2002]. </sbc.ucdavis.edu/outreach/resource/gm_food_safety.htm>.

Messina, Lynn, ed. *Biotechnology.* New York: H.W. Wilson, 2000.

Nottingham, Stephen. *Eat Your Genes: How Genetically Modified Food is Entering Our Diet.* New York: Palgrave Press, 1998.

One World Guide to Biotechnology [cited July 16, 2002]. <www.oneworld,org/guides/biotech/index.htm>.

Pinstrup-Anderson, Per, and Ebbe Schioler. *Seeds of Contention: World Hunger and the Global Controversy over GM Crops.* Baltimore, MD: Johns Hopkins University Press, 2000.

Pollan, Michael. "Playing God in the Garden." *The New York Times Sunday Magazine* (October 25, 1998).

Rampton, Sheldon. *Trust Us, We're Experts: How Industry Manipulates Science and Gambles with Your Future.* Los Angeles: JP Tarcher, 2000.

"Real Food Campaign." Friends of the Earth [cited July 16, 2002]. <www.foe.co.uk/campaigns/real_food/>.

Roberts, Cynthia. *The Food Safety Information Handbook.* Westport, CT: Oryx Press, 2001.

"Special Report: GM Foods." New Scientist [cited July 16, 2002]. <www.newscientist.com/hottopics/gm/>.

Tai, Wendy. "Plant Biotechnology: A New Agricultural Tool Emerges as the World Seeks to Vanquish Chronic Hunger." *In Focus* 1 (April 2002): 1.

Teitel, Martin. *Genetically Engineered Food: Changing the Nature of Nature.* Rochester, VA: Inner Traditions, 2001.

Ticciati, Laura. *Genetically Engineered Foods.* New York: Contemporary Books, 1998.

Ticciati, Laura, and Robin Ticciati. *Genetically Engineered Foods: Are They Safe? You Decide.* New Canaan, CT: Keats Publishing, 1998.

Von Wartburg, Walter, and Julian Liew. *Gene Technology and Social Acceptance.* New York: University Press of America, 1999.

Yount, Laura. *Biotechnology and Genetic Engineering.* New York: Facts on File, 2000.

In order to combat potential threats from hostile governments and terrorist groups, should the United States resume research and development programs on biological weapons?

Viewpoint: Yes, biological weapons are a genie already out of the bottle and awaiting a master. Research and development on biological weapons should be resumed.

Viewpoint: No, resumption of research on and development of biological weapons is a medically and scientifically unneeded, inefficient, and unethical response to threats of bioterrorism.

Since formally terminating its biological weapons program in 1969, the United States has condemned the spread of biological weapons programs and lobbied for an international treaty to ban such weapons. Those who have closely followed the American biological weapons programs point out that even though research dedicated to the weaponization of biological agents supposedly ended in 1969, research that addressed the identification of and reaction to the use of biological weapons by hostile agents continued. From World War II until 1969 the United States operated a major biological weapons programs that conducted secret open-air tests of biological agents on American soil. Moreover, the project had weaponized lethal biological agents, toxins, and incapacitating agents and had stockpiled biological bombs for potential battlefield use.

In the United States, biological warfare research began during World War II in the Army Chemical Warfare Service. Most of the early work focused on anthrax and botulism, but biological agents for use against people and plants also were studied. After the war, the United States established research and development facilities at Fort Detrick, Maryland, test sites in Mississippi and Utah, and a manufacturing plant near Terre Haute, Indiana, that were dedicated to biological warfare projects. During the cold war era, the Federal Civil Defense Administration and other governmental agencies issued various warnings to the general public about the threat of biological warfare. Such warnings including a 1950 manual titled *Health Services and Special Weapons Defense* and pamphlets such as "What You Should Know about Biological Warfare." The public was informed that deadly biological organisms and toxins could be transmitted by air, food, and water supplies.

Research and field trials had clearly demonstrated that biological agents could be used effectively as offensive weapons by the time President Richard M. Nixon decided to terminate the American biological warfare program. In 1969 Nixon announced that "the United States of America will renounce the use of any form of deadly biological weapons that either kill or incapacitate." He also ordered the destruction of existing stocks of bacteriological weapons. In the future, Nixon asserted, all American bacteriological programs would be "confined to research in biological defense, on techniques of immunization, and on measures of controlling and preventing the spread of disease."

Although the first White House statement referred to only biological and bacteriological agents, Nixon later extended the ban to include toxins and toxic agents.

In 1972 the United States was a leading supporter of a new Biological Weapons Convention (BWC), formally known as the Convention on the Prohibition of the Development, Production and Stockpiling of Bacteriological (Biological) and Toxin Weapons and on Their Destruction. Article I stated that states that accepted the convention would never "develop, produce, stockpile or otherwise acquire or retain" biological agents or toxins that "have no justification for prophylactic, protective or other peaceful purposes." The convention also banned "weapons, equipment or means of delivery designed to use such agents or toxins for hostile purposes or in armed conflict." Article II called on all parties to "destroy, or to divert to peaceful purposes . . . all agents, toxins, weapons, equipment and means of delivery specified in article I."

The BWC treaty was originally signed by 79 nations, including the United States, the Soviet Union, Great Britain, and Canada. However, the agreement was not binding on any signatory until that nation's government formally ratified the document. In 1974 the Senate ratified the 1972 convention, and President Gerald R. Ford signed it in January 1975. Critics argued that the convention had two major loopholes. The treaty did not define "defensive" research, nor did it provide a means of detecting and punishing noncompliance.

The 1972 BWC was not the first attempt to secure a global ban on germ warfare. After World War I, the Geneva disarmament convention included a "Protocol for the Prohibition of the Use in War of Asphyxiating, Poisonous or other Gases, and of Bacteriological Methods of Warfare." Although the United States and 28 other nations in 1925 signed the Geneva protocol against bacteriological warfare, the U.S. Senate did not ratify the document. Nevertheless, because the Geneva protocol banned only the use of biological warfare, nations conducting germ warfare research and development were technically in compliance.

Since 1969 the United States has been a leading advocate for efforts to enforce the ban on biological weapons. By the 1990s, there was considerable evidence that several states that had signed and ratified the 1972 convention were not in compliance. The Soviet Union in particular had expanded its program to develop, test, and stockpile biological weapons. American scientists suspected this in 1979 when reports of an anthrax outbreak in Sverdlovsk began to circulate. In 1992 Russian president Boris Yeltsin finally admitted that military biowarfare research had been responsible for the Sverdlovsk outbreak.

Thus, in the 1990s, on the basis of fears that rogue states or terrorist groups would use biological agents, the United States embarked on a secret project to make and test biological bombs. The Pentagon sponsored the assembly of a germ factory in the Nevada desert in order to determine how easy it would be for terrorists and rogue nations to build plants that could produce significant quantities of deadly germs. Such weapons have often been referred to as the poor man's nuclear bomb. Critics argued that such tests were banned by the 1972 BWC. The Bush administration as well as the Clinton administration argued that the convention allowed research on microbial agents and delivery systems for defensive purposes. Government officials said that the goal of this research was to mimic, in order to understand the threat, the steps that terrorists or hostile nations would take to create a biological arsenal. Iraq, for example, in the 1980s apparently began a program to produce and stockpile biological weapons.

Administration officials said the need to keep such projects secret was a significant reason for President Bush's rejection of a draft agreement to strengthen the 1972 treaty. The new provisions called for international inspections of facilities that could be conducting germ warfare research and development. The Bush administration said that it supported strengthening the treaty but was interested in finding alternative ways to enforce compliance.

In the wake of the September 11, 2001, attacks on the World Trade Center and the Pentagon, policy analysts began asking whether the United States should conform to the BWC or abandon treaty obligations and openly develop and stockpile biological weapons. Some critics of the 1972 convention argue that the United States could pursue research on weaponization without stockpiling biological weapons and that this would not be a serious breach of treaty obligations. They also argue that treaties are unlikely to have any effect on terrorist groups and rogue nations, such as Iraq, Iran, Libya, Syria, and North Korea, that are interested in attacking American citizens. Other analysts contend that if the United States insists on its own right to conduct research on biological weapons, it would be setting a dangerous precedent that might stimulate the global development and dissemination of weapons of mass destruction. Those who oppose a return to deliberate research on weaponization argue that defensive research would provide the information needed to prevent or respond to existing or novel biological agents. Moreover, an emphasis on weapons research would deplete resources and talent from other areas of biomedical research. Others argue that there is no real distinction between research on defensive and research on aggressive biological agents. Thus any potentially valuable research in this area should not be artificially obstructed. Opponents argue that weapons research is likely to be very secretive and highly classified. Thus useful scientific developments would not contribute to advances in nonmilitary biomedical science. —LOIS N. MAGNER

Viewpoint:

Yes, biological weapons are a genie already out of the bottle and awaiting a master. Research and development on biological weapons should be resumed.

A contentious issue that forces us to confront humankind's deepest fears, the debate over whether the United States should resume research and development programs on biological weapons often disregards the fact that although the United States biological weapons programs formally ended in 1969, by President Richard M. Nixon's executive order, research on potential biological weapons has never stopped. Only formal research dedicated to the weaponization of agents has stopped. As abhorrent as the argument may be on a personal level, there are valid strategic reasons for the United States to abandon its current BWC policy and treaty obligations in order to openly resume formal research and development programs on the actual weaponization of biological agents. Renewed research on and development of biological weapons do not demand that the United States begin to accumulate or stockpile such weapons, and this essay makes no such argument. Regardless, the evolution of political realities in the last half of the twentieth century clearly points toward the probability that within the first half of the twenty-first century, biological weapons will surpass nuclear and chemical weapons as a threat to the citizens of the United States. An effective defense against biologic agents—and the development of strategic weapons that will deter attack on the United States—can be obtained only through limited but deliberate biological weapons research.

With regard to the strategic long-term interests of the United States, the arguments in favor of resuming biological weapons programs essentially depend on assessments and assertions made regarding the potential threat, the potential enemies who might use similar weapons, and the scientific and/or strategic goals that might be achieved by such programs.

It is difficult to overstate the threat from effective biological weapons. An airborne weapon in the form of a virus or bacteria with a 100% lethality on exposure that is also stable at a range of atmospheric conditions, easily transmittable, and releasable from multiple sites could drive human population numbers low enough to threaten extinction and/or so isolate populations (i.e., gene pools) as to effectively induce speciation or subspeciation that would signal the formal extinction of *Homo sapiens.* Some life might continue to exist, but in a very real sense, it would not be human as we currently understand and use the term. In addition, rapid advances in knowledge regarding the human genomic sequence and advances in the use of virus vectors to insert specific sequences of DNA into target host sequences—procedures based on gene therapies that medical scientists currently argue may yield a bounty of beneficial medical treatments—can, with similar technological tools and training, be twisted into specifically tailored biological weapons targeted at groups with specific genetic traits. Once a horror of science fiction, recent advances in human genomics make these types of weapons both feasible and obtainable.

Between the extremes of extinction-level-event biological weapons and targeted biological weapons lie a multitude of threat scenarios. Weapons of far less lethality and/or transmissibility can be used to collapse a nation or society by overburdening its medical and economic infrastructure. An effective biological weapon could have no effect on humans but induce famine and economic ruin if targeted at livestock or agricultural production. The ultimate biological weapon would, of course, be one devastating to a specifically targeted enemy for which effective countermeasures (e.g., an appropriate vaccine) could be developed to protect the aggressor and chosen allies. Such a weapon would carry to an extreme—and far surpass—the intended strategic usefulness of the high-radiation neutron bomb designed to maximize lethal radiation exposure with a minimal blast damage (i.e., a bomb intended to kill populations with a minimal impact on the physical, nonbiological infrastructure of a target).

Although fears of terrorist use of weapons of mass destruction, including nuclear, biological, and chemical weapons, against the United States increased in the aftermath of the September 11, 2001, terrorist attacks on the World Trade Center and Pentagon, the threat before the attacks was deemed significant enough that in 1996 the U.S. Congress passed the Defense Against Weapons of Mass Destruction Act. The legislation provides millions of dollars in annual financing to train emergency responders in the treatment of nuclear, chemical, and biological injuries.

The list of potential enemies who might effectively use biological agents against the

United States is growing. The development of easily obtainable global transportation and communication also has facilitated the globalization of terrorism. Historically, nation-states are self-preserving in that they take actions in their interest that do not threaten their own continuation. Akin to biological self-interest in preservation, this characteristic of political entities was the psychological underpinning of the development and implementation of a strategic policy based on mutually assured destruction (MAD) that guided research and development of nuclear weapons programs throughout the cold war between the Soviet Union and the United States. Although force reductions and specific target profiles may soon alter the protocols for the use of small tactical nuclear weapons, the MAD doctrine remains the governing policy in formulating a response by the United States to the use of nuclear weapons against it. Moreover, in response to the threat of Iraqi chemical and biological weapons during the 1991 Gulf War, the United States asserted that an attack on its forces by an enemy using chemical and/or biological weapons would be considered grounds to respond with any and all weapons, not excluding nuclear weapons, in the U.S. arsenal.

Essentially, in lieu of a MAD-like biological weapons program response to a use of biological weapons, the United States simply attempted to extend the MAD nuclear shield to protect against attack by biological weapons. Although there are other arguments for and against the viability of this policy, it is undisputed that the use of nuclear weapons requires a highly identifiable and localizable enemy target (e.g., capitol, industry, military bases). Accordingly, except in extreme cases in which terrorist groups can be linked to a sponsoring government (as the United States linked the al-Qaeda terrorist group to the former Taliban-led government of Afghanistan) or be localized in training facilities, the extended nuclear shield offers little to no deterrence. Moreover, terrorist groups, especially those based on religious zealotry in which adherents of a particular faith focus on supernatural rewards rather than earthly political and/or military consequences, may act without regard to self-preservation. Terrorist groups organized into semiautonomous cells that operate without the formal permission or cooperation of host governments or within nations traditionally allied with the United States also are not targetable with strategic nuclear weapons.

Tangible and potential terrorist enemies of the United States are attracted to the use of biological weapons because of the potential to produce agents economically at multiple sites. In addition, these sites do not offer the more identifiable and fixed-target profile of a nuclear or chemical weapons facility. Moreover, because the medical science training required to develop biological agents is not as technically demanding as the advanced physics and chemistry training required to engineer effective nuclear and chemical weapons—and is more readily obtainable under the guise of medical science training for legitimate purposes, it is arguably easier to acquire the professional expertise to develop at least low-grade biologic weapons.

If one of the goals of renewed research on biological weapons is to counter terrorist threats, there is an admitted weakness in any MAD-like strategic defense in which biological weapons are used. As with nuclear weapons, such strategic weapons would be of little use against cells of terrorist groups. However, because terrorist groups must often work in conjunction with self-preserving nation states and because terrorist groups often evolve into self-preserving entities or political forces, it is arguable that biological weapons with built-in attenuation (self-killing factors) might provide a

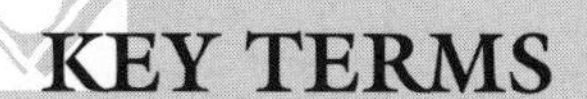

KEY TERMS

BIOLOGICAL WEAPONS CONVENTION (BWC): An international treaty prohibiting development and use of biological weapons. The BWC describes biological weapons as "repugnant to the conscience of mankind." As of 2002, more than 159 countries had signed and 141 had ratified this 1972 agreement that prohibits the development, production, and stockpiling of biological weapons. The BWC specifies and directs that biological toxins and pathogens that have no established protective (prophylactic) ability, beneficial industrial use, or use in medical treatment not be developed. Biological weapons stockpiles were to have been destroyed by 1975.

MOBILIZATION CAPACITY: The capacity to turn peacefully produced biological agents into biological weapons.

NBC: A common abbreviation for "nuclear, biological, and chemical."

WEAPONIZATION: The production of biological agents (e.g., anthrax spores) of such nature and quality that they can be used to effectively induce illness or death. Weaponization includes concentrating a strain of bacterium or virus, altering the size of aggregations of biologic agents (e.g., refining clumps of anthrax spores to sizes small enough to be effectively and widely carried in air currents), producing strains capable of withstanding normally adverse environmental conditions (e.g., disbursement weapons blast), and the manipulation of a number of other factors that make biologic agents effective weapons.

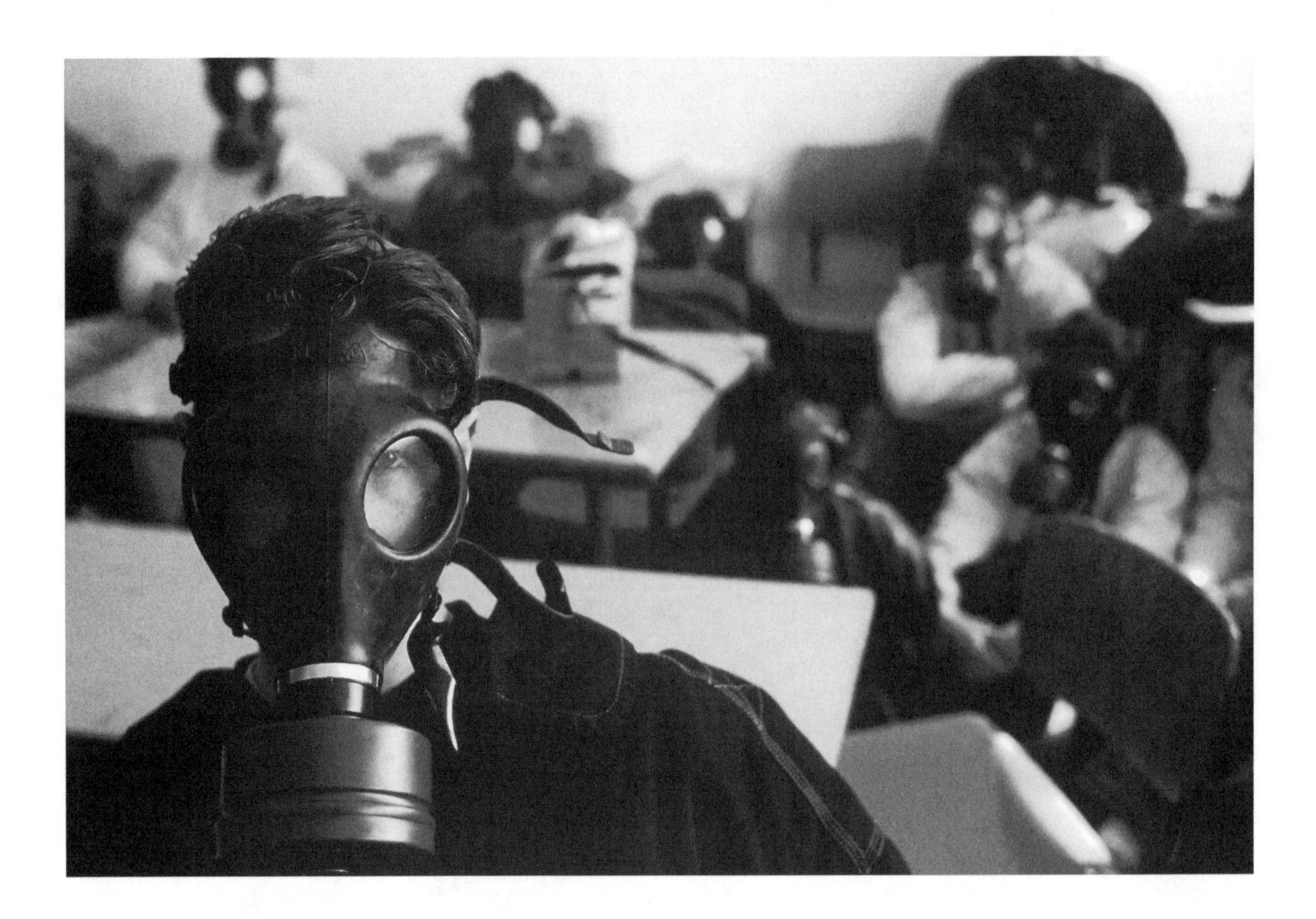

Israeli children wearing gas masks during the Gulf War in 1991. *(© Corbis. Reproduced by permission.)*

more usable and targetable strategic deterrent than would more difficult to contain and longer contaminating radiological weapons.

In the aftermath of the September 11 terrorist attacks and the subsequent deliberate spread of spores of the anthrax bacterium, *Bacillus anthracis,* via the U.S. mail system (events that as of this writing have not yet been conclusively linked), a number of federal agencies already conducting active and vigorous research into defenses to potential biological weapons established new programs and upgraded facilities to counter an increased terrorist threat. For example, partially in response to terrorist attacks on the United States, in February 2002 the U.S. National Institute of Allergy and Infectious Diseases established a study group of experts to evaluate potential changes in research to address terrorist threats more effectively. The group was charged with the specific task of recommending changes in research programs to conduct more effective and expeditious research into smallpox, anthrax, botulism, plague, tularemia, and viral hemorrhagic fever.

Advances in effective therapeutic treatments (e.g., the development of antitoxins and antibodies) are fundamentally dependent on advances in the basic biologic and pathological mechanisms of microorganisms. Research on pathogenic agents is essential because research challenges are similar for combating the introduction of a pathogenic organism into a population whether it arrives as a naturally emerging infectious disease or a deliberately spread weapon of terrorism. The natural evolution of microorganisms and the development of increasingly resistant strains demand that medical technology advance—not only to counter deliberate terrorism but also to serve as a mechanism to keep pace with natural threats.

Although the emphasis of current research is to understand and develop defensive countermeasures against potential biological threats, one could characterize any research into emerging infectious agents as research with unintended benefit to potential biological weapons programs. Excepting the deliberate development and production of new strains and types of offensive biological weapons, in calling for a continued moratorium on research on biological weapons, opponents of such programs are hard-pressed to define and draw the boundary lines of prudent and responsible basic research that, intent aside, is purely defensive in effect. It is a rather modest technological and procedural leap to make the transition from genetic engineering research with vaccine agents (e.g., attenuated agents used for smallpox vaccination), ectromelia (mouse pox), and/or monkey pox to direct manipulation of factors to increase virulence or to create chimera viruses (a chimera is an organism from multiple genetic sources).

Accordingly, the argument over whether the United States should resume research and development programs on biological weapons becomes one of assessing benefit from conducting formal research into developing and weaponizing offensive agents.

Without active research into weaponization, the defensive programs of the United States are put in a blind and reactive posture. In addition to exploring defenses against potential biological threats, it is only by exploring the possibilities of actual weaponization that effective counterintelligence measures can be developed to monitor the potential development of such agents by hostile groups. In essence, unless the United States conducts active weaponization research, its medical and intelligence personnel are left to guess at defenses to mount or what to look for in terms of threats. In addition, without research on weaponization, the United States cannot formulate plans to assure adequate mobilization capacity to respond to a strategic or tactical biological weapons threat.

Potential strategic biological agents do not have to be lethal (e.g., smallpox, respiratory anthrax, plague). Biological weapons can be developed that with subsequent treatment are potentially only incapacitating (e.g., tularemia, Venezuelan equine encephalomyelitis). In addition, antibiotic-resistant bacteria or immunosuppressive agents could also be turned into weapons. Accordingly, if research and limited development of biological weapons should resume, the U.S. program would not necessarily be obliged to blindly develop weapons of mass destruction. A more prudent policy would involve the reimplementation of former safety polices that dictated the development of biological weapons proceed only when a vaccine or treatment was available for the weaponized agent.

A reliance on treaty obligations with regard to terrorist organizations is inane and with regard to formal states has proved ineffective. Although it was a signatory party to the 1972 BWC, the Soviet Union maintained a well-funded and high-intensity biological weapons program throughout the 1970s and 1980s. Tens of thousands of scientists worked to produce and stockpile biological weapons, including anthrax and smallpox agents. U.S. intelligence agencies have raised doubt about whether successor Russian biological weapons programs have been completely dismantled. As of 2002, intelligence estimates compiled from various agencies provide indications that more than two dozen countries are actively involved in the development of biological weapons. The U.S. Office of Technology Assessment at various U.S. Senate committee hearings has identified a list of potential enemy states developing biological weapons. Such potentially hostile nations include Iran, Iraq, Libya, Syria, North Korea, and China.

The United States is in what former Secretary of Defense William S. Cohen described as a "superpower paradox." As the United States increasingly becomes the world's dominant superpower, potential enemies are more likely to mount unconventional or asymmetrical threats using biological weapons. Only by resuming aggressive research and development programs on biological weapons can the United States exploit its scientific and technological advantages in defense against biological weapons of mass destruction. —K. LEE LERNER

Viewpoint:

No, resumption of research on and development of biological weapons is a medically and scientifically unneeded, inefficient, and unethical response to threats of bioterrorism.

The United States should not resume research and development programs on biological weapons. Such resumption—in addition to being a violation of current treaties and conventions prohibiting the development of biological weapons and other weapons of mass destruction—is unneeded from a medical and scientific standpoint. The United States can obtain the highest levels of both deterrence and defense against an attack by a hostile force using biological weapons without further research formally dedicated to biological weapons. Moreover, the drain of talent and resources demanded by such programs would actually weaken the defenses of the United States.

Should biological weapons programs resume and biological weapons subsequently be developed and stockpiled as part of a strategic defense initiative to deter biological attack by a hostile force, the very same problems and issues confronting an aging nuclear force will confront any strategic biological weapons arsenal. The problems of weapons aging, deterioration, and disposal of outmoded weapons that plague the nuclear weapons programs will find new life in the biological warfare arena. In addition, hazards unique to use of living organisms will make the maintenance of such a force more hazardous than maintenance of nuclear and chemical weapons. There is also the historical precedent that the existence of a biological weapons stockpile will raise calls for advanced testing protocols conducted under the justification of "safeguarding the stockpile." Given the secrecy of such strategic programs, it is difficult to envision the wide input of the medical and scientific community needed to develop protocols for safe storage and testing.

From both strategic and tactical standpoints, the development of biological weapons is unneeded. Although a nightmare scenario, current military options available against identifiable bioterrorist threats are far more likely to eliminate the development or capacity to develop biological weapons by hostile forces. The current arsenal contains precision-guided conventional thermal fuel-air bombs and low-grade nuclear weapons that are more than adequate to the task of destroying even hardened enemy biological research and/or storage facilities. Moreover, should the need arise for a strike against a biological weapons stockpile, it would be most effective to use weapons capable of destroying the biological agents.

The resumption of a biological weapons program is of even less utility against nonlocalizable hostile forces (e.g., terrorist cells) that operate domestically or within the borders of other nations.

Last with regard to strategic or tactical consideration, any resumption of research on and development of biological weapons sends a negative message to developing third world nations that such programs provide a usable strategic defense. Because of lower technical demands and lower weaponization costs, the development of biological weapons is an already too attractive alternative to the development of nuclear or chemical weapons. A resumption of formal biological weapons research by the United States would almost certainly result in a global biological arms race.

Reinstitution of a formal research and development program for biological weapons will also shift financial resources away from a balanced, effective, and flexible response to biological, chemical, and radiological threats. Financial resources would be better spent, for example, on further development of the National Pharmaceutical Stockpile Program (NPS) as part of a strategy to mount a rapid response to an attack with biological weapons. A strategic stockpile of antibiotics, vaccines, and other countermeasures that can be rapidly deployed to target areas is the most tangible and viable defense to the most likely biological attack scenario—the limited use of biological agents by terrorists.

The financial and scientific talent costs of resuming biological weapons research would impede the development of defense strategies at a time when their development is most critical. Testimony before the U.S. Congress by federal agencies involved in coordinating a response to a biological weapons attack clearly asserts that the U.S. public health infrastructure is currently not yet in the optimal position to either detect or respond to an attack with biological weapons. Although additional funds have been allocated to train emergency responders—and this is an ongoing annual budgetary commitment—more funds are needed to improve hospital and laboratory facilities.

Even with the most advanced biological weapons program, it would be impossible to anticipate every agent or form of biological attack that might confront the United States. It is therefore more imperative that scientific and medical responses be improved for maximum scientific flexibility and response to future threats. The Centers for Disease Control and Prevention (CDC), based in Atlanta, has already established a bioterrorism response program. The program, by increasing both basic science testing and treatment capacities, is the model and the nucleus for the most effective response to a bioterrorist attack. The CDC strategy calls for the further development of laboratories equipped to detect and study biological agents, renewed emphasis on epidemiological detection and surveillance, and the development of a public heath infrastructure capable of providing information and treatment guidance. By strengthening both preparedness and response, the CDC program enhances the national defense far more than does clandestine research on biological weapons, which has only intangible benefit.

The CDC openly identifies potential biological threats and publishes on its Web pages a list of the biological agents most likely to be used by hostile forces. Along with basic scientific information, the CDC and National Institutes of Health (NIH)–funded agencies provide information designed to inform the public about threats and appropriate medical responses. Criticism of such openness and calls for clandestine research programs are misguided. Experts testifying before Congress have asserted that it takes less than $10,000 worth of equipment to make crude biological weapons. Accordingly, because the total prevention of the development of crude biological weapons by terrorists is not feasible, the only rational and effective response is to openly prepare for such an attack. Arguably, an effective defense preparedness program might actually deter attacks by all but the most suicidal extremist groups. Little gain could be anticipated through the use of the most common agents, and the actual damage done in such attacks would be minimal.

In contrast to weapons research programs, increased knowledge and communication between medical and nonmedical personnel concerning the symptoms of exposure to pathogens strengthen the detection and epidemiological tracking capabilities essential to an early and effective response. Programs such as the developing Health Alert Network (HAN), National Electronic Disease Surveillance System (NEDSS), and Epidemic Information Exchange

(Epi-X) are designed to coordinate and enhance information exchange.

The most effective response against the use of new strains—or against more exotic biological agents—is mounted by the CDC Rapid Response and Advanced Technology laboratory and the tiered system of laboratory facilities designed for early and accurate detection and identification of pathogenic agents. After the terrorist attacks on the United States on September 11, 2001, additional funds were quickly allocated to enhance the U.S. Department of Health and Human Services 1999 bioterrorism initiative. One of the key elements of the Bioterrorism Preparedness and Response Program (BPRP) was to develop and make operational a national laboratory response network. The development of this testing capacity has double value because it facilitates identification and information sharing regarding emerging natural diseases. Even before the latest expansion, the additional laboratory capacity already demonstrated its merit with the detection of several viral encephalitis strains along the east coast of the United States.

At the treatment end of the response spectrum, it is difficult to envision a plan capable of adequately preparing any local medical facility for a full response to a biological weapons attack. Regardless, specific response plans to bioterrorism events are now a part of the latest accreditation requirements of the Joint Commission on Accreditation of Healthcare Organizations. The minimal benefit derived from extensive scientific study of biologic agent weaponization potential (the alteration of biological agents and disbursement mechanisms to make biological weapons more effective) pales in comparison with the very real defense returned by investment of financial and scientific talent resources in rapid reinforcement of local response capability by federal resources (e.g., the availability of antibiotics, vaccines, and medicines stockpiled in the NPS to assure deployment to target areas with 12 hours).

Instead of a renewed research emphasis on biological weapons, the United States has thus far maintained a more tangible defense response, primarily through increased funding of basic science research programs by the NIH. In December 2001, Congress more than doubled the previous funding for bioterrorism research. Current funding levels, estimated at approximately one billion dollars, are expected to increase over the next several years. Because of security issues, a resumption of a biological weapons program would, however, cloud research-funding issues and almost certainly involve the duplication of much basic biological research in both open and clandestine programs. This becomes particularly problematic because many scientists fear that the increased emphasis on funding research on potential agents of bioterrorism may divert resources from the existing problems of malaria, flu, tuberculosis, and other bacterial or viral agents that already kill thousands of Americans and millions people worldwide each year.

A U.S. marine wears protective biological gear during a training exercise, November 2001. *(© Corbis. Reproduced by permission.)*

Any balanced response to bioterrorism in terms of allocation of research resources must balance the tragic deaths of five innocent Americans caused by the deliberate spread of spores of *B. anthracis* via the U.S. mail system in the fall of 2001 (as of this writing the number of deaths so attributed) against the estimated 12,000 to 18,000 deaths of flu each year in the United States and the millions of deaths worldwide of diarrhea caused by bacterial infection.

In 1969, five years before ratifying the BWC, the United States renounced first use of biological weapons and restricted future weapons research programs to issues concerning defensive responses (e.g., immunization, detection). There is no current evidence or argument that clandestine research programs would enhance public safety. In fact, such programs would be in stark contrast to the public bioterrorist defense initiatives previously outlined.

At present, the CDC identifies approximately 36 microbes (e.g., Ebola virus variants, plague bacterium) that might be used by bioterrorists. Instead of renewed attempts to develop new agents and improve weaponization of existing agents, basic science research on existing microbes will not only facilitate the development of treatment protocols but also, within the boundaries of the BWC, allow the identification of additional agents or genetic variants that could be used by bioterrorists. Most experts who argue before the U.S. Congress insist that instead of adding the layers of secrecy inherent in a biological weapons programs, the long-term interests of the United States would be better served by the strengthening of BWC verification procedures (currently the ad hoc VEREX commission of scientists).

Arguments that somehow the resumption of biological weapons programs will enhance the ability of intelligence services to detect biological weapons production are nebulous. To be sure, existing efforts at BWC verification are made more difficult by the nature of biological weapons (e.g., the small quantities needed, the ability to conceal or camouflage agents), but no coherent argument is raised about how research into weaponization will enhance these detection efforts. Accordingly, any resumption of research and development on biological weapons is a medically and scientifically unneeded, inefficient, and unethical response to threats of bioterrorism. — BRENDA WILMOTH LERNER

Further Reading

"Biological Diseases/Agents Listing." Centers for Disease Control and Prevention. April 5, 2002 [cited January 23, 2002]. <http://www.bt.cdc.gov/Agent/Agentlist.asp>.

Cole, Leonard A. *The Eleventh Plague: The Politics of Biological and Chemical Warfare.* New York: WH Freeman and Company, 1996.

Dando, Malcolm. *Biological Warfare in the 21st Century.* New York: Macmillan, 1994.

Dennis, D. T., et al. "Tularemia as a Biological Weapon." *Journal of the American Medical Association* 285, no. 21 (June 6, 2001): 2763–73.

Dire, D. J., and T. W. McGovern. "CBRNE: Biological Warfare Agents." *eMedicine Journal* 3, no. 4 (April 2002): 1–39.

Fleming, D. O., and D. L. Hunt. *Biological Safety: Principles and Practices.* 3rd ed. Washington, DC: American Society for Microbiology, 2000.

Does logging of the fire-ravaged Bitterroot National Forest pose a threat to the environment?

Viewpoint: Yes, logging of recently fire-ravaged forests in the Bitterroot Mountains poses numerous threats to the recovery of this land, as well as to the animal and plant populations that reside there.

Viewpoint: No, plans by the United States Forest Service to allow logging of dense forests in the Bitterroot Mountains would not cause lasting damage to the soil and rob the forest of natural systems of recovery; on the contrary, such plans would reverse a century's worth of mismanagement that allowed the national forests to become so dense that raging fires were inevitable.

For almost six weeks during the summer of 2000, a confluence of intense wildfires devastated about 365,000 acres of forests in southwestern Montana. About 20% of the acreage affected was in the Bitterroot National Forest. In addition to the fires in Montana that August, forest fires also occurred Idaho, Oregon, Washington, Utah, South Dakota, and Texas. At the federal level, the cost of fighting the forest fires that burned 8.4 million acres during the summer of 2000 amounted to about $1.4 billion.

When the fires finally ended, the United States Forest Service announced plans to cut and sell trees from about 46,000 acres in the Bitterroot National Forest, including tracts of undeveloped, roadless forest. The proposal called for salvaging 176 million board feet of ponderosa pine and other trees. Many environmental groups, scientists, and concerned citizens vigorously opposed the plan and warned that if it succeeded it would serve as a precedent for the logging of other fire-damaged national forests.

Opponents of the proposed logging plan argued that scientific studies demonstrated that logging of recently burned land inhibits the natural recovery process and causes long term damage to the complex forest environment. To counter opposition by environmentalists, the Forest Service argued that careful logging was an essential aspect of recovery and protection for the fire-ravaged forest. Although the Forest Service is officially charged with overseeing the health of national forests and selling timber, representatives of the Service were also eager to point out that the plan would create thousands of jobs and raise money that could be used to help replant trees. When environmentalists challenged the sale on a procedural issue, a federal judge agreed and stopped the project. However, the Forest Service immediately announced its intention to appeal the decision and asked the judge to allow some sales to proceed on an emergency basis. The case was finally settled in February 2002, when the Forest Service agreed to reduce the logging area to 14,700 acres.

Opponents of logging the Bitterroot Mountain area argue that after a fire, dead trees play a vital role in protecting the forest floor and local waterways from the effects of direct sunlight and erosion caused by wind and water. Removing dead trees exacerbates damage to the soil and causes a further loss of habitat and shelter for plants and animals. Dead trees, whether they are still standing or have fallen to the forest floor, provide shade and help the soil retain moisture. Fallen trees act as natural dams to stop erosion on steep slopes and enrich the soil as they decay. Soil scientists note that logging is detrimental to soil structure and health, particularly to the topsoil layers in which the seeds of pine grass, huckleberry, and beargrass would sprout. Logging scrapes away the essential upper layers of soil and surface litter and compacts the remaining soil. This removes the seeds of plants that normally sprout after a fire and the nutrients needed for plant growth. Forest Service officials said that damage to the soil would be minimized by allowing logging only on frozen and snow-covered ground and by techniques that would restrict the use of heavy equipment in vulnerable areas.

Although advocates of logging believe that thinning the fire-damaged forests would remove the fuel that might promote future wildfires and encourage infestations by insect pests, their critics argue that attempts to demonstrate the value of this procedure have been negative or, at best, inconclusive. Some studies suggest that very intense forest fires and outbreaks of timber diseases are more likely to occur in areas that have been logged than in natural forests. The piles of twigs and debris left behind by loggers may have contributed to fires in forests subjected to thinning.

Nevertheless, advocates of logging argue that the Forest Service plan would actually benefit the forest by reversing misguided management policies that allowed overly dense national forests to become increasingly vulnerable to devastating wildfires. Critics of existing fire suppression policies argue that this approach leads to overly dense forests where pine needles, logs, and debris provide masses of kindling that fuel dangerous infernos. Forest Service officials claim that because of previous mismanagement some national forests now support 700 trees per acre rather than the 100 trees per acre that would occur under more natural conditions. Therefore, thinning and controlled burns would prevent fires and promote healthier forests.

Environmentalists and loggers disagree about the best way to encourage restoration of the fire-damaged Bitterroot Mountain forests, but they do agree that the region faces a high probability of extremely destructive fires in the not-too-distant future. Both sides in this controversy urge further studies in order to develop sound fire-management policies. —LOIS N. MAGNER

Viewpoint:

Yes, logging of recently fire-ravaged forests in the Bitterroot Mountains poses numerous threats to the recovery of this land, as well as to the animal and plant populations that reside there.

The one-two punch of wildfire and logging can cause long-term harm by removing the dead trees that shield direct sunlight from the forest floor, streams, and other shallow waterways, help disrupt wind and water erosion, and protect the soil from the drying effects of the sun. This combination can affect the growth of replacement trees and make the forest susceptible to future wildfires. It can also have a devastating impact on both the terrestrial and aquatic animals.

The controversy over whether to log the Bitterroot Mountains began in the summer of 2000 when wildfire swept through approximately 300,000 acres of dense forests. By fall, the U.S. Forest Service had proposed logging more than 40,000 acres—about 14%—of the burned area, including 16,800 undeveloped, roadless acres. American Wildlands described the proposed harvest of 176 million board feet as "over 2.5 times the amount of timber sold from Bitterroot National Forest between 1988 and 1999—enough to fill 56,000 log trucks lined up end to end for 475 miles." Concerned citizens and numerous environmental groups fought the proposal, eventually filing legal appeals. A February 8, 2002, settlement reduced the logging area to 14,700 acres.

Those opposed to the Forest Service proposal pointed to opinions from forestry experts, biologists, and others, as well as numerous scientific studies demonstrating that the logging of recently burned land disrupts the natural recovery process and has lasting negative effects on forests.

Soil's Part in Recovery Soil, particularly its uppermost layers, plays a vital role in forest recovery following a fire. These layers provide the microhabitat that allows the rejuvenative greening of the charred forest floor. The uppermost surface litter includes newly fallen organic matter that not only will eventually decompose into life-supporting nutrients for sprouting plants, but also will provide a blanket of sorts to protect tender shoots from exposure to harsh winds and help maintain the moisture content within the topsoil and subsoil. The topsoil, which is where seeds initially sprout, must retain sufficient moisture, contain proper nutrients, and provide sufficient porosity (spaces between soil particles) for the movement of air, water movement, and growing roots.

Logging has been shown to impair soil health. According to Forest Service soil scientist Ken McBride, who is a specialist on the Bitterroot Mountains, logging on this site would damage the soil and sidetrack the forest's recovery, regardless of the logging methods used. He argued against the forest service proposal as a private citizen rather than as an employee, stating that even the most careful logging procedures would cause harm by: 1) scraping off any remaining surface litter and the underlying topsoil; and 2) compacting the soil. The scraping of surface litter and topsoil would remove the nutrients plants require, as well as the seeds for plants that normally begin post-fire forest rejuvenation. It would also make the upper soil vulnerable to the drying effects of the wind and direct sunlight, potentially generating unsuitable conditions for seed germination and seedling growth. Soil compaction would decrease soil porosity and make it much more difficult for any existing or newly transported seeds to sprout successfully.

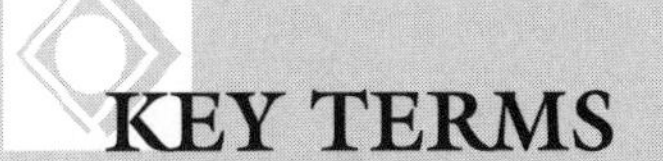

KEY TERMS

FUEL-REDUCTION: In forestry, a term for efforts directed toward the cutting of trees and underbrush that might otherwise provide fuel to future wildfires.

SALVAGE LOGGING: A logging practice that involves the removal of dead trees.

SUBSOIL: Plant roots often extend down to this level of soil, which lies beneath the topsoil layer.

SURFACE LITTER: Recently deposited organic matter, such as leaves, dead plants, or small twigs that cover the soil beneath.

TOPSOIL: The level of soil immediately beneath the surface litter. It is mainly humus (organic matter undergoing decomposition) and is the site of many plant roots.

Role of Dead, Standing Trees Removal of dead, charred trees can likewise hinder a forest's recovery. Standing dead trees continue to provide cover for animals still living in the forest. They also provide homes as well as food for burrowing and other insects. Pine bark beetles, for example, will converge on burned ponderosa pine forests. The presence of insects, in turn, draws insectivorous birds back into the forests. Returning birds can have a major impact on forest rejuvenation by eating seeds from surrounding areas, flying into the burned area and dispersing them. The dead, standing trees also provide homes for many of the cavity-dwelling birds, as well as nesting areas, hunting perches, and resting sites.

A team of researchers from the Pacific Northwest Research Station's Forestry and Range Sciences Laboratory in La Grande, Oregon, conducted an expansive study of a forest ecosystem's overall health. The three researchers—an entomologist, a wildlife biologist, and a plant physiologist—found that standing and fallen dead trees, which they call snags and logs, respectively, are critical components of the ecosystem. Snags, particularly those that had become hollow, were significant bird roosts. Although ornithologists already knew that woodpeckers used dead trees for roosts, the research team learned that woodpeckers actually may use several roosts in different snags during the year. Woodpeckers also create their own holes in trees by pounding through the bark in search of insects. For instance, the pileated woodpecker (*Dryocopus pileatus*), a foot-and-a-half-long, red, black, and white bird, uses its chisel-like beak to make holes large enough to serve as nests and homes for bats, flying squirrels, and other birds.

The team also reported that these snag-using animals opted for entrances that were about 33–98 ft (10–30 m) above the ground, again accentuating the importance of leaving standing, dead trees in a forest. One of the researchers noted that the team's data, and other research reports, suggest that forest managers actually increase the number of snags to promote forest vigor.

In addition, these standing trees provide some welcome shade to an area that was likely quite dry even before the wildfire raced through. Like the soil surface litter, the shade helps retain moisture left at ground level, and gives any new rainfall more time to penetrate the soil than if the trees were logged and the expanse was left unshielded from the sunlight.

In riparian areas, dead trees shade streams and other waterways, helping to block direct sunlight and to keep the water at the lower temperatures that fish and other aquatic species demand.

Positive Effect of Fallen Trees Fallen trees also play a role in a forest's recovery. Just as the ground is moister under a rock or a lying log than in the open, the soil beneath a fallen tree trunk retains moisture better. The trees also protect the soil from wind and water erosion by disturbing, and thus slowing, air and water flow. As a result, they minimize the loss of the organic top layers of the soil. Trees are particularly beneficial in areas with steep inclines, such as stream banks. After a fire, streams need some time to recover from the influx of ash and other post-fire debris, and dead trees along the shores can at least divert or slow some of that surge.

In addition, fallen trees create pockets of more stable soil, perhaps along the trunk or

The dense forests of the Bitterroot Mountains were damaged by a severe wildfire in 2000. *(Photograph by Don Emmert. AFP/Corbis-Bettmann. Reproduced by permission.)*

between a web of branches. Here, seeds find less disturbance from wind or water flow, germinate more readily, and encounter greater moisture to promote successful growth.

In its study of the importance of dead and dying trees, the Pacific Northwest Research Station research team also found that undeveloped forests typically had a much greater diversity and number—50 to 140 per acre—of fallen trees than was previously realized. They also documented the use of logs by a variety of animals, including black bear, and other smaller animals, including insects. In particular, the study's entomologist noted that 11 species of ants live in dead wood and eat spruce budworms, one of the organisms that kills trees. The wildlife biologist on the team followed up by examining woodpecker scat and finding ants. In just this one example, they were able to describe a small circle of life in the forest between ants, budworms, woodpeckers, logs, and snags. Without the snags or logs, the circle is incomplete.

The team's wildlife biologist, Evelyn Bull, commented in *PNW Science Findings*, "Our ongoing research brings out the direct conflict between retaining deadwood for wildlife and reducing fuels for wildfire."

Thinning as an Option In the Bitterroot Mountains and many other northwest U.S. forests, proposals to thin the forests or undertake salvage logging as preventive measures against future wildfire, and timber-disease outbreaks, are common. The idea is that the removal of fuel, including burned trees, will remove the dried, fire-prone understory, and decrease the number of dead or damaged trees that might invite insect disease vectors. Neither prevention method, however, has been shown to have a positive effect on thwarting future wildfires.

For example, several studies have provided evidence that the opposite is true: Forest fires and disease outbreaks are more prevalent in areas that have been logged. Research has repeatedly shown that logging is strongly correlated with the occurrence of intense fires in the Pacific Northwest. Studies of the Sierra Nevada demonstrated that thinned forests are much more susceptible to hot, damaging fires than unmanaged forests. Likewise, research into Oregon and Washington forests indicate that logging begets more intense fires, because logging practices frequently leave behind piles of highly flammable twigs, branches, and other loggers' refuse. The resulting conditions are ripe for an intensely hot fire that even the flame-resilient, thick bark on older-growth trees may not be able to withstand. Other reports indicate that even when flammable brush is removed, the forest remains vulnerable to fire because of the drying effects of the sun and the wind on the forest floor.

Numerous Forest Service reports have associated logging with fire propensity. One study examined the five-year period beginning in 1986, and reported that private lands, which are typically the most heavily logged, experienced a 20% increase in the number of trees that died from fire or disease. During the same time period, those areas with the least logging, such as national parks, experienced a decrease in tree mortality of 9%. Other Forest Service research corroborates those findings, and indicates that salvage logging compounds the effects of the original fire and should not be considered an option, especially in roadless, undeveloped, and previously unlogged areas.

Forest Restoration Alternatives If salvage logging is not the answer, can anything be done to assist the restoration of the Bitterroot Mountain forests? Environmentalists, scientists, and loggers agree that the potential for future, highly destructive wildfires is enormous. The current situation in the drier forests of Montana and the Northwest has its primary root in the fire-suppression policies of the twentieth century. With the support of politicians and public opinion, the United States spent decades keeping fires to a minimum. Before the policy, a Northwest forest would commonly go up in a blaze every decade—and occasionally every year. The fire proceeded quickly and with little strength, effectively eliminating the understory, yet leaving the larger trees virtually unscathed. The forest did not accumulate excess, flammable fuel at ground level. The incinerated understory yielded nutrients for the soil. The forests thrived. After the fire-suppression policy, the undergrowth had time to take hold. Twigs, limbs, and dry brush collected for decades. In these conditions, a spark can trigger a massive inferno that can burn hot and long enough to ignite flame-resistant bark and climb older growth trees. These so-called "crown fires" move through the forest as tall walls of fire, and roar through thousands of acres.

More studies are needed on the best fire-management policy. The bulk of existing evidence, however, clearly reveals that salvage logging is incompatible with forest recovery or fire/disease prevention, and salvage logging on public lands should not be an option, especially in national parks. Even the most cautiously applied logging techniques cause damage to the forest. Logging can cause:

- removal and compaction of vital soil layers
- wind and water erosion to the forest floor
- wind- and sun-caused drying of the soil
- the removal of dead trees that would have provided shelter and food for innumerable animal species
- a significant increase in soil erosion
- an elimination of snags and logs that would have assisted moisture retention at ground level, and shielded direct sunlight from the forest floor, streams, and other shallow waterways

Says Evan Frost in a report for the World Wildlife Fund, "Although thinning within the context of intensive forestry is not new, its efficacy as a tool for fire hazard reduction at the landscape scale is controversial, largely unsubstantiated and fundamentally experimental in nature."

Rob Smith, director of the Sierra Club's southwest office, remarked in *Sierra*, "Wilderness is the last place—not the first place—you mess around with forestry experiments." —LESLIE MERTZ

Viewpoint:

No, plans by the United States Forest Service to allow logging of dense forests in the Bitterroot Mountains would not cause lasting damage to the soil and rob the forest of natural systems of recovery; on the contrary, such plans would reverse a century's worth of mismanagement that allowed the national forests to become so dense that raging fires were inevitable.

For many who lived through the fires that ravaged the western United States in the summer of 2000, the *USA Today* headline from August 28 of that year most likely calls back powerful, painful memories: "Montana Fires Merge, Creating Mega-Inferno." The Bitterroot Valley, south of Missoula in the southwestern corner of Montana, had erupted with wildfire that ultimately destroyed 365,000 acres, approximately 20% of it in the Bitterroot National Forest. More than 1,500 people had to be evacuated, though fortunately no one was killed, and some 70 homes were destroyed. Only after approximately six weeks did the smoke, hanging heavy throughout the valley, finally lift, letting in the sunlight.

But as the headline suggested, the Bitterroot inferno was not an isolated instance. Fire had broken out along the Continental Divide, and by August the fires had merged; nor were the Montana conflagrations the extent of the destruction that raged throughout the western United States during the summer of 2000. In all, more than 79 fires were burning across 1.6 million acres (647,500 hectares) in Montana, Idaho, Washington, Oregon, Utah, Texas, and South Dakota during just one weekend in August. By the time the summer was over, fire had claimed 8.4 million acres nationwide, and the federal government had spent $1.4 billion fighting it.

Searching for Causes and Solutions Even as the fires burned, Montana's governor, Marc Racicot, blamed the administration of President Bill Clinton for the fires. "The conditions of our forests," Racicot told *USA Today*, "are such that each administration that is charged with dealing with them has to, on their watch, make absolutely certain that they do all they can to maintain forest health. They [the Clinton administration] have not done that." In response, Interior Secretary Bruce Babbitt and Agriculture Secretary Dan Glickman blamed the forces of nature, combined with past mismanagement. In particular, they cited a government policy, in effect for about a century, that prevented the clearing of the trees and underbrush. As a result, with the summer heat and dry weather as a spark, the heavy undergrowth of the forests provided fuel for the fires.

While the fires and the politicians raged (Babbitt claimed that complaints from Racicot, a Republican and supporter of then-presidential candidate George W. Bush, were politically motivated), people searched for ways to prevent

A burned tree awaits sale as salvage timber. Does logging of such fire-damaged areas rob the forest of natural sources of recovery?
(© Raymond Gehman/Corbis. Reproduced by permission.)

another such fire from happening. Thus was commenced a broad public discussion that included not only scientists and government officials but the populace as a whole. Groups concerned with the forests, across a wide spectrum of interests that ranged from those of environmental activists to those of logging companies, each sought a solution. As Scott Baldauf reported in the *Christian Science Monitor,* "more than a decade of increasingly dangerous and devastating fires out West appears to be giving activists on both sides a common goal—healthy forests—and a growing belief that fire is a natural process that is here to stay."

Whereas the Native Americans who once controlled what is now the Western United States had used fires to clear forests, Baldauf wrote, white settlers "saw fire as an implacable enemy, and began a century of strict fire suppression. With nothing but the saw to threaten them, trees today grow closer together and pine needles pile up like so much kindling." Furthermore, as Baldauf noted, restrictions on the harvesting of timber, imposed for the purpose of protecting endangered species, have helped create a situation of incredibly dense forests. Contrary to popular belief, there are actually about twice as many trees per acre (0.404 hectares) in today's forests than there were in the forests of 1900. Baldauf concluded by noting that "Maintaining the long-term health of forests will clearly take a change of methods, including thinning and, in some cases, controlled burns. The latter will likely face resistance from the growing number of those who are building homes at forest's edge."

Bitter Fight over Bitterroot Actually, the fiercest opposition involved not the controlled burns but the thinning, and the most outspoken opponents were not local homeowners, but environmentalists. The cause for this opposition came when, in 2001, the Forest Service of the new Bush administration announced plans to allow the commercial salvaging of some 46,000 acres (18,616 hectares) of burned timber in Bitterroot National Forest. Environmentalists, fearful that the policy would set a precedent for the logging of forests damaged by fires, managed to stall action by the government, such that 16 months passed without the cutting of a tree. During this time, natural conditions threatened to erode the commercial viability of the wood, which after 18–24 months would be useless as lumber due to cracking.

By February 2002, the 18-month mark had been reached, and still the only activity that had taken place was in court. On February 8, after three days of legal arguments, the U.S. Forest Service agreed to reduce the planned area of logging to just 14,000 acres (5,666 hectares). Though Rodd Richardson, supervisor of the Bitterroot forest, told the *New York Times* that "everybody got some of what they wanted," the newspaper report did not indicate any gains made by the Forest Service in the negotiations. As for the environmentalists, their attitude was one of triumph tempered only by a desire for greater victory and a promise that the fight had only begun. Tim Preso, attorney for the environmental organization EarthJustice, told the *Times,* "This was a test case on the nature of the appropriate response to a large-scale wildfire." Having forced the government to give up plans for logging in areas without roads, he announced that "Roadless areas are off the table as far as we're concerned," and promised that "we will fight for every acre on these wildlands."

A three-year-old child, or a visitor from another planet, might be forgiven for assuming that environmentalist strong-arm groups are purely interested in protecting the environment. The reality, however, is that the organizations (if not their members) are at least as concerned with political power. This must be the case, because otherwise there is no way to make sense of their opposition to a policy that would clean up much of the destruction left by the 2000 Bitterroot fires, prevent future conflagrations, and even pay back some of the money spent on fighting those fires. Judging by the actions of environmentalists in the Bitterroot fight, the danger of a company making a profit from the cleanup is more grave than the danger of future fires.

Is Profit Wrong? In the aftermath of the fires, David Foster of the *Los Angeles Times* profiled Bitterroot locals who had profited from the fires. For example, homemaker Jennifer Sain "put her kids in day care and rented her GMC Suburban to the Forest Service for $80 a day, then hired herself out at $11.63 an hour to drive it, ferrying fire officials and reporters around the forest." According to Sain, she made $10,000 for two months' work. However, "Some of her government-wary ranching relatives didn't appreciate her working for the Forest Service, she says. But economics prevailed. 'I wouldn't have done it if it wasn't good money,' Sain says."

Discussing these and other examples of the economics involved in attacking the fires, Foster wrote, "Most people here will tell you the fires were awful. At the same time, however, the government was spending $74 million to fight the Bitterroot fires, and a good portion of that money trickled into the local economy." Regarding the matter of locals profiting, Bitterroot National Forest public information officer Dixie Dies told Foster, "It isn't discussed. There was so much hurt, people don't want to talk about who did well." Yet Arizona State University professor Stephen Pyne, author of 11 books on the subject of wildfires in the West, explained that "Profiting from wildfire is part of Western history." According to Pyne, during the 1930s, residents of the Pacific Northwest set fires to the woods around them, knowing that President Franklin D. Roosevelt's New Deal programs would provide them with jobs putting out those very fires.

Where the fires of the 1930s are concerned, the moral terrain is not clearly mapped, since the people were struggling with the effects of the Great Depression, and they had to suffer the long-term effects of the fires set in their own home areas. On the other hand, it is clearly immoral for people to profit unfairly in the face of tragedy by price-gouging. But people *do* profit from tragedy, and not all such profits are immoral: few people would say, for instance, that Jennifer Sain should have provided her services for free.

Similarly, there is absolutely no basis, on moral grounds, for opposing fuel-reduction projects—that is, the cutting of trees that would provide fuel to future conflagrations—by logging companies. Only by such cutting can the forests be protected from fires that would destroy the trees and underbrush, thus removing sources of organic material to enrich the soil, and virtually ensuring erosion and permanent deforestation. Perhaps the environmentalists would be more willing to permit such logging if it were done by the government at great expense, and if the wood were not used commercially. It is this same mentality that condemned the United States for profiting, by ensuring cheaper oil supplies, from the Gulf War of 1991. The federal government so rarely manages to do anything profitable; why fault it in those rare instances when it does?

Because of their outspoken opposition to business, capitalism, and profit, environmentalists in the Bitterroot case seem willing to sacrifice the best interests of the environment for the interests of politics. As a result, much of Bitterroot will remain as poorly managed as it has always been, with densely overgrown forests just waiting for another hot, dry summer. Perhaps the next fires will cause more severe damage, and then the environmentalists will have succeeded: no logging company will be interested in the trees of Bitterroot National Forest. —JUDSON KNIGHT

Further Reading

Baldauf, Scott. "Reviving Forests After the Flames: Even Before the Smoke Clears, Scientists Are Assessing How to Rejuvenate Burned Areas." *Christian Science Monitor* (August 24, 2000): 2.

Bull, E. L., C. G. Parks, and T. R. Torgersen. "Trees and Logs Important to Wildlife in the Interior Columbia River Basin." USDA Forest Service, General Technical Report PNW-GTR-391, 1997.

"Conservationists and U.S. Forest Service Reach Agreement on Controversial Bitterroot Salvage Project." American Wildlands Website [cited July 16, 2002]. <http://www.wildlands.org/publands/bitterroot.html>.

Duncan, S. "Dead and Dying Trees: Essential for Life in the Forest." *PNW Science Findings* 20 (November 1999): 1–6.

Foster, David. "In Wildfire's Wake, There's Money to Burn: Montana's Bitterroot Valley Is an Example." *Los Angeles Times* (July 29, 2001): B-1.

Frost, E. "The Scientific Basis for Managing Fire and Fuels in National Forest Roadless Areas." Prepared for World Wildlife Fund as supplementary comments on the Notice of Intent regarding National Forest System Roadless Areas (CFR: 64 No 201, 10/19/99). December 16, 1999 [cited July 16, 2002]. <http://www.fire-ecology.org/science/roadless_area_fire_managem.htm>.

Gray, Gerry. "Deciphering Bitterroot." *American Forests* 108, no. 1 (spring 2002): 16.

"In Plan to Help a Forest Heal, A Wider Rift Opens." *New York Times* (December 9, 2001): A-55.

Kris, Margaret. "Environmentalists Are Howling Again." *National Journal* 34, no. 1 (January 5, 2002): 50–51.

Paige, Sean. "Fight Rages over Fate of Deadwood and Timber Sales." *Insight on the News* 18, no. 2 (January 14–21, 2002): 47.

Parker, Laura. "Montana Fires Merge, Creating Mega-Inferno." *USA Today* (August 28, 2000): 3-A.

Robbins, J. "Is Logging Bane or Balm? Plan Stirs Debate." *New York Times* (January 29, 2002).

———. "Forest Service and Environmentalists Settle Logging Dispute." *New York Times* (February 8, 2002): A-14.

USDA Forest Service. "Initial Review of Silvicultural Treatments and Fire Effects of Tyee Fire." Appendix A, *Environmental Assessment for the Bear-Potato Analysis Area of the Tyee Fire.* Wenatchee, WA: Chelanand Entiat Ranger Districts, Wenatchee National Forest, 1995.

Weatherspoon, C. P., and C. N. Skinner. "An Assessment of Factors Associated with Damage to Tree Crowns from the 1987 Wildfire in Northern California." *Forest Science* 41 (1995): 430–51.

Was Rep. John Dingell's investigation of scientific fraud unjustified in the "Baltimore case"?

Viewpoint: Yes, the Dingell investigation was an abuse of congressional power that hindered the objectivity of a scientific investigation.

Viewpoint: No, the Dingell investigation of scientific fraud was not unjustified because peer review and self-regulation cannot work alone in maintaining scientific integrity.

Although scientists agree that they should be accountable to the federal government for the funds they receive, they object to the concept that members of Congress can evaluate the merits of scientific experiments, or settle debates about scientific concepts and the interpretation of research results. In an attempt to pass judgment on science, members of Congress could use their power to determine the course of scientific research or direct research funds to their own constituents. Moreover, press coverage of congressional hearings puts scientists at a distinct disadvantage. The dispute known as the "Baltimore Case" is the best known example of a recent confrontation between scientists and a congressman who believed that he had evidence of scientific fraud. The case revolved around a paper published in the journal *Cell* in 1986.

Many observers believe that the scientists who faced Congressman John Dingell (D-Michigan), chairman of the House Energy and Commerce Committee and the House Subcommittee on Oversight and Investigations, were systematically bullied, intimidated, threatened, and smeared in the press by committee members who leaked confidential and incomplete information to reporters. Critics of the Dingell investigation called it a witch-hunt and offered the Congressman's own words as evidence. When David Baltimore failed to demonstrate what the Congressman considered proper humility in the face of questioning by his committee, Dingell promised his colleagues that he would "get that son of a bitch . . . and string him up high." Given the fact that the "son of a bitch" in question was a Nobel Laureate and a highly respected molecular biologist, many scientists interpreted the long, drawn-out investigation as an attempt to intimidate all scientists who might become involved in scientific disputes.

David Baltimore was elected to the National Academy of Sciences in 1974 and shared the 1975 Nobel Prize in Physiology or Medicine with Renato Dulbecco, and Howard Temin, for "discoveries concerning the interaction between tumor viruses and the genetic material of the cell." Temin and Baltimore independently discovered an enzyme called reverse transcriptase that allows genetic information to flow from RNA to DNA. The discovery of reverse transcriptase has had a tremendous impact on molecular biology and cancer research. In 1968 Baltimore joined the faculty of the Massachusetts Institute of Technology. He was appointed director of the new Whitehead Institute for Biomedical Research in 1984. Four years after the publication of the controversial *Cell* paper, Baltimore was appointed president of Rockefeller University. Because of the problems caused by the investigation of fraud charges brought against coauthor Dr. Thereza Imanishi-Kari, Baltimore resigned from the presidency in 1991. Three years later Baltimore returned to MIT as the

Ivan R. Cottrell Professor of Molecular Biology and Immunology. Reflecting on the factors that had guided his career, Baltimore said: "My life is dedicated to increasing knowledge. We need no more justification for scientific research than that. I work because I want to understand."

In the wake of the Baltimore Case, many scientists worried that the definition of scientific misconduct adopted by the Federal government could inhibit innovative scientists and stifle cutting-edge research. Federal rules define scientific misconduct as "fabrication, falsification, plagiarism or other practices that seriously deviate from those that are commonly accepted within the scientific community. . . ." Scientists fear that politicians could exploit the rather ill-defined and amorphous concept of "deviant" to force scientists to accept officially sanctioned research goals and punish those who pursue innovative ideas.

The history of medicine and science provide many examples of research claims that seriously deviated from accepted norms of the time. For example, based on accepted norms, a peer-reviewed journal in England rejected Dr. Edward Jenner's 1796 report on the use of cowpox to provide immunity against smallpox. Similarly, Nobel Prize–worthy reports on innovative research on the hepatitis B virus and a radioimmunoassay technique that can detect trace amounts of substances in the body were rejected by prominent peer-reviewed journals before they were published in other journals.

Observers with legal training, which includes many government officials, call attention to the difference between fraud in the civil tort law and fraud or misconduct in science. There are also differences between the rules that govern scientific misconduct and criminal statutes. The kinds of behavior that are seen as scientific misconduct tend to change with time. Now that the biomedical sciences have become the driving engine of biotechnology, the theft or unauthorized removal of materials from the laboratory has become a matter of increasing concern. Yet under the rules established by the government for grant oversight, theft is not defined as misconduct. The federal government regulations concerning misconduct are limited to fabrication, falsification, and plagiarism. Officials at the Office of Research Integrity (ORI), which serves as the watchdog of science at the U.S. Department of Health and Human Services, explain that there are already criminal statutes against theft, but the criminal code does not deal with plagiarism, fabrication, and falsification.

However, after reviewing a series of cases involving scientific fraud and misconduct, some observers conclude that science cannot be trusted to police itself. Therefore, outside investigators, such as the Office of Scientific Integrity (now the ORI), or Congressman Dingell's subcommittee, are needed to reveal misconduct and enforce proper standards of conduct among those conducting scientific research with public funds. Because science is done by fallible human beings, critics argue that Congress should play a role in investigating charges of fraud or misconduct. Moreover, they believe that the threat of such investigations might deter misconduct by other scientists. Enforcing high standards of scientific integrity is essential to prevent the publication and dissemination of false and misleading data and conclusions by unscrupulous researchers.

Those who investigate scientific misconduct admit that the cases may be very complex, but they argue that peer review and self-regulation are not sufficiently rigorous in detecting and prosecuting scientific fraud and misconduct. The Baltimore Case does support the argument that investigating charges of scientific fraud is a difficult task; pursuing the allegations of misconduct took 10 years, five investigations, and three congressional hearings. Finally, in 1996 a federal appeals panel overturned Dingell's findings and concluded that the evidence he had used to arrive at a verdict of fraud in 1994 was unreliable and based on "unwarranted assumptions." —LOIS N. MAGNER

Viewpoint:

Yes, the Dingell investigation was an abuse of congressional power that hindered the objectivity of a scientific investigation.

"An ambitious congressman using his power to decide what is true or false in science." This was Anthony Lewis's summation of the "Baltimore Case," in a *New York Times* Op-Ed dated June 24, 1996. If ever there was a scary picture of science, Lewis has painted it perfectly with this one sentence.

Science isn't decided in a committee, it's only funded there. Should scientists be accountable for the money they receive? Absolutely. But the people who fund the science aren't necessarily the ones who can decide whether that science is "good." If that were the case, it would be Congress, not the NIH, reviewing grant applications for scientific projects. Furthermore, conducting congressional hearings to settle scientific disputes sets us on a slippery slope towards government-controlled science (where only "politically correct" research is allowed), and, indeed, government-controlled thinking, reminiscent of George Orwell's *1984*.

Politicians and bureaucrats view the world quite differently than scientists. Because they are not versed in the language of science, nor familiar with the culture of scientific research, they are ill equipped to make determinations on scientific disputes. Unfortunately, what the politicians lack in understanding they make up in power and privilege, and they have an attentive audience in the press that few scientists enjoy. This is a dangerous mix. The Baltimore Case was judged in the press, with committee members leaking confidential, incomplete information to chosen reporters. David Baltimore himself first heard about the imminent hearing from the press, who called for his reaction.

A Difference in Views of Science The controversy concerning Federal definition of serious scientific misconduct is a perfect illustration of the differences between scientists and bureaucrats. The rules define science misconduct as ". . . fabrication, falsification, plagiarism or other practices that seriously deviate from those that are commonly accepted within the scientific community for proposing, conducting and reporting research." The "practices that seriously deviate" part makes sense to a Federal agency trying to protect public funds from wasteful wild-goose chases. But to scientists, this phrase sets a dangerous precedent of a government coercing scientists to think and work under a universally accepted formula, where innovations could bring charges of misconduct. Dr. David Goodstein, Vice Provost of CalTech, wrote: "To me it seems poor public policy to create a government bureaucracy mandated to root out transgressions that are not specified in advance."

Goodstein also points out the difference between fraud in the civil tort law, which most politicians would be familiar with, and fraud in science. In civil cases fraud must be proven to have caused actual damage. In science the intentional misrepresentation of procedures or results is considered fraud, regardless of whether this misrepresentation harmed anyone, or whether the conclusions of the research were right or wrong. Science, therefore, takes a much stricter view of fraud than civil tort law.

The Baltimore Case Revisited Most people forget (or don't realize) that when Margot O'Toole first expressed doubts about the *Cell* paper, she felt there were errors in the paper, but was adamant that she did *not* suspect fraud. The issues she raised were typical of scientific disputes—doubts about the experiments, data interpretation, and the conclusions drawn from the data. Like many scientific disputes, this one could have been settled with further research. In fact, in the years since the publication of the paper, other researchers were able to verify the central claim of the paper, the very one O'Toole called into question. It is entirely acceptable for researchers to disagree with experimental interpretations. Many scientific journals contain dialog regarding a research paper, in the form of letters to the editors and responses from the original investigators. In fact, David Baltimore suggested such an exchange to O'Toole regarding the paper in question. She claimed she feared his written response would deter the *Cell* editors from publishing her critique, and she refused to engage in this exchange. In all the hearings at M.I.T. and Tufts, O'Toole denied the possibility of fraud. Her story started to change when "fraud busters" Stuart and Feder from the NIH got involved. Stuart and Feder were not highly regarded in the NIH. They got involved with the *Cell* paper after receiving calls from Charles Maplethorpe, a disgruntled graduate student who worked in Thereza Imanishi-Kari's lab. From their involvement, the case exploded into accusations of fraud.

Congressman John Dingell shown in a 1983 photograph. *(© Bettmann/Corbis. Reproduced by permission.)*

The Danger of Congressional Power After David Baltimore, Nobel laureate and coauthor of the infamous *Cell* paper, locked horns with Congressman Dingell while defending Imanishi-Kari, Dingell vowed, "I'm going to get that son of a bitch. I'm going to get him and string him up high." From this scientific vantage point, Dingell embarked on a witch-hunt that many later compared to the workings of McCarthy in his time. Like McCarthy, he had the power of the govern-

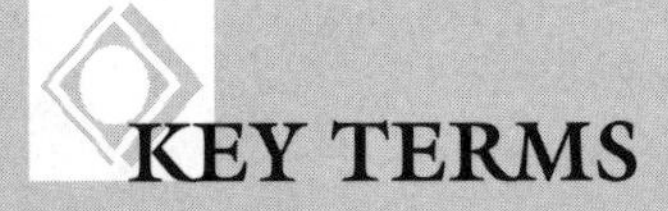

KEY TERMS

FABRICATION: Making up false data.
FALSIFICATION: Altering or distorting data.
INTERFERENCE: Purposefully damaging or taking material related to another scientist's research.
MISAPPROPRIATION: Deliberately plagiarizing or using information about other scientists' work when reviewing grant applications or manuscripts.
MISREPRESENTATION: Intentionally leaving out important facts or lying about research.
PEER REVIEW: The process by which a scientist's research is examined by other scientists before being accepted for publication in a scientific journal.
PLAGIARISM: Stealing other people's ideas or words without giving due credit.
WHISTLE-BLOWER: A person who informs the authorities or the public about the wrong actions of another person or institution.

ment at his side, and immunity as a member of Congress.

This meant that Dingell could circumvent the law and access information that was personal and confidential, and completely irrelevant to the case he was "investigating." In one instance, according to Dr. Bernadine Healy, director of NIH between 1991 and 1993, Dingell's committee accessed medical information in the personnel records of an accused scientist.

Dingell used his congressional power to call in the Secret Service, to run forensic tests on Imanishi-Kari's lab notebooks. He hoped to prove the parts of the notebooks called into question by O'Toole were fabricated. Not surprisingly, the Secret Service concluded the parts in question were indeed created later than they were supposed to have been created. Also not surprisingly, the appeals board that exonerated Imanishi-Kari noted that the forensic analysis "provided no independent or convincing evidence that the data or documents were not authentic or could not have been produced during the time in question." Ironically, the appeals panel found that the Secret Service's report contained the same flaws that cast suspicion over Imanishi-Kari—results were altered and data omitted. The U.S. attorney in Maryland, to whom Dingell sent the information he accumulated in his hearings, declined to prosecute Imanishi-Kari. One reason was that an independent forensic analysis of the notebooks found the Secret Service conclusions to be "erroneous."

When Dingell set up the Office of Scientific Integrity (OSI), whose purpose was to investigate scientific misconduct, he did not let the office run an impartial inquiry. Dr. Brian Kimes, the OSI's first director, recalled, "The N.I.H. was trying to separate from Dingell. Our leadership wanted us to make our own independent decision. But that was almost an impossibility." The Dingell subcommittee, holding its hearing in parallel to the OSI's investigation, bullied OSI staff members into supplying them with documents and confidential information. The OSI investigators conducted their hearings in fear of being second-guessed by Dingell. When the appeals panel was allowed to conduct its own scientific investigation, without the political pressure from Congressman Dingell, Imanishi-Kari was fully exonerated of all charges against her. During the appeals process, unlike the previous hearings, Imanishi-Kari was allowed to see the allegations against her, call witnesses, and examine the evidence. She was therefore able to respond to the accusations leveled at her. Prior to her appeal, she was denied due process. In *The Baltimore Case*, David Kevles asserts that Dingell's committee treated her like a defense contractor, not a scientist.

It is interesting to contrast the treatment Baltimore and Imanishi-Kari received at the hands of Dingell and his associates with the treatment NASA received after the *Challenger* explosion. With the exception of legendary physicist Richard Feynman, who was strongly critical of NASA and its methods, the Rogers Commissions treated the agency with kid gloves. The fact that NASA knew of the design flaw in the Solid Rocket Booster and, to paraphrase Feynman's words, continued to play Russian roulette with the lives of the astronauts, was swept under the rug. Congressman Rogers declined to press criminal negligence charges against responsible parties at NASA because it would "not be in the national interest." Yet Dingell contacted the U.S. attorney in Maryland, suggesting Imanishi-Kari be prosecuted because, in his mind, she was guilty of fraud. The damage to the image of science that was caused by the Dingell committee hearings was also not in the national interest.

Can Science Police Itself? Most of the time, yes, science can police itself. Most claims of fraud are investigated quietly and handled well. On the other hand, many of the ORI's (the former OSI) major decisions were overturned over the years by the federal appeals board that exonerated Imanishi-Kari. Obviously, science can police itself better than the government can, most of the time. One of the problems we face in our society is the elevation of the scientist to a god-like status, where he or she can't be human, and in fact is expected to be more than human. Mistakes in science happen all the time. Mistakes are part of the scientific method, one might say, but the public reacts with fury when they come to light. Fraud is unacceptable to most scientists, but like all human beings some will be tempted to commit fraud. And because those who supervise scientists are also human, they want to believe there is nothing wrong in their realm. Do whistle-blowers get punished? Unfortunately yes, sometimes. Do inquiries gloss over misconduct? Unfortunately yes, sometimes. But instead of holding a kangaroo court and wasting taxpayers' money trying to settle scientific disputes, the government would be better off holding hearings that investigate the pressures on scientists today.

Science used to be about ideas as much as results. Published papers showed a work in progress. Diminished funding and, therefore, the need for concrete results drive science today. As Goodstein points out, monetary gain is rarely a motive in science fraud. But the need to produce results in order to receive your next grant is a powerful motive. Scientists want to stay in science, and the only way to do so is by receiving grants, which are in short supply. By understanding how science today differs from science 30 years ago, for example, the government, and the public, might be able to engage in a constructive dialog with the scientific community, and perhaps devise a way to ease the burden on scientists, so that all they have to concentrate on is the science, not the politics of funding.

The argument above isn't meant to excuse scientific fraud—outright fraud is inexcusable. And the government can request that academic institutions receiving federal funding put in place better procedures to handle investigations of fraud and protection of whistle-blowers. In fact, the government has done so. And as Goodstein points out, at least at CalTech and the universities that copied CalTech, the procedures he instituted require a purely scientific investigation, not a judicial process. And their record, at the time the article was written, was better than the government's record, in terms of having the decisions upheld on appeal.

In addition to the above, not all allegations of misconduct or fraud are easy to judge. Dr. Bernadine Healy brought an example of an investigation at the Cleveland Clinic, where a three-month extensive investigation into allegations of misconduct by a scientist resulted in "no conclusive evidence" of misconduct. The scientist was penalized for sloppy techniques, but nothing else. The ORI spent two years on its own investigation, reaching the same conclusion in April 1992 and reversing itself in July that same year. A year later the appeals board overturned the ORI's decision, exonerating the scientist of the charges of misconduct, as the Cleveland Clinic did in the beginning.

Science can police itself. It isn't perfect, and never will be, just like every other human endeavor. But while scientists are trying to police themselves, and doing a better job at it, who polices the lawmakers? Congressional power should never be abused the way it was during Dingell's witch-hunt. The congressman would have been wise to remember the words of Senator Margaret Chase Smith, who in 1950 stated: ". . . I am not proud of the way we smear outsiders from the floor of the Senate, hide behind the cloak of congressional immunity and still place ourselves beyond criticism."

What happens when Congress interferes with science? The best summary may be the federal appeals panel's final decision in the Baltimore Case. Dingell's brain-child, the OSI/ORI, which charged Imanishi-Kari with fraud in its 1994 verdict, presented evidence that was "irrelevant, had limited probative value, was internally inconsistent, lacked reliability or foundation, was not credible or not corroborated, or was based on unwarranted assumptions." —ADI R. FERRARA

Viewpoint:

No, the Dingell investigation of scientific fraud was not unjustified, because peer review and self-regulation cannot work alone in maintaining scientific integrity.

In 1996, Dr. Thereza Imanishi-Kari was acquitted of 19 charges of scientific misconduct in what had become known simply as the "Baltimore Case." The fact that it had taken 10 years, five investigations, and three congressional hearings to get to this "truth" is often cited as proof that the federal inquiries headed by Representative John Dingell (a Michigan Democrat and chairman of the House Energy and Commerce Committee) were unjustified and that Imanishi-Kari should not have been investigated for fraud in the first place. However, many people—scientists and non-scientists alike—believe that there is a positive role for Congress today in investigating scientific fraud.

Science as Part of Society Scientists and the science they engage in are a fundamental part of society. In 1995, the National Academy of Scientists (NAS) issued a booklet for practicing scientists entitled *On Being a Scientist: Responsible Conduct in Research*, which states that "scientific knowledge obviously emerges from a

David Baltimore, photographed in 2000.
(© Richard Schulman/Corbis. Reproduced by permission.)

process that is intensely human, a process indelibly shaped by human virtues, values, and limitations and by societal contexts." Many scientists do not dispute that this is so. For example, neurologist Dr. Oliver Sacks describes science as "a human enterprise through and through, an organic, evolving, human growth."

By contrast, some scientists like to claim that their work is value-free and completely objective, and that science is somehow on a higher plane than the rest of society. However, the NAS booklet suggests that "science offers only one window on human experience. While upholding the honor of their profession, scientists must seek to avoid putting scientific knowledge on a pedestal above knowledge obtained through other means." So, if science is a human activity practiced by highly intelligent and knowledgeable—but nonetheless, ordinary—people, it follows that this endeavor is vulnerable to the same kinds of abuses as activities in other areas of society.

As human beings, scientists are not immune from making mistakes. In fact, the scientific process itself depends on researchers continually examining evidence, trying to improve on earlier work, coming up with alternative interpretations of data, and eliminating error from the established scientific record. However, that process can only work if scientists are able to examine and question the data presented by their peers and replicate the original experimental results in their own laboratories. Even in the absence of any intention of fraud, the possibility of a future federal investigation might make some researchers keep better records.

Imanishi-Kari was the subject of several investigations, including those by the Office of Scientific Integrity (OSI; later the Office of Research Integrity, or ORI), Rep. Dingell's Oversight and Investigations Subcommittee, and even the Secret Service. Eventually, the appeals board of the Health and Human Services (HHS) judged her to have been guilty of nothing more than being "aberrant in her data recording and rounding patterns" when working on the experiments that form the basis of the flawed 1986 *Cell* paper. But this was enough for post-doctoral fellow Margot O'Toole to have difficulty in reproducing the results and to claim that the paper was "riddled with errors." In 1989, three of the coauthors—Imanishi-Kari, D. Weaver, and Baltimore—corrected some of these errors in another *Cell* paper, although they contend that these mistakes were not very important scientifically. In 1991, the original paper was retracted completely by its coauthors.

Unfortunately, misconduct, like errors, is found in science as much as in other social contexts, and it can be difficult to tell the difference between inadvertent errors and deliberate fraud without a full and proper investigation. In science, misconduct is traditionally defined as fabrication, falsification, and plagiarism (together known as FFP), and other departures from standard scientific practice, although the Commission on Research Integrity (CRI) replaced these with misappropriation, interference, and misrepresentation as part of its new definition of scientific misconduct in 1995 (see Key Terms).

The Importance of Scientific Integrity Not only are scientific researchers human like the rest of us, in many cases the work they do also has an important impact on how we all live our lives. This is particularly true in the fields of medicine and genetics. False or misleading information can hold up scientific progress and—in the worst-case scenario—even delay and seriously hinder the development of medical cures or provoke health scares. In turn, this feeds public fear and skepticism about science and gives ammunition to those that would attack science endeavor.

These days, a large amount of public money is used to fund scientific research, and in return scientific establishments need to be open and accountable. For this reason in particular, research misconduct is no different than fraud in other areas of society. Rep. Dingell states that ". . . Congress authorizes approximately $8 billion annually for [the National Institutes of Health (NIH)] alone, and it is the subcommittee's responsibility to make sure that this money is spent properly and that the research institutions, including the NIH, that receive these federal funds behave properly."

Scientists are also hired and promoted on the basis of how well they can attract federal funds to their departments, and the reputations (and hence financial power) of researchers, as well as their colleagues and associated institutions, rest firmly on their incorruptibility. Since 1990, federal regulations have stipulated that institutions receiving training grants from the NIH must include training in "principles of scientific integrity" and have written procedures to follow when complaints of scientific misconduct are reported.

It is highly unlikely that fraud can ever be stamped out completely. Misconduct will occur wherever large numbers of people are involved in a social system, so it is essential that institutions have adequate machinery in place for dealing quickly and efficiently with cases of suspected scientific fraud. Some people believe that if the scientific community cannot be trusted to do this alone, Congress should intervene and ensure that research integrity is maintained.

Peer Review and Self-Regulation There have been many critics of increased congressional involvement in the sciences, not least David Baltimore himself. They argue that the process of peer review and self-regulation are enough to keep scientific honor intact. This is true, in theory, most of the time, but practice has occasionally proved otherwise.

In recent years, the peer review system itself has even been shown to be vulnerable to corruption. The problem lies in the fact that while researchers in the same field are often in direct competition with each other, they are generally the best people to understand, review, and judge the validity of each other's work. Implicit in this system is the concept of trust. Scientists must be able to trust each other not to steal each other's ideas and words or to mislead each other when reviewing pre-publication papers, presenting posters or giving conference lectures, and even just talking about their work with other researchers.

Once a scientist suspects a colleague of committing errors, or even scientific misconduct, it is essential to the atmosphere of trust within research institutions that there be a way for a person to voice his or her concerns without fear of reprisals. If there was such measure in place, the scientific community could certainly be left alone to regulate itself. However, many people question whether this is the case and claim that scientific establishments have done a very poor self-regulatory job in the past. Allegations of misconduct can have grave consequences for both the accused and the whistle-blower and should not be undertaken, or dealt with, lightly. If the coauthors of the 1986 *Cell* paper had taken Margot O'Toole's claim of scientific error seriously and re-examined their findings and conclusions in the beginning, science would have been advanced and several careers would have been saved earlier. Rep. Dingell believes that research institutions are too slow to react to fraud allegations and are usually more concerned with keeping the situation away from the public eye than with examining the accusations and dealing with them fairly. He asserts that ". . . science is essentially a quest for truth. The refusal to investigate concerns raised by fellow scientists, and to correct known errors, is antithetical to science. . . ."

Involvement of Congress Some scientists are worried that congressional intervention in scientific affairs will stifle creativity and the kind of novel thinking that sometimes leads to huge advances in science. There has even been talk of the "science police"—a threat that has not really materialized. Rep. Dingell and his associates believe that federal rules and regulations merely serve to help universities and other research institutions act quickly and decisively when allegations of scientific misconduct surface. This gives researchers the freedom to concentrate on being experts in their own field—science, rather than law.

Another claim is that federal investigations favor the whistle-blower over the accused scientist and are unfair because they do not allow researchers under investigation to find out what charges and evidence are being used against them. However, this was rectified in 1992, when the HHS appeals board was set up for this purpose. The existence of such an appeals procedure might also have the advantage of deterring researchers from deliberately making false accusations against their colleagues and competitors, in itself a form of scientific misconduct.

In conclusion, science and scientists do not stand outside society but are an integral part of it. Modern scientific activity is so dependent on the public and the government for money and good will that it must be seen as accountable and beyond reproach. The peer review process and self-regulation are extremely important to scientific progress, but they are not sufficient for maintaining research integrity. Although Rep. Dingell has been criticized for his heavy-handed approach to the investigations of the "Baltimore Case," the fact remains that an outside body needs to ensure that allegations of scientific misconduct are investigated thoroughly and that researchers complaining about the behavior of other scientists are taken seriously and protected from retaliation. In an ideal situation, peer review, self-regulation, and federal investigation will work hand-in-hand to preserve scientific integrity. —AMANDA J. HARMAN

Further Reading

Baltimore, David. Correspondence in "Shattuck Lecture—Misconduct in Medical Research." *The New England Journal of Medicine* 329 (September 2, 1993): 732–34.

Bulger, Ruth Ellen, Elizabeth Heitman, and Stanley Joel Reiser, eds. *The Ethical Dimensions of the Biological Sciences.* Cambridge, England: Cambridge University Press, 1993.

Goodman, Billy. "HHS panel issues proposals for implementing misconduct report." *The Scientist* [cited July 16, 2002]. <http://www.the-scientist.com/yr1996/july/miscond>.

———. "Multiple Investigations." *The Scientist* [cited July 16, 2002]. <http:// www.the-scientist.com/yr1996/ august/aftermath>.

Goodstein, David. "Conduct and Misconduct in Science" [cited July 16, 2002]. <www.its.caltech.edu/~dg/conduct.html>.

Guston, David H. "Integrity, Responsibility, and Democracy in Science." *Scipolicy* 1 (Spring 2001): 167–340.

Healy, Bernadine. "The Dingell Hearings on Scientific Misconduct—Blunt Instruments Indeed." *The New England Journal of Medicine* 329 (September 2, 1993): 725–27.

———. "The Dangers of Trial by Dingell." *New York Times* (July 3, 1996).

Imanishi-Kari, T., D. Weaver, and D. Baltimore. *Cell* 57 (1989): 515–16.

Kevles, Daniel J. *The Baltimore Case: A Trial of Politics, Science and Character.* New York: W.W. Norton & Company, 1998.

Lewis, Anthony. "Abroad at Home; Tale of a Bully." *New York Times* (June 24, 1996).

National Academy of Sciences. *On Being a Scientist: Responsible Conduct in Research.* Washington: National Academy Press, 1995.

"Noted Finding of Science Fraud Is Overturned by Federal Panel." *New York Times* (June 22, 1996).

Sacks, Oliver. Introduction to *Hidden Histories of Science* by R. B. Silvers. London: Granta Books, 1997.

Weaver, D., et al. "Altered Repertoire of Endogenous Immunoglobulin Gene Expression in Transgenic Mice Containing a Rearranged mu Heavy Chain Gene." *Cell* 45 (1986): 247–59.

"What Can We Learn from the Investigation of Misconduct?" *The Scientist* [cited July 16, 2002]. <http://www.the-scientist.com/yr1989/jun/opin>.

MATHEMATICS AND COMPUTER SCIENCE

Does catastrophe theory represent a major development in mathematics?

Viewpoint: Yes, catastrophe theory is a major development in mathematics.

Viewpoint: No, catastrophe theory does not represent a major development in mathematics: not only is it a close cousin of chaos theory, but both are products of a twentieth-century intellectual environment that introduced quantum mechanics and other challenges to a traditional "common sense" worldview.

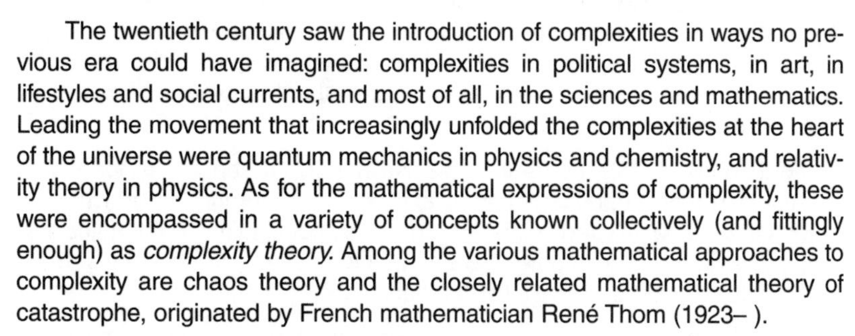

The twentieth century saw the introduction of complexities in ways no previous era could have imagined: complexities in political systems, in art, in lifestyles and social currents, and most of all, in the sciences and mathematics. Leading the movement that increasingly unfolded the complexities at the heart of the universe were quantum mechanics in physics and chemistry, and relativity theory in physics. As for the mathematical expressions of complexity, these were encompassed in a variety of concepts known collectively (and fittingly enough) as *complexity theory.* Among the various mathematical approaches to complexity are chaos theory and the closely related mathematical theory of catastrophe, originated by French mathematician René Thom (1923–).

There are several varieties of misunderstanding surrounding the ideas of complexity, chaos, and catastrophe. First of all, the umbrella term *complexity* seems to suggest complexity for complexity's sake, or an exclusive focus on things that are normally understood as complex, and nothing could be further from the truth. In fact, complexity theory—particularly in the form known as *catastrophe theory*—is concerned with events as apparently simple as the path a leaf makes in falling to the ground. As complexity theory shows, however, there is nothing simple at all about such events; rather, they involve endless variables, which collectively ensure that no two leaves will fall in exactly the same way.

There is nothing random in the way that the leaf falls; indeed, quite the contrary is the case, and this points up another set up misconceptions that arise from the names for various types of complexity theory. Chaos theory is not about chaos in the sense that people normally understand that term; in fact, it is about exactly the opposite, constituting a search for order within what appears at first to be disorder. In attempting to find an underlying order in the seemingly random shapes of clouds, for instance, chaos theory is naturally concerned with complexities of a magnitude seldom comprehended by ordinary mathematics.

Likewise, catastrophe theory does not necessarily involve "catastrophes" in the sense that most people understand that word, though many of the real-world applications of the idea are catastrophic in nature. In the essays that follow, catastrophe theory is defined in several different ways, and represented by numerous and diverse examples. At heart, all of these come back to a basic idea of sudden, discontinuous change that comes on the heels of regular, continuous processes. An example—and one of the many "catastrophic" situations to which catastrophe theory is applied—would be the sudden capsizing of a ship that has continued steadily to list (tilt) in a certain direction.

At issue in the essays that follow is the question of whether catastrophe theory represents a major development in mathematics. To an extent, this is a controversy that, like many in the realm of mathematics or science, takes place "above the heads" of ordinary people. Most of us are in the position of an unschooled English yeoman of the eighteenth century, poking his head through a window of Parliament to hear members of the House of Lords conducting a debate concerning affairs in India—a country the farmer has never seen and scarcely can imagine.

Yet just as England's position vis-à-vis India *did* ultimately affect the farmer, so our concern with catastrophe mathematics represents more than an academic interest. If catastrophe theory does represent a major new movement, then it opens up all sorts of new venues that will have practical applications for transportation, civil engineering, and other areas of endeavor that significantly affect the life of an ordinary person. On the other hand, if catastrophe theory is not a significant new development, then this is another blow against complexity theory and in favor of the relatively straightforward ideas that dominated mathematical thought prior to the twentieth century.

Catastrophe theory has a long and distinguished history, with roots in the sixteenth-century investigations of Leonardo da Vinci (1452–1519) into the nature of geometric patterns created by reflected and refracted light. Serious foundational work can be found in the bifurcation and dynamical systems theory of French mathematician Jules Henri Poincaré (1854–1912). Yet the heyday of catastrophe as a mathematical theory was confined to a period of just a few years in the third quarter of the twentieth century: thus, by the time its originator had reached the mere age of 55, even supporters of his theory were already proclaiming catastrophe a dead issue. Thom himself, in a famous statement quoted by both sides in the controversy at hand, stated that "catastrophe theory is dead."

If both sides concede that catastrophe theory is dead as an exciting mathematical discipline, then one might wonder what question remains to be decided. But it is precisely the meaning of this "death" that is open to question. Seeds die, after all, when they are planted, and the position of catastrophe supporters in this argument is that catastrophe is dead only in a formal sense; far from being truly dead, it flourishes under other names. Detractors, on the other hand, deride catastrophe theory as a mere gimmick, a sophisticated parlor trick that has been misapplied in sometimes dangerous ways to analyses of social science.

Both sides, in fact, agree that the use of catastrophe theory for analyzing social problems is questionable both in principle and in application. Adherents of catastrophe theory as a bold new mathematical idea are particularly strong on this point, taking great pains to differentiate themselves from those who at various times have sought to use catastrophe as a sort of skeleton key to unlocking underlying secrets of mass human behavior.

Again, however, there is a great deal of misunderstanding among the general public with regard to the ways mathematical theories are used in analyzing social problems. The mere use of mathematical models in such situations may strike the uninitiated as a bit intrusive, but in fact governments regularly depend on less controversial mathematical disciplines, such as statistics, in polling the populace and planning for the future.

A final area of agreement between both sides in the catastrophe dispute is a shared admiration for the genius of Thom and for the work of other mathematicians—most notably Dutch-born British mathematician Christopher Zeeman (1925–)—who helped develop the theory as a coherent system. When it comes to catastrophe theory, evidence does not support the assertion that "to know it is to love it." However, even those who would dismiss catastrophe as mere mathematical trickery still admit that it is an extraordinarily clever form of trickery. —JUDSON KNIGHT

Viewpoint:

Yes, catastrophe theory is a major development in mathematics.

Not only is catastrophe theory a major development in mathematics, it also has many uses in other fields as diverse as economics and quantum physics. Catastrophe theory is itself a development of topology, and builds on the work of earlier mathematical concepts. The only controversy surrounding the status of catastrophe theory is not a mathematical one, but rather over some of the non-mathematical uses to which it has been applied.

Catastrophe theory attempts to describe situations where sudden discontinuous results follow from smooth, continuous processes. This puts it outside the framework of traditional calculus, which studies only smooth results. The real world is full of discontinuous jumps, such as when water boils, ice melts, the swarming of locusts, or a stock market collapse. Catastrophe

KEY TERMS

CATASTROPHE THEORY: An area of mathematical study devoted to analyzing the ways that a system undergoes dramatic changes in behavior following continuous changes on the part of one or more variables.

CAUSTICS: The word means "burning" and usually refers to light rays, and the intense, sharp, bright curves that form on certain surfaces; for example, the patterns of light in a cup of coffee, an empty saucepan, or a rainbow. They can be thought of as maximum points of intensity of the light.

CHAOS THEORY: An area of mathematical study concerned with non-linear dynamic systems, or forms of activity that cannot easily be represented with graphs or other ordinary mathematical operations.

COMPLEXITY: A mathematical theory based on the idea that a system may exhibit phenomena that cannot be explained by reference to any of the parts that make up that system.

DYNAMICAL SYSTEMS: A theory of mathematics concerned with addressing entities as systems moving through a particular state.

MORPHOGENESIS: The shaping of an organism by embryological processes of differentiation of cells, tissues, and organs. Also used in a broader sense as any complex system-environment interaction that alters a system's form and structure. The most common example is the growth of an animal from a fertilized egg, where seemingly identical embryonic cells differentiate into the specialized parts of the body.

SINGULARITY THEORY: Generalization of the study of mathematical functions at the maximum and minimum points. Hassler Whitney introduced the topological notion of mappings, and concluded that only two kinds of singularities are stable, folds and cusps.

SYSTEM: In mathematics and the sciences, a system is any set of interactions that can be separated from the rest of the universe.

TOPOLOGY: The mathematics of geometrical structures. It is concerned with the qualities, not the specific quantities, of such structures, such as how many holes an object has. For example, topologically speaking a donut and a coffee mug are the same, as they have one hole. Another example is the London Underground Map, which does not have accurate directions and distances, but does show how all the lines and stations connect. Topology often deals with structures in multiple dimensions.

theory is an attempt to offer a geometric model for such behaviors.

The theory can be described in a number of ways: as the study of the loss of stability in a dynamic system, the analysis of discontinuous change, a special subset of topological forms, or even a controversial way of thinking about transformations. However, the most accurate and unambiguous way of describing catastrophe theory is by using mathematical terminology. Unfortunately, even the most basic mathematical understanding requires knowledge of calculus in several variables and some linear algebra. Yet, by virtue of its initial popularity, many books have been written on catastrophe theory for the non-mathematician. While a good number of these do a wonderful job of describing the nature of catastrophe theory, giving easy-to-understand examples and infusing the reader with enthusiasm for the topic, they also, by their very nature, simplify the subject. They tend to stress the non-mathematical applications of the theory, and that is when the controversy begins.

Catastrophe Theory and the Media The media does not usually take notice of the release of a mathematics text. Yet the English translation of René Thom's work *Structural Stability and Morphogenesis: An Essay on the General Theory of Models*, published originally in French in 1972, was widely reviewed. It not only spawned dozens of articles in the scientific press, but also in such mainstream publications as *Newsweek*. The theory's name was partly to blame. The word "catastrophe" evoked images of nuclear war and collapsing bridges, and the media suggested the theory could help explain, predict, and avoid such disasters.

Yet the aim of the theory was altogether different, if almost as ambitious. Thom had a long track record of pure mathematical work in topology, including characteristic classes, cobordism

Chaos theory is also known as the *butterfly effect*, because of the idea that a butterfly beating its wings in China may ultimately affect the weather in New York City.
(Robert J. Huffman/Field Mark Publications.)

theory, and the Thom transversality theorem (for which he was awarded a Fields medal in 1958). However, it was while thinking about the biological phenomena of morphogenesis that he began to see some topological curiosities in a broader light. Biological morphogenesis is the astonishing way in which the small collection of embryonic cells suddenly specialize to form arms, legs, organs, and all the other parts of the body. Such sudden changes fascinated Thom, and he began to see such transformations in form everywhere, from the freezing of water into ice, to the breaking of waves. He began formulating a mathematical language for these processes, and with the help of other mathematicians, such as Bernard Malgrange, he reached a remarkable conclusion. In 1965 Thom discovered seven elementary topological forms, which he labeled "catastrophes" for the manner in which small changes in the parameters could produce sudden large results. These elementary catastrophes, Thom argued, were like the regular solids and polygons in Greek geometry. Despite the infinite number of sides a polygon can have, only three regular polygons—the triangle, square, and hexagon— can be placed in edges to fill a surface such as a tiled floor. Also, there are only five regular solids that can be made by using regular polygons as the sides. These forms recur again and again in nature, from honeycombs to crystals, precisely because they are the only ways to fill a plane surface, or form a regular shape. Thom suggested that the seven elementary catastrophes also recur in nature in the same way. This offers a new way of thinking about change, whether it be a course of events or a dynamic system's behavior, from the downfall of an empire to the dancing patterns of sunlight on the bottom of a swimming pool.

Thom left much of the mathematical "dirty work" to others, and so it was John Mather who showed that the seven abstract forms Thom called elementary catastrophes existed, were unique, and were structurally stable. Indeed, Thom's methods were one of the reasons catastrophe theory was initially criticized. Thom's mathematical work has been liken to that of a trailblazer, hacking a rough path through a mathematical jungle, as opposed to other, more rigorous mathematicians, who are slowly constructing a six-lane highway through the same jungle. Even one of Thom's most ardent followers noted that "the meaning of Thom's words becomes clear only after inserting 99 lines of your own between every two of Thom's."

Wider Applications However obscurely phrased, Thom's ideas quickly found an audience. In a few years catastrophe theory had been used for a wide variety of applications, from chemical reactions to psychological crises. It was the work of E. Christopher Zeeman, in particular, that was most provocatively reported. Zeeman had attempted to apply the theory in its broadest sense to many fields outside of mathematics. Zeeman modeled all kinds of real-world actions he argued had the form of catastrophes, from the stability of ships to the aggressive behavior of dogs. Perhaps his most controversial model was that of a prison riot at Gartree prison in England. The study was challenged on the grounds that it was an attempt to control people in the manner of Big Brother, and some media reports incorrectly stated that hidden cameras were being used to invade the privacy of the test subjects. In reality the modeling used data obtained before the riot by conventional means that was available in the public domain.

While new supporters of the theory quickly applied it to an ever-growing list of disciplines, from physics to psychology, others argued that catastrophe theory offered nothing of value, used *ad hoc* assumptions, and was at best a curiosity, rather than a universal theory of forms as Thom had suggested. Indeed, some saw it not just as insignificant, but as a danger. Hector Sussmann and Raphael Zalher criticized it as "one of the many attempts that have been made to deduce the world by thought alone," and Marc Kac went so far as to charge that catastrophe theory was "the height of scientific irresponsibility."

Some of the applications were easy targets for the critics. For example, political revolutions were modeled by Alexander Woodcock using the two-variable cusp elementary catastrophe. The variables chosen were political control and popular involvement. The "results" of mapping flow lines onto the topological surface of the cusp catastrophe suggested some controversial conclusions. High popular involvement in politics would always seem to bring a dictatorship to a catastrophe point, yet many historians would argue that sometimes high popular involvement helps stabilize a dictatorship, such as the early years of the Nazi regime. Some critics suggested that the chosen variables were not as important as other factors, for example, economic performance, or media perception.

A New Tool, Not a Replacement Yet the criticisms, while valid, miss the point of such models. They are not to be seen as complete explanations of real-world events, but merely attempts to offer new insights, however incomplete they might be. Competing models, using other variables, can be compared, and more sophisticated, higher-dimensional catastrophes can be used, although such four, five, and higher dimensional models lose their visual representation, being impossible to draw. Catastrophe theory is not a replacement for old methods, but rather a new supplement to test and compare other methods against. Even if the models have no validity at all in the long term, they still have the power to challenge conventional wisdom and offer new and exciting avenues of inquiry.

Part of the problem with the media perception of catastrophe theory is the difficulty of relating the mathematical details of the theory to a non-mathematical audience. Mathematically catastrophe theory is a special branch of dynamical systems theory, using differential topology. It attempts to describe situations in which gradually changing forces lead to abrupt changes. This is impossible using differential calculus, but that does not mean that catastrophe is a replacement for calculus, rather it is a new tool to use in situations previously unexplored by traditional mathematics. For example, calculus cannot predict the path of a leaf, or where it will land, as there are too many variables. However, there are patterns that can be deduced. No two falling leaves follow the same quantitative path, but they have a shared qualitative behavior that can be studied.

Based on a Solid Mathematical Foundation Catastrophe theory was not a mathematical revolution, but a development from the work of previous mathematicians. It builds on the theory of caustics, which are the bright geometric patterns created by reflected and refracted light, such as a rainbow or the bright cusp of light in a cup of coffee. Early work on caustics was done by Leonardo da Vinci (1452–1519), and their name was given by Ehrenfried Tschirnhaus (1651–1708). It also uses bifurcation theory, which was developed by Henri Poincaré (1854–1912) and others, and examines the forking or splitting of values. It also draws heavily on Hassler Whitney's (1907–1989) work on singularities. Indeed, without Whitney's developments the theory could not have been conceived. Thom's own earlier work on transversality was also important. The calculus of variations, which has a long and distinguished mathematical history, is also important to the theory. Catastrophe theory's development from such diverse fields shows that it is not something out of left field, but a novel way of combining many previous mathematical ideas and some new work on topological surfaces. In this way catastrophe theory is a major development in mathematics, even without considering any wider applications.

Catastrophe theory's basis in topology means that in general it will be concerned with qualities, not quantities. The elementary catastrophes are like maps with a scale. Models using them can only be approximate, and stress properties, not specific results. This means that such models cannot be used to predict events. Thom himself has noted that the scientific applications of his theory are secondary to its formal beauty and power, and its scientific status rests on its "internal, mathematical consistency," not the questionable models it has been used for.

The Death of Catastrophe Theory? Another criticism that has been leveled against catastrophe theory is that it quickly vanished. In just a few years, after the initial flurry of papers and attempted real-world applications the theory seemed to disappear from sight. Indeed, as early as 1978 supporters of catastrophe theory were already noting its demise. Tim Poston and Ian Stewart, in their book *Catastrophe Theory and its Applications*, remarked that "Catastrophe theory is already beginning to disappear. That is, 'catastrophe theory' as a cohesive body of knowledge with a mutually acquainted group of experts working on its problems, is slipping into the past, as its techniques become more firmly embedded in the consciousness of the scientific community." Thom himself stated, "It is a fact that catastrophe theory is dead. But one could say that it died of its own success." Rather than disappearing, catastrophe theory was absorbed into other disciplines and called by other, less controversial, names such as bifurcations and singularities. Catastrophe theory and its descendants have been particularly successful in fields already using complex mathematical analysis. From quantum physics to economics, catastrophe theory continues to be used in more and more innovative ways.

While catastrophe theory is an example of the normal development of mathematics, its importance should not be downplayed. Even the critics of catastrophe theory admit that the mathematics behind catastrophe theory is elegant and exciting. Their opposition to the theory is not a mathematical one, but rather with the non-mathematical uses the theory has been applied to. Like fractals and chaos theory, the ideas of catastrophe theory briefly captured popular attention. However, while the media may have suggested that catastrophe theory had the power to predict everything from stock market crashes to earthquakes, the reality of the theory is such that casual uses offer nothing more than interesting oddities. Even the most rigorous mathematical uses of catastrophe theory may return very little in terms of modeling the real world. However, the disappointments that the theory could not live up to the initial hype and exaggerations in no way diminishes the mathematical development that the theory represents. —DAVID TULLOCH

Viewpoint:

No, catastrophe theory does not represent a major development in mathematics: not only is it a close cousin of chaos theory, but both are products of a twentieth-century intellectual environment that introduced quantum mechanics and other challenges to a traditional "common sense" worldview.

The mathematical principles known under the intriguing, though somewhat misleading, title of "catastrophe theory" are certainly not lacking in what an advertiser might call "sex appeal." Like the more well-known ideas of relativity theory or quantum mechanics, or even chaos theory (to which catastrophe theory is closely related), the precepts of catastrophe mathematics are at once intellectually forbidding and deceptively inviting to the uninitiated.

It is easy to succumb to the intoxicating belief that catastrophe theory is a major development in mathematics—that it not only signals a significant shift in the direction of mathematical study, but that it provides a new key to understanding the underlying structure of reality. This, however, is not the case. Catastrophe theory is just one of many intriguing ideas that emerged from the twentieth-century intellectual environment that introduced quantum mechanics and relativity, two theories that truly changed the way we view the world.

Relativity and quantum mechanics, of course, were and continue to be major developments in modern thought: almost from the time of their introduction, it was clear that both (and particularly the second of these) presented a compelling new theory of the universe that overturned the old Newtonian model. These ideas—and again, particularly quantum mechanics—served to call into question all the old precepts, and for the first time scientists and mathematicians had to confront the possibility that some of their most basic ideas were subject to challenge. The resulting paradigm shift, even as it signaled a turn away from the old verities of Newtonian mechanics, also heralded the beginnings of an intellectual revolution.

Chaos and its close relative, catastrophe theory, are the result of extraordinary insights on the part of a few creative geniuses working at the edges of mathematical discovery. But does either area represent a major development in mathematics, the opening of a new frontier? If either of the two were likely to hold that status, it would be the more broad-reaching idea of chaos theory, which Robert Pool analyzed in *Science* in 1989 with an article entitled "Chaos Theory: How Big An Advance?" The dubious tone of the title suggests his conclusion: paraphrasing Steven Toulmin, a philosopher of science at Northwestern University, Pool wrote, "Chaos theory gives us extra intellectual weapons, but not an entirely new worldview."

As for catastrophe theory, there is this pronouncement: "It is a fact that catastrophe theory is dead. But one could say that it died of its own success. . . . For as soon as it became clear that the theory did not permit quantitative prediction, all good minds . . . decided it was of no value." These words belong to French mathematician René Thom (1923–), originator of catastrophe theory, quoted by the theory's chief popularizer, Dutch-born British mathematician Christopher Zeeman (1925–).

Complexity Theory Though they arose independently, chaos and catastrophe theories can ultimately be traced back to the theory of dynamical systems, originated by French mathematician Jules Henri Poincaré (1854–1912). Whereas ordinary geometry and science are concerned with the variables that differentiate a particular entity, dynamical systems is devoted purely to addressing entities as systems moving through a particular state. Thus if the "clouds" of cream in a cup of coffee exhibit behavior similar to that of a hurricane, then this relationship is of interest in dynamical systems, even though a vast array of physical, chemical, and mathematical variables differentiate the two entities.

These three—catastrophe, chaos, and dynamical systems—are part of a larger phenomenon, which came into its own during the twentieth century, known as complexity theory. Also concerned with systems, complexity is based on the idea that a system may exhibit phenomena that cannot be explained by reference to any of the parts that make up that system. Complexity theory in general, and chaos theory in particular, involve extremely intricate systems in which a small change may yield a large change later. Weather patterns are an example of such a system, and in fact one of the important figures in the development of chaos theory was a meteorologist, Edward Lorenz (1917–).

Specifically, chaos theory can be defined as an area of mathematical study concerned with non-linear dynamic systems, or forms of activity that cannot easily be represented with graphs or other ordinary mathematical operations. The name "chaos," which may have originated with University of Maryland mathematician James Yorke (1941–)—a leading proponent of the theory—is something of a misnomer. The systems addressed by chaos theory are the opposite of chaotic; rather, chaos theory attempts to find the underlying order in apparently random behavior.

In his infamous 1995 "manifesto," Theodore Kaczynski, also known as the "Unabomber," referred to the writings of mathematician Christopher Zeeman.
(Photograph by John Youngbear. AP/Wide World Photos. Reproduced by permission.)

How Far Should We Take Catastrophe Theory? Similarly, catastrophe theory is not really about catastrophes at all; rather, it is an area of mathematical study devoted to analyzing the ways that a system undergoes dramatic changes in behavior following continuous changes on the part of one or more variables. It is applied, for instance, in studying the way that a bridge or other support deforms as the load on it increases, until the point at which it collapses. Such a collapse is, in ordinary terms, a "catastrophe," but here the term refers to the breakdown of apparently ordered behavior into apparently random behavior. Again, however, there is the underlying principle that even when a system appears to be behaving erratically, it is actually following an order that can be discerned only by recourse to exceptionally elegant mathematical models, such as the catastrophe theory.

Catastrophe theory has proven useful in discussing such particulars as the reflection or refraction of light in moving water, or in studying the response of ships at sea to capsizing waves; however, its results are more dubious when applied to the social sciences. In the latter vein, it has been utilized in attempts to explain animals' fight-or-flight responses, or to analyze prison riots. One mathematician (though he will be remembered in quite different terms) even saw in catastrophe theory a model for the breakdown of Western society. His name: Theodore Kaczynski, a.k.a. the Unabomber, who in his infamous 1995 "manifesto" referred to the writings of Zeeman.

Catastrophe theory may be useful for analyzing aspects of the universe, but it is not a key to the *entire* universe. Even relativity and quantum mechanics, much more far-reaching theories, only apply in circumstances that no one has directly experienced—i.e., traveling at the speed of light, or operating at subatomic level. In the everyday world of relatively big objects in which humans operate, old-fashioned Newtonian mechanics, and the mathematics necessary to describe it, still remain supreme. This suggests that neither catastrophe theory nor any other esoteric system of mathematical complexity is truly a major development in our understanding of the world. —JUDSON KNIGHT

Further Reading

Arnol'd, V. I. *Catastrophe Theory.* Berlin: Springer-Verlag, 1992.

Ekeland, Ivar. *Mathematics and the Unexpected.* Chicago: University of Chicago Press, 1988.

Hendrick, Bill. "Clues in Academia: Manifesto Steered Probe to 'Catastrophe Theory.'" *Journal-Constitution* (Atlanta) (April 6, 1996).

"Indexes of Biographies." School of Mathematics and Statistics, University of St. Andrews, Scotland. May 31, 2002 [cited July 16, 2002]. <http://www-groups.dcs.st-and.ac.uk/~history/BiogIndex.html>.

Pool, Robert. "Chaos Theory: How Big an Advance?" *Science* 245, no. 4913 (July 7, 1989): 26–28.

Poston, Tim, and Ian Stewart. *Catastrophe Theory and Its Applications.* London: Pitman, 1978.

Saunders, Peter Timothy. *An Introduction to Catastrophe Theory.* Cambridge, England: Cambridge University Press, 1980.

"Turbulent Landscapes: The Natural Forces That Shape Our World." *Exploratorium.* May 31, 2002 [cited July 16, 2002]. <http://www.exploratorium.edu/complexity/>.

Zeeman, Eric Christopher. *Catastrophe Theory—Selected Papers 1972–1977.* Reading, MA: Addison-Wesley, 1977.

Would most first-year college students be better off taking a statistics course rather than calculus?

Viewpoint: Yes, most first-year college students would be better off taking a statistics course rather than calculus because statistics offers a greater variety of practical applications for a non-scientific career path.

Viewpoint: No, first-year college students would not be better off taking a statistics course rather than calculus because calculus is wider in application and teaches an intellectual rigor that students will find useful throughout their lives.

Statistics is a branch of mathematics devoted to the collection and analysis of numerical data. Calculus, on the other hand, is a mathematical discipline that deals with rates of change and motion. The two, it would seem, have little in common, yet they are linked inasmuch as they are at the center of an educational debate, here considered from two perspectives, as to which course of study is more appropriate for first-year college students.

The two disciplines are also historically linked, both being products of the early Enlightenment. They emerged in an intellectually fruitful period from the middle of the seventeenth to the beginning of the eighteenth centuries, and collectively they are the achievement of four of the greatest minds the world has ever produced. Each discipline benefited from the contributions of two major thinkers, but in the case of statistics and probability theory, French mathematicians Pierre de Fermat (1601–1665) and Blaise Pascal (1623–1662) worked together. By contrast, English physicist Sir Isaac Newton (1642–1727) and German philosopher Gottfried Wilhelm von Leibniz (1646–1716) presented two competing versions of calculus, and in later years, their respective adherents engaged in a bitter rivalry.

While statistics is the branch of mathematics concerned with the collection and analysis of numerical data relating to particular topics, probability theory is devoted to predicting the likelihood that a particular event will occur. The two are closely related, because a proper statistical database is necessary before probabilities can be accurately calculated. Furthermore, they emerged as areas of study together, and only later became differentiated.

Though statistics and probability theory can be, and are, applied in a number of serious pursuits, they also play a part in gambling and games of chance. It was a fascination with such games—albeit a professional rather than personal fascination—that led Fermat and Pascal to undertake the first important work in probability theory and statistics. The two disciplines remained closely linked until the end of the eighteenth century, by which time statistics emerged as an independent discipline.

By the beginning of the twentieth century, several factors contributed to growing respect for statistics and probability theory. In physics and chemistry, quantum mechanics—which maintains that atoms have specific energy levels governed by the distance between the electron and the nucleus—became the prevailing theory of matter. Since quantum mechanics also holds that it is impossible to predict the location of an electron with certainty, probability calculations became necessary to the work of physicists and chemists.

The early twentieth century also saw the rise of the social sciences, which rely heavily on statistical research. In order to earn a graduate degree today, a prospective psychologist or sociologist must undergo an intensive course of study in statistics, which aids in the interpretation of patterns involving complex variables—for instance, the relationship between schoolchildren's diet and their test scores, or between the age at which a person marries and his or her income.

Another factor behind the growing importance of probability and statistics was the emergence of the insurance industry. Insurance is an essential element of economically advanced societies, which literally could not exist if individuals and businesses had no way of protecting themselves financially against potential losses. In the world of insurance, statistics and probability theory are so important that an entire field of study, known as actuarial science, is devoted to analyzing the likelihood that certain events will or will not happen. Without actuarial analysis, which makes it possible to calculate risks, an insurance company could not make a profit and still charge reasonable prices for coverage.

Though the mathematics involved in calculating statistics go far beyond the education of the average person, the discipline itself is rooted in phenomena easily understandable to all. By contrast, calculus is likely to be more intellectually intimidating. Actually, there are two kinds of calculus: differential calculus, which is concerned with finding the instantaneous rate at which one quantity, called a *derivative,* changes with respect to another, and integral calculus, which deals with the inverse or opposite of a derivative. On the surface, this sounds rather abstract, though in fact calculus has numerous practical applications in areas ranging from rocket science to business.

The roots of calculus go back to ancient Greece, and the discipline owes much to the work of the medieval French mathematician Nicole d'Oresme (1323–1382). He advanced the study of motion, or kinematics, with his work on uniform acceleration. Galileo Galilei (1564–1642), building on Oresme's findings, proved that falling bodies experience uniform acceleration, and this in turn paved the way for Newton's own studies of gravitation and motion. In order to quantify those studies, Newton required a new form of mathematics, and therefore devised calculus.

One of the great controversies in the history of mathematics is the question of whether Newton or Leibniz deserves credit for the creation of calculus. In fact, both developed it independently, and though it appears that Newton finished his work first, Leibniz was the first to publish his. Over the years that followed, mathematicians and scientists would take sides along national lines: British scholars tended to support Newton's claims, and those of Germany and central Europe backed Leibniz. Even today, European mathematicians emphasize the form of calculus developed by Leibniz. (As noted below, Leibniz's calculus emphasizes integrals, and Newton's derivatives.)

During the eighteenth century, even as mathematicians debated the relative contributions of Newton and Leibniz—and some even questioned the validity of calculus as a mathematical discipline—calculus gained wide application in physics, astronomy, and engineering. This fact is illustrated by the work of the Bernoulli family. Influenced by Leibniz, a personal friend, Jakob Bernoulli (1645–1705) applied calculus to probability and statistics, and ultimately to physics, while his brother Johann (1667–1748) appears to have supplied much of the material in the first calculus textbook, by French mathematician Guillaume de L'Hôpital (1661–1704). Johann's sons all made contributions to calculus. Most notable among them was Daniel (1700–1782), who applied it in developing the principles of fluid dynamics, an area of physics that later made possible the invention of the airplane. In addition, the Bernoulli brothers' distinguished pupil, Leonhard Euler (1707–1783), used calculus in a number of ways that would gain application in physical studies of energy, wave motion, and other properties and phenomena.

The first-year college student attempting to choose between the study of statistics and calculus, then, has a tough choice. Each has merits to recommend it, as the essays that follow show. Statistics may seem more rooted in everyday life, while calculus may appear to draw the student into a lofty conversation with great mathematical minds; yet there is plenty of practical application for calculus, and statistics certainly brings the student into contact with great thinkers and great ideas. —JUDSON KNIGHT

Viewpoint:

Yes, most first-year college students would be better off taking a statistics course rather than calculus because statistics offers a greater variety of practical applications for a non-scientific career path.

In today's highly competitive world, it can easily be said that a college education should prepare a student for the world of work. Certainly most students enter college today with the expectation of getting a job after graduation. They often express the desire to be well-rounded and marketable. Therefore, they need to be practical when deciding what courses to take. With the exception of math majors, most first-year college students would be far better off taking a statistics course rather than a calculus course. Why? The answer is simple: statistics are part of everyday life. A student who understands statistics has an advantage over those who do not. Understanding statistics can be useful in college, employment, and in life.

The Scholastic Advantage Taking a statistics course encourages a student to flex his or her intellectual muscles, but at the same time, is more practical than a calculus course. This is especially true for a first-year college student who may be uncertain about his or her career path. If someone plans to be a math major, then taking calculus makes perfect sense, but for someone pursuing another career, taking calculus in the first year of college could be a waste of time and money. Not all professionals will need a calculus background, but most professionals will need at least some knowledge of statistics. In fact, studying statistics helps first-year college students hone their critical and analytical thinking skills—skills they can draw upon throughout their college career regardless of what major they choose.

In the *Columbia Electronic Encyclopedia,* statistics is defined as "the science of collecting and classifying a group of facts according to their relative number and determining certain values that represent characteristics of the group." With that definition in mind, it's easy to see how understanding statistics would help a first-year college student. The main aim of most college-level introductory statistics courses is to get the student comfortable with such principles as statistical modeling, confidence intervals, and hypothesis testing, concepts that will help students become better consumers of information.

Most college students are required to write at least one paper in their first year, which often involves going to the library and combing through stacks of research material. Understanding statistical measures, such as mean and standard deviation, can certainly help them make sense of what they find. Students who truly understand what the numbers represent will not just be regurgitating facts and figures reported by others; instead, they will be able to come to their own conclusions based on the data presented. In essence, they will have the edge, because they will be better able to identify the strengths and weaknesses of another's research with the skill set that knowledge in statistics provides. An added bonus is that students who are able to think clinically and delve deeper into a particular subject are often rewarded with better grades.

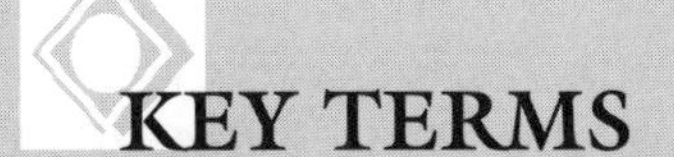

KEY TERMS

CALCULUS: A branch of mathematics that deals with rates of change and motion.

DIFFERENTIAL CALCULUS: A branch of calculus concerned with finding the instantaneous rate at which one quantity, called a derivative, changes with respect to another.

INTEGRAL CALCULUS: A branch of calculus that deals with integrals, which identify the position of a changing object when its rate of change is known. An integral is the inverse of a derivative.

MEAN: An average value for a group of numerical observations.

STANDARD DEVIATION: A measure of how much the individual observations differ from the mean.

STATISTICS: A branch of mathematics concerned with the collection and analysis of numerical data.

The Employment Advantage Even before graduation many first-year college students find themselves looking for work to help pay for college expenses. A statistics course, coupled with one in business math, is going to be far more helpful in most job situations than calculus. Many college students get a job in a retail setting, for example, where calculus has little value, but being able to understand and interpret a survey could prove helpful to them not only as employees, but as consumers, too.

An article by Margaret Menzin and Robert Goldman for the Association for Women in Mathematics states that the United States Department of Labor forecasts a faster rate of job growth in the area of statistics than in computer science. Statisticians, they remind us, are utilized in a variety of fields from archeology to zoology. For example, statisticians are employed by many finance and insurance companies throughout the United States. Having a knowledge of statistics, whether a person pursues a career in statistics or not, will enhance a person's marketability and will most likely help him or her perform better on the job.

The Life Advantage Indeed, understanding statistics leads to a better understanding of the world around us. Certainly, if one wants to be an informed consumer, the study of statistics is valuable. Here is how a working knowledge of statistics can help. Say, for example, that someone hears on a television advertisement that four out of five doctors polled preferred XYZ medicine for tooth pain. The statement sounds impressive, but it does not mean that most doctors prefer that particular brand of pain reliever or that someone

should choose it over another brand. How many doctors were actually polled? The statement does not mean very much if only five doctors were polled. Even if 1,000 doctors were polled, what types of doctors were they? Their specialty could influence their response. Knowing the particulars about the sample size and type are also important. Were they dentists, orthodontists, anesthesiologists, or a combination of these and other specialists? Asking questions about sampling is just one way to analyze a statistic. There are many other ways as well. A consumer-savvy person armed with some knowledge about statistics is less likely to be misled by clever advertising. After all, the marketing campaign might influence someone who does not know statistics to buy the product, but someone who knows statistics will know to ask the right questions.

Also, understanding statistical procedures can help students understand why certain tasks are done a certain way, which in some cases will help reduce stress. Understanding probability, for example, helps someone determine the likelihood that a certain event will occur. In fact, there are several everyday situations that will seem much less stressful and much more logical if someone understands statistical principles such as the queuing theory, for instance. Menzin and Goldman define the queuing theory as a theory that predicts what happens as people or things get in line. Queuing theory can be used to predict how traffic bunches up in toll booths or how long your report will take to print out of a shared printer. Say, for example, that a student is attempting to register for classes. When he or she gets to the registration hall and sees one long line that feeds into several windows, his or her first reaction might be to get angry and say, "Why don't they form several lines?" But someone who understands the queuing theory will realize that one line that feeds to several windows is actually more efficient and will reduce the wait time for everyone.

Conclusion Clearly, a first-year college student can benefit greatly from taking courses that have practical applications. Statistics is certainly one of those courses and can be beneficial to all students, not just math majors. Knowledge of statistics allows a student to intelligently question information he or she receives in the classroom, on the job, and in life. —LEE A. PARADISE

Viewpoint:

No, first-year college students would not be better off taking a statistics course rather than calculus because calculus is wider in application and teaches an intellectual rigor that students will find useful throughout their lives.

In a November 28, 2001, *New York Times* piece, Richard Rothstein, a research associate of the Economic Policy Institute, made a compelling case for a renewed emphasis on statistics—as opposed to calculus—in the classroom. Though he was discussing mathematics teaching in high school rather than college, Rothstein's argument could as easily be applied to college education. Among the reasons he offered for teaching statistics rather than calculus was that few jobs make use of the latter, whereas statistics will aid students in understanding much of the data that bombards them from the news media.

While Rothstein's points are arguable, he makes an undeniably compelling case; however, his very emphasis on practicality undermines the assertion that statistics education is more important than calculus. On one level at least, it is precisely because calculus is not obviously a part of practical, everyday knowledge that it should be taught. Whereas one is assaulted with statistics at every turn—in newspapers and magazines, on television, and in public debates over issues such as health care or social security—calculus seems a more esoteric pursuit reserved for scientists and mathematicians. And this, ironically, makes it all the more appropriate for a young college student preparing to embark on a four-year education.

To some extent, teaching a first-year college student statistics is like teaching him or her to listen to popular music; the subject hardly needs to be emphasized, because the ambient media culture already does the job so well. Admittedly, there is considerable merit in studying statistics, not least because such study allows the individual to see through the sometimes deceptive claims made under the veil of supposedly unassailable statistical data. This deceptiveness is encapsulated in President Harry S. Truman's famous declaration that there are three kinds of untruth: "lies, damn lies, and statistics."

But while statistics will have practical application in the student's later life, calculus offers a degree of intellectual rigor that will equip the student to think about all sorts of issues in much greater depth. Teaching a first-year college student calculus is like teaching him or her the Greek classics, or logic, or the appreciation of Mozart and Beethoven—riches of the mind that the student might not seek out on his or her own, but which he or she will treasure for life.

Calculus in Brief Of all the subjects a student is likely to encounter, mathematical or otherwise, calculus is among the most intimidating.

College students studying in a math lab.
(© Bob Rowan; Progressive Image/Corbis. Reproduced by permission.)

Only the most scholastically advanced high-school students ever take calculus, and then only as seniors, after four years of study that include algebra, geometry, algebra II, and trigonometry. It might be comforting to learn that calculus is really very easy to understand—it *might* be, that is, if such were the case. In fact calculus is extremely challenging, much more so than statistics, and the student who emerges from a course in it is like an athlete who has undergone an extraordinarily rigorous form of physical training.

The branch of mathematics that deals with rates of change and motion, calculus is divided into two parts, differential calculus and integral calculus. Differential calculus is concerned with finding the instantaneous rate at which one quantity, called a derivative, changes with respect to another. Integral calculus deals with integrals, the inverse of derivatives.

What does all this mean? Suppose a car is going around a curve. Differential calculus would be used to determine its rate of change (that is, its speed) at any given point, but if its speed were already known, then integral calculus could be used to show its position at any given time. Imagine any object following a curved path: this is the territory of calculus, which measures very small changes.

Applications of Calculus Despite its apparently esoteric quality, the applications of calculus are myriad. Physicists working in the area of rocket science use it to predict the trajectory of a projectile when fired, while nuclear scientists apply calculus in designing particle accelerators and in studying the behavior of electrons. Calculus has also been used to test scientific theories regarding the origins of the universe and the formation of tornadoes and hurricanes.

Indeed, calculus is the language of science and engineering, a means whereby laws of the natural world can be expressed in mathematical form. It makes possible the analysis and evaluation of those physical laws, predictions regarding the behavior of systems subject to those principles, and even the discovery of new laws. Yet the importance of calculus extends far beyond the realm of science, engineering, and mathematics.

Though electricians apply scientific knowledge, their work is obviously much more practical and everyday in its emphasis than that of scientists, yet as Martha Groves reported in the *Los Angeles Times,* electrical workers' unions require apprentices to study calculus. Businesses also use calculus widely, to increase efficiency by maximizing production while minimizing costs, and for other purposes.

Calculus, Change, and Infinitesimal Matter

A journal of the paint manufacturing industry might seem an odd place in which to make a case for the study of calculus, but that is pre-

cisely what Joseph B. Scheller of Silberline Manufacturing Company of Tamaqua, Pennsylvania, did in a 1993 editorial for *American Paint & Coatings Journal.* Students, he maintained, should learn about the concept of change, an essential aspect of life, and particularly modern life. "What do I think we should teach young children?" he wrote near the conclusion of his editorial. "I think they should learn the concepts of physics and calculus . . . because they tell us about ourselves and the world in which we live."

The entire universe is characterized by constant movement—change in position over time—at every conceivable level. Galaxies are moving away from one another at a breathtaking rate, while our own planet spins on its axis, revolves around the Sun, and participates in the larger movements of our Solar System and galaxy. At the opposite end of the spectrum in terms of size, all matter is moving at the molecular, atomic, and subatomic levels.

Molecules and atoms move with respect to one another, while from the frame of reference of the atom's interior, electrons are moving at incredible rates of speed around the nucleus. Literally *everything* is moving and thereby generating heat; if it were not, it would be completely frozen. Hence the temperature of absolute zero (-459.67°F or -273.15°C) is defined as the point at which all molecular and atomic motion ceases—yet precisely because the cessation of that motion is impossible, absolute zero is impossible to reach.

What does this have to do with calculus? Two important things. First of all, calculus is all about movement and change—not merely movement through space, but movement through time, as measured by rates of change. Secondly, calculus addresses movement at almost inconceivably small intervals, and by opening up scientists' minds to the idea that space can be divided into extremely small parts, it paved the way for understanding the atomic structure of matter. At the same time, it is perhaps no accident that calculus—with its awareness of things too small to be seen with the naked eye—came into being around the same time that the microscope opened biologists to the world of microbes.

Joining a Great Conversation To study calculus is to join a great conversation involving some of the world's finest minds—a conversation that goes back to the fifth century B.C. and continues today. In his famous paradoxes, the Greek philosopher Zeno of Elea (c. 495–c. 430 B.C.) was perhaps the first to envision the idea of subdividing space into ever smaller increments. A century later, Eudoxus of Cnidus (c. 390–c. 342 B.C.) developed the so-called method of exhaustion as a means, for instance, of squaring the circle—that is, finding a way of creating a circle with the same area as a given square. Archimedes (287–212 B.C.) applied the method of exhaustion to squaring the circle by inscribing in a circle a number of boxes that approached the infinite.

The Greeks, however, were extremely uncomfortable with the ideas of infinite numbers and spaces on the one hand, or of infinitesimal ones on the other. Thus calculus did not make its appearance until the late seventeenth century, when Gottfried Wilhelm von Leibniz (1646–1716) and Sir Isaac Newton (1642–1727) developed it independently. There followed a bitter debate as to which man should properly be called the father of calculus, though in fact each staked out somewhat separate territories.

In general, Newton's work emphasized differentials and Leibniz's integrals, differences that resulted from each man's purposes and interests. As a physicist, Newton required calculus to help him explain the rates of change experienced by an accelerating object. Leibniz, on the other hand, was a philosopher, and therefore his motivations are much harder to explain scientifically. To put it in greatly simplified terms, Leibniz believed that the idea of space between objects is actually an illusion, and therefore he set out to show that physical space is actually filled with an infinite set of infinitely small objects.

Challenging the Mind Mathematicians did not immediately accept the field of calculus. Critics such as the Irish philosopher George Berkeley (1685–1753) claimed that the discipline was based on fundamental errors. Interestingly, critiques by Berkeley and others, which forced mathematicians to develop a firmer foundation for calculus, actually strengthened it as a field of study.

As calculus gained in strength over the following centuries, it would prove to be of immeasurable significance in a vast array of discoveries concerning the nature of matter, energy, motion, and the universe. In this sense, too, the study of calculus invites the student into a great conversation involving the most brilliant minds the world has produced. Admission to that conversation has nothing to do with gender, race, religion, or national origin; all that is required is a willingness to surrender old ways of thinking for newer, more subtle ones.

Calculus is a challenging subject, but it is an equal-opportunity challenger. For this reason, as reported in *Black Issues in Higher Education,* the College Board (which administers the SAT, or Scholastic Achievement Test) announced in 2002 that it would revise the test to include greater emphasis on calculus and algebra II. These subjects would be emphasized in place of vocabulary and analogy questions, which some critics have maintained are racially biased.

At the same time, as noted by Sid Kolpas in a review of Susan L. Ganter's *Calculus Renewal* for *Mathematics Teacher,* calculus teaching itself is entering a new and exciting phase. "Spurred by the mathematical needs of other subject areas, dramatic advances in cognitive science and brain research, and rapidly changing technology and workplace demands," Kolpas wrote, "calculus renewal seeks a transformation of the intimidating, traditionally taught calculus course. It seeks a brave new calculus with more real-world applications, hands-on activities, cooperative learning, discovery, writing, appropriate use of technological tools, and higher expectations for concept mastery." In these words, one glimpses a hint of the great intellectual challenge that awaits the student of calculus. Certainly there is nothing wrong with a student undergoing a course in statistics, but it is hard to imagine how that course could awaken the mind to the degree that calculus does. —JUDSON KNIGHT

Further Reading

Berlinski, David. *A Tour of Calculus.* New York: Pantheon Books, 1995.

"College Board Proposes SAT Overhaul." *Black Issues in Higher Education* 19, no. 4: 17.

The Columbia Electronic Encyclopedia. "Statistics" [cited July 22, 2002]. <http://www .inforplease.com/ce6/sci/A0846567.html>.

Ganter, Susan L., ed. *Calculus Renewal: Issues for Undergraduate Mathematics Education in the Next Decade.* New York: Kluwer Academic/Plenum Publishers, 2000.

Groves, Martha. "More Rigorous High School Study Urged: Panel Finds Graduating Seniors Are Often Ill-Prepared for College." *Los Angeles Times* (October 5, 2001): A-27.

Kolpas, Sid. "Calculus Renewal: Issues for Undergraduate Mathematics Education in the Next Decade." *Mathematics Teacher* 95, no. 1: 72.

Menzin, Margaret, and Robert Goldman. "Careers in Mathematics" [cited July 22, 2002]. <http://math.usask.ca/document/netinfo/careers.html>.

Rothstein, Richard. "Statistics, a Tool for Life, Is Getting Short Shrift." *New York Times* (November 28, 2001): D-10.

Scheller, Joseph B. "A World of Change: There's Nothing Static About Modern Life, and Education Should Deliver That Message." *American Paint & Coatings Journal* 78, no. 21 (October 29, 1993): 20–21.

Should mathematics be pursued for its own sake, not for its social utility?

Viewpoint: Yes, mathematics should be pursued for its own sake, as doing so will eventually present practical benefits to society.

Viewpoint: No, mathematics should first be pursued by everyone for its societal utility, with the mathematically gifted being encouraged to continue the pursuit of math for math's sake at advanced levels.

One of the great problems in mathematics is not a "math problem" at all; rather, it is a question of the means by which the study of mathematics should be pursued. This may be characterized as a debate between "pure" mathematics, or mathematics for its own sake, and applied mathematics, the use of mathematics for a specific purpose, as in business or engineering. This is a concern that goes back to the historical beginnings of mathematical inquiry.

The earliest mathematicians, those of Mesopotamia, were concerned almost entirely with what would today be called applied mathematics. The Sumerians, for instance, developed the first mathematical systems not for the purposes of abstract inquiry, but rather to serve extremely practical needs. They were businesspeople, and their need for mathematical techniques was much the same as that of a modern-day corporation: merchants needed to keep track of their funds, their inventory, their debts, and the monies owed to them.

Today, accountants and mathematicians employed in business use concepts unknown to the Sumerians, utilizing mathematics for such esoteric pursuits as calculating risks and benefits or optimizing delivery schedules. The techniques involved originated from abstract studies in mathematics, meaning that the thinkers who developed them usually were not driven by immediate needs, but today the application of those methods is decidedly down-to-earth.

The other great mathematicians of ancient Mesopotamia were the Babylonians, who likewise used numbers for highly practical purposes: the study of the heavens and the development of calendars. Astronomy arose from ancient astrology, which first came into mature form in Babylonia during the second and first millennia before Christ. Today, astrology hardly seems like a "practical pursuit," but the situation was quite different in ancient times, when rulers relied on astrological charts the way modern politicians consult polls and focus-group results.

To a high-school student confronted with the vagaries of trigonometry, the discipline might seem quite abstract indeed. Trigonometry analyzes the relationships between the various sides of a right triangle, as well as the properties of points on a circle, and requires the student to understand concepts and values such as sine, cosine, tangent, and so forth. Yet trigonometry did not originate from the imagination of some abstracted mathematical genius; rather, it had its origins in the astrological and astronomical studies of the ancient Egyptians and Babylonians.

For centuries, in fact, trigonometry had the status not of a mathematical discipline, but of an aid to astronomy. Only in the late Middle Ages, on the brink of the voyages of discovery that would put trigonometry to use in navigation, would it be fully separated, as a mathematical discipline, from its many applications. This separation, however, began nearly 2,000 years earlier, and is one of many legacies for which the world of mathematics is indebted to the ancient Greeks.

Despite the advances made previously by the Babylonians, and to a lesser extent the Sumerians and Egyptians, the Greeks are almost universally recognized as the originators of mathematics as a formal discipline. This is true even when one takes into account the advances made by the Chinese, of which the Greeks were of course unaware, because Chinese mathematicians were likewise primarily concerned with the application of mathematics to practical, everyday problems. Only the Hindu mathematicians of early medieval India, who developed the number system in use today (including zero, a number unknown to the Greeks and Romans), may be regarded as mathematical innovators on anything like a par with the Greeks.

What was different about the Greek mathematicians? Their interest in mathematicians for its own sake. Whereas the Sumerians had envisioned numbers in purely concrete terms, so much so that they had no concept of 5, for instance (only "five goats" or "five fingers" and so on), the Greeks recognized the abstract properties of numbers. Likewise they pursued geometry, not for the many practical applications it would eventually yield, but rather as a great puzzle or game in which the attempt to find solutions was its own reward.

Greek mathematicians became fascinated with problems such as the squaring of the circle (finding a circle with an area equal to that of a given square using only a compass and straight edge) or determining prime numbers. Many of these pursuits might have seemed, to an outside observer, like so much useless theoretical speculation, but in the centuries to come, they would yield a vast wealth of mathematical knowledge that find application in everything from architecture and engineering to computers and rocket science.

The two essays that follow examine the question of applied vs. pure mathematics, not as a dilemma that has only one possible answer, but as a question that offers several answers of differing relative worth. Thus, for the defender of pure mathematics, it is implicit that the pursuit of math for math's sake will eventually present practical benefits to society. Likewise, for the partisans of applied mathematics, there is no censure attached to pure mathematics; rather, it is a question of which pursuit is most appropriate for the largest number of students. —JUDSON KNIGHT

Viewpoint:

Yes, mathematics should be pursued for its own sake, as doing so will eventually present practical benefits to society.

Social utility may be defined as whatever contributes toward making life easier, safer, longer, more comfortable, harmonious, invigorating, fulfilling, and enjoyable. Every human endeavor ought to promote social utility somehow or to some degree, but not necessarily in a direct or primary way.

The Nature and Role of Mathematics Mathematics promotes social utility by enabling scientific advances that may ultimately result in new products, services, systems, or insights that make life better. The old slogan of E. I. du Pont de Nemours was "Better Things for Better Living Through Chemistry." The company that created nylon, Teflon, Lucite, and Freon could easily say that, but it must also be remembered that chemistry depends on mathematics. So does physics, biology, technology, and even medicine. The "better things" constitute scientific utility, or the products of applied science, and the "better living" is the social utility that emerges from applied science and ultimately from pure science, including mathematics.

Since the late seventeenth century, statistics, algebra, and calculus have all been components of epidemiology, especially since the pioneering theory of deterministic epidemiological models by William H. Hamer in 1906 and Sir Ronald Ross in 1911 began to make biometrics a rigorous mathematical science. The subsequent mathematical understanding of epidemics has saved countless lives. Hamer and Ross were deliberately working toward practical medical goals, but Sir Isaac Newton (1642–1727) and Gottfried Wilhelm Freiherr von Leibniz (1646–1716), who invented the calculus that biometricians now use, were each working on purely theoretical levels. The practical applications of their mathematical discoveries came later.

Mathematics is the epitome of theoretical thought, even more so than philosophy, which is grounded in some kind of perceived reality. Pure mathematics is the epitome of apparently impractical thought, even more so than speculative philosophy, which typically has an object in view. Mathematicians are interested in determining what fits, or what coheres to make each

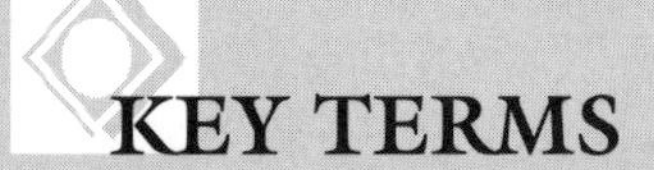

KEY TERMS

ABACUS: An ancient counting instrument characterized by sliding counting beads on rods.

ACADEMIC EDUCATION: Study and teaching intended to widen and deepen the students' grasp of concepts, without regard for the practical value of the curriculum. *Library of Congress Subject Headings* calls it "critical thinking."

APPLIED (OR PRACTICAL) MATHEMATICS: Mathematics as put to practical use in the physical world, to solve either specific problems *ad hoc* or classes of recurring problems. These problems may be very simple, such as measuring household gardens, or very complex, such as programming missile guidance systems.

POIÊSIS: Aristotle's technical philosophical term meaning "production" or "creativity," i.e., theory as applied to practice, or the union of *theôria* and *praxis*. The English word "poetry" derives from it.

PRACTICAL (OR APPLIED) SCIENCE: Any area of knowledge or research pursued specifically and solely for the sake of its immediate or obvious practical applications.

PRAXIS: Standard philosophical term from the ancient Greek meaning "practice" or "human action." It exploits *theôria* and its adherents live the "active life." Marx valued it above *theôria*.

PURE (OR THEORETICAL) MATHEMATICS: Mathematics with no apparent practical purpose.

SLIDE RULE: A calculating instrument consisting of a ruler with an inserted sliding rule in the middle, and the rules marked with numbers in a logarithmic scale.

THEORETICAL (OR PURE) SCIENCE: Any area of knowledge or research pursued "for its own sake," or without any immediate practical application in view.

THEÔRIA: Standard philosophical term from the ancient Greek meaning "theory" or "human thought." It scrutinizes *praxis* and its adherents live the "contemplative life." Most non-Marxist and non-Utilitarian philosophers value it above *praxis*.

VOCATIONAL EDUCATION: Study and teaching intended to give students a "trade," without regard for the students' wider or deeper grasp of concepts outside the necessary skills of that trade. *Library of Congress Subject Headings* calls it "occupational training."

whole theoretical system work. Other thinkers may discover how to apply these coherent deductive systems, such as calculus, trigonometry, Lobachevskian geometry, set theory, etc., to everyday problems in the real world. If mathematicians had to consider the practical ends of their thought as well as its theoretical coherence, very few of them would have the intellectual talent, resources, or stamina to achieve any worthwhile results, either practical or theoretical. Society is better served by dividing the intellectual labor between the theoreticians and the practical thinkers.

The ancient Babylonians and Egyptians developed the theoretical science of trigonometry out of astronomy, to explain quantitatively their observations of the movements of stars and planets. Many centuries later the ancient Greeks found practical applications for trigonometry in surveying, military tactics, and navigation. Similarly, elaborate theoretical developments regarding Hilbert space, topology, probability, statistics, game theory, model theory, computer science, and many other areas of mathematical inquiry may not yet have realized their potential in the practical business of improving society, but that does not make them any less valuable, either as mental exercises or as cultural products.

Paradoxically, pursuing mathematics as a whole for its own sake promotes social utility more reliably than pursuing it specifically for the sake of either social utility or any other ulterior purpose. However, since utility is more likely to be served by unrestricted mathematical investigations than by mathematical projects designed with specific, practical ends in view, even the most theoretically inclined mathematicians should remain aware that their results may ultimately have practical uses. By thinking "outside the box," as it were, these mathematicians actually aim at such practical uses, but not directly.

Ancient Greek philosophy was not distinct from the natural sciences and included mathematics. The philosopher Thales (ca. 624–ca. 547 B.C.) studied mostly geometry, astronomy, navigation, meteorology, and ontology. Once he inferred from his meteorological calculations that the next year's olive crop was likely to be abundant. Therefore he bought as many shares as he could afford in all the olive presses in the area. When his prediction proved correct, he made a fortune by selling his shares at whatever price he named. The point is that Thales did not study mathematics and science in order to become rich, but when a practical application of his self-acquired knowledge presented itself, he saw his opportunity and took it. This story, whether true or not, illustrates that to pursue mathematics and science only for their utility limits their scope and hinders mathematicians and scientists, but to pursue them for their own sake does not preclude finding practical uses for them.

Theoretical results usually have broader and more varied application than practical results. Practical applications of theoretical results are sometimes recognized only much later, and sometimes even by accident or serendipity. Results in pure mathematics may sit idle long after their discovery until someone finds practical uses for them. Such was the case with the equations for determining plane functions from line integrals. In 1917 Austrian mathematician Johann Radon discovered these equations, which turned out to be necessary for computerized axial tomography (CAT) scans, but South African physicist Allan Macleod Cormack only learned of them in 1972, nine years after he had independently solved that problem and successfully applied it to computerized composite imaging. Cormack's work at first almost completely escaped the notice of physicians. Not until another physicist, Godfrey Newbold Hounsfield, unaware of both Radon and Cormack, led a team of British radiologists in the construction and installation of the first clinical CAT scanner in 1971 did the medical community show interest. CAT scans have proved to be the greatest single advance in radiology since x rays. Thus, not from ingratitude but from ignorance, Hounsfield failed to acknowledge either Radon or Cormack in the Nobel Prize–winning article which introduced CAT scans to the world. The Nobel Committee, however, allowed Cormack to share the prize with him.

Theoretical researchers sometimes serve practical or social purposes despite themselves. In 1994, mathematicians Julius L. Shaneson of the University of Pennsylvania and Sylvain E. Cappell of New York University, building on the earlier work of University of Cambridge mathematician Louis Joel Mordell, published a result in theoretical algebra that they believed had no practical value. They were not seeking any practical value when they were working on the problem. Nevertheless, economists recognized almost instantly the importance of the Mordell-Shaneson-Cappell method for certain kinds of business problems and began using it successfully.

John Forbes Nash's results in game theory, especially his discovery of mutual gain for all players in non-cooperative games, first published as "Equilibrium Points in N-Person Games" in the 1950 *Proceedings of the National Academy of Sciences*, were later applied to practical economics in ways that he had not foreseen, yet for this work in pure mathematics he shared the 1994 Nobel Prize in economic sciences.

Philosophical Underpinnings The question of the relative importance of theoretical interest versus practical value is at least as old as ancient Greek philosophy and can be applied to any area of knowledge or research. Aristotle (384–322 B.C.) distinguished between pure and applied mathematics, claiming that the practical was justified by the theoretical, not by its relevance to the physical world. He also systematically discussed *theôria* ("human thought"), *praxis* ("human action"), and *poiêsis* ("human creativity or production"). Logically, but not always chronologically, *theôria* precedes *praxis. Poiêsis* emerges from the cooperation of the two.

For Aristotle, *poiêsis* not only opposes *praxis*, but also enlists *theôria* to supersede *praxis* and create new products. *Praxis* and *theôria* are each primary and irreducible, and although *poiêsis* is secondary to both *praxis* and *theôria*, ultimately both *poiêsis* and *praxis* depend on *theôria* for their impetus and advancement.

Karl Marx (1818–1883) had little use for pure theory, declaring in his eleventh thesis on Feuerbach that the point of philosophy is not to understand the world, but to change it, and implying in the eighth of these theses that *theôria* is misled by the kind of thinking that has no practical aim or design, and can be rescued from its airy wandering only by *praxis*, which essentially is the same as social utility or common life. For Marx, the purpose of *theôria* is to understand and assist *praxis*, not to ruminate on vague, impractical mysteries such as God, pure science, or *theôria* itself. His attitude directly contrasts Aristotle's, for whom the highest form of thinking is contemplative thought about nothing but itself.

Martin Heidegger (1889–1976) distinguished between "meditative thinking" (*besinnliches Denken*) and "calculative thinking" (*rechnendes Denken*). This distinction is similar to Aristotle's *theôria* and *praxis*, except that, for Heidegger, the gulf is wider and the extremes are more sharply defined. While Aristotelian *theôria* may eventually have a practical end, Heideggerian meditative thinking has none, and would be compromised and denigrated if any practical use were ever found for it. In this regard, Heidegger resembles the Taoists.

The subtle power of pure thinking has long been known in Chinese wisdom. Lao-Tse (604?–531? B.C.), the legendary founder of Taoism, called it *wu wei*, "non-action." He wrote in the *Tao-te Ching*, Chapter 37, "The Tao never takes any action, yet nothing is ever left undone." The relation between Taoism and the other great Chinese religious system, Confucianism, is analogous to that between *theôria* and *praxis*, the former centered in individual contemplation, intellectual advancement, and mental tranquillity, the latter on social action, political organization, and worldly progress.

Heidegger and Taoism (pro-*theôria*) and Marxism and Confucianism (pro-*praxis*) represent the two extremes. Aristotelianism takes a middle ground by regarding *theôria* and *praxis* as opposite and equally essential elements of human life, which together find their greatest manifestation as *poiêsis*, the practical yet creative production of the substance of human culture and civilization.

Serving Practical Needs The Radon/Cormack/Hounsfield case illustrates the need for better communication among pure mathematicians and those scientists who may be seeking practical applications of mathematical discoveries. Corollary to that would be a subsequent need for better cooperation between academic and vocational educators, to emphasize the *poiêsis*, or productive power, generated by reciprocity between *theôria* and *praxis.* But this is not to say that pure mathematicians should not remain free to let their highest thoughts proceed wherever they will, without the impediment of practical considerations.

Applied mathematics is taught not only in secondary schools on non-college-preparatory tracks, but also at the Massachusetts Institute of Technology (MIT), where it constitutes a separate department. In 1959 at the University of Cambridge, England, George Batchelor founded the Department of Applied Mathematics and Theoretical Physics. At first blush, such a name might seem incongruous, but not when we consider that it does not refer to "applied mathematics" in the usual sense. Rather, it means mathematics applied to theoretical physics, that is, mathematics as the basis of theoretical physics.

Immanuel Kant (1724–1804) famously wrote in the *Critique of Pure Reason*, "Percepts without concepts are empty; concepts without percepts are blind." By this he meant that merely being in the world, experiencing, perceiving, working, and living day-to-day is meaningless if we do not think about, reflect upon, or criticize what we experience; and by the same token that merely thinking, ruminating, and contemplating is misguided if we do not also somehow act in the world, or put our thoughts to use.

Drawing the lines between concepts and percepts, theory and practice, *theôria* and *praxis*, theoretical and practical science, pure and applied mathematics, or academic and vocational education, is difficult. The lines are gray. In general, though, we can say that concepts, theory, *theôria*, theoretical science, pure mathematics, and academic education are characterized by being open-ended and making progress in the sense of achieving a Kantian overarching reason (*Vernunft*), while, on the other hand, percepts, practice, *praxis*, practical science, applied mathematics, and vocational education are characterized by not progressing beyond the narrow focus of the solution to each particular defined problem, and thus achieving only a Kantian understanding (*Verstand*) of circumstances. But the union or harmony of all these dichotomous factors would be *poiêsis*, which is indeed a social good.

To concentrate primarily on present particular problems instead of the broader theoretical elements of mathematics prevents the development and accumulation of the very knowledge that will be needed to solve future particular problems. The attitude of those who prefer to focus narrowly on *ad hoc* puzzles rather than ponder the general truths or overarching unity of mathematics recalls the closing lines of "Ode on a Distant Prospect of Eton College" by Thomas Gray: "Thought would destroy their paradise. / No more; — where ignorance is bliss / 'Tis folly to be wise." —ERIC V.D. LUFT

Viewpoint:

No, mathematics should first be pursued by everyone for its societal utility, with the mathematically gifted being encouraged to continue the pursuit of math for math's sake at advanced levels.

There is widespread agreement that the pursuit of mathematics starts at an early age. Mathematics is a part of the curriculum from kindergarten through grade 12 in essentially all public education today. The Executive Summary in *Before It's Too Late: A Report to the Nation from the National Commission on Mathematics and Science Teaching for the 21st Century* identifies "enduring reasons" for children to achieve competency in mathematics. Among them are the widespread workplace demands for mathematics-related knowledge and abilities; the need for mathematics in everyday decision-making; and for what the authors describe as "the deeper, intrinsic value of mathematical knowledge that shapes and defines our common life, history, and culture." Except for the last and quite esoteric reason, the Commission is squarely on the side of studying mathematics for its practical utility in society.

Disputes abound among educators when it comes to mathematics. Although educators generally agree on the importance of pursuing mathematics, and that it should be promoted for its utility, there are heated differences on how it should be presented, and how the results should be measured. The extent of the use of computers and calculators is another point of contention among some educators.

There are many theories on how humans learn. One of the most influential in education, Jean Piaget's, provides some clues on how students process abstract concepts at various stages of development. According to students of Piaget's theory, the percentage of the population that will ever reach the level of learning needed to pursue math for math's sake is not large. Academic enrollments at high-tech universities such as MIT support that theory.

Why Teach/Learn Math? Before going further in this discussion, it must be pointed out that teaching anything does not guarantee that learning will result. That can be a problem. There is no doubt that both teaching and learning mathematics is important to the future of our children, our economy, our progress in technology, and even our success on the global scene. Just how to accomplish this is the big question.

In the forward to the *Before It's too Late* report, John Glenn, Commission Chairman, says, "at the daybreak of this new century and millennium, the Commission is convinced that the future well-being of our nation and people depends on just how well we educate our children generally, but on how well we educate them in mathematics and science specifically." Since each and every science has quantitative components, that makes the teaching, and learning, of mathematics even more important.

Glenn continues, "From mathematics and the sciences will come the products, services, standards of living, and economic and military security that will sustain us at home and around the world. From them will come the technological creativity American companies need to compete effectively in the global marketplace."

The Commission report expresses strong concern that in the United States, the education systems as a whole are not teaching children to understand and use the ideas of science and mathematics, with the least success coming at the higher grades. Studying mathematics for math's sake is most likely to begin at the higher grades when students are beginning to follow their interests and aptitudes. Unfortunately, the number of students going on to major in math make up a very small percentage of those pursuing higher education.

How to Teach Math: The Math "Wars" The debate concerning how to teach mathematics has been described as a "war" by some, among them David Ross, a mathematician at Kodak Research Labs. The National Council of Teachers of Mathematics (NCTM), a nonprofit, nonpartisan education association with 100,000 members dedicated to improving mathematics teaching and learning, takes the position that every student will enter a world that is so different from the one that existed a few years ago that modifications in school mathematics programs are essential. They point out that every student will assume the responsibilities of citizenship in a much more quantitatively driven society, and that every student has the right to be mathematically prepared. They believe that quality mathematics education is not just for those who want to study mathematics and science in college—it is for everyone.

Jean Piaget

(© Bettmann/Corbis. Reproduced by permission.)

NCTM states that students must learn basic computation facts and know how to compute, and that practice is important, but practice without understanding is a waste of time. Although the opponents of the teaching techniques promoted by NCTM do not see much merit in what they are doing, NCTM does include basic skill mastery in their goals. NCTM just takes a more flexible approach to teaching mathematical concepts. NCTM reasons that the old method did not work very well and did little to promote students' problem-solving skills. Citing from the NCTM *Standards 2000*:

"Many adults are quick to admit that they are unable to remember much of the mathematics they learned in school. In their schooling, mathematics was presented primarily as a collection of facts and skills to be memorized. The fact that a student was able to provide correct answers shortly after a topic was generally taken as evidence of understanding. Students' ability to provide correct answers is not always an indicator of a high level of conceptual understanding."

The NCTM emphasizes giving the teacher freedom to determine how topics should be approached in any given lesson with an emphasis on problem-solving skills and real-life applications. This all sounds very idealistic. Opponents call it "fuzzy math" and take a firm stand

A student uses a calculator to solve math problems.
(© Bob Rowan, Progressive Image/Corbis. Reproduced by permission.)

that students have to be well grounded in fundamentals. They do not see all the old rote learning as bad. Opponents to the NCTM approach definitely stand by the old saying that practice makes perfect.

The Place for Calculators The dispute on how best to achieve math literacy includes debates on how, when, and where calculators and computers should be used. The NCTM points out that students live in a fast-paced world of TV and video games. They describe today's students as "Internet savvy, smoothly juggling multiple images and inputs" in one of their "Setting the Record Straight about Changes in Mathematics Education" bulletins. They describe calculators as one of many tools to enhance students' learning. NCTM believes that calculators support the learning of mathematics and knowing how to use calculators intelligently is part of being prepared for an increasingly technological world. NCTM recommends calculators be made available for use by students at every grade level from kindergarten through college. Various other organizations agree, according to research compiled at Rice University for the Urban Systemic Initiative/Comprehensive Partnership for Mathematics and Science Achievement (1997).

The Rice University survey suggests that by reducing the time spent on tedious paper-and-pencil arithmetic and algebraic algorithms, calculators allow students to spend more time developing mathematical understanding, reasoning, number sense, and applications. Quite understandably, those on the other side of the math "wars" do not see the time spent on paper-and-pencil calculations as time wasted. They see calculators as useful tools, but feel it is important to learn the fundamentals before relying on them.

The use of "tools" in the teaching, learning, and application of mathematics has been going on since the first person looked at his or her fingers to count. The Chinese have been using an abacus for at least 700 years. The National Museum of American History at the Smithsonian has a Web site titled "Slates, Slide Rules, and Software," where one can learn about the various devices students have used to master abstract mathematical concepts.

The debates and disputes on how to get students interested in learning mathematics, for either its utility or for its own sake, are ongoing. According to a report to the President of MIT, only 3.3% of the students enrolled at the school during the 1999–2000 academic year were math majors. Since so few students pursue mathematics as a discipline, it make sense to encourage the study of math for its social utility. —M. C. NAGEL

Further Reading

Aristotle. *Basic Works*, edited by Richard McKeon. New York: Random House, 1941.

Bailey, Norman T. J. *The Mathematical Theory of Infectious Diseases and Its Applications.* London: Charles Griffin, 1975.

Ball, W.W. Rouse. *A Short Account of the History of Mathematics.* New York: Dover, 1960.

Before It's Too Late: A Report to the Nation from the National Commission on Mathematics and Science Teaching for the 21st Century [cited July 29, 2002]. <http://www.ed.gov/americounts/glenn/toolate-execsum.html>.

Calinger, Ronald. *A Contextual History of Mathematics: To Euler.* Upper Saddle River, NJ: Prentice Hall, 1999.

Cappell, Sylvain E., and Julius L. Shaneson. "Genera of Algebraic Varieties and Counting Lattice Points." *Bulletin of the American Mathematical Society* 30, no. 1 (January 1994): 62–69.

"Du Pont Heritage" [cited July 25, 2002]. <http://heritage.dupont.com/>.

"The Du Pont Story" [cited July 25, 2002]. <http://www.dupont.com/corp/overview/anniversary/story.html>.

Grattan-Guinness, Ivor. *The Norton History of the Mathematical Sciences: The Rainbow of Mathematics.* New York: Norton, 1998.

Hamer, William H. "The Milroy Lectures on Epidemic Disease in England: The Evidence of Variability and Persistence of Type." *Lancet* 1 (1906): 733–39.

Haven, Kendall F. *Marvels of Math: Fascinating Reads and Awesome Activities.* Englewood, CO: Teacher Ideas Press, 1998.

Heidegger, Martin. *Discourse on Thinking*, translated by John M. Anderson and E. Hans Freund. New York: Harper and Row, 1969.

Hogben, Lancelot. *Mathematics in the Making.* London: Macdonald, 1960.

Katz, Victor J., ed. *Using History to Teach Mathematics: An International Perspective.* Washington, DC: Mathematical Association of America, 2000.

Kincheloe, Joe L. *Toil and Trouble: Good Work, Smart Workers, and the Integration of Academic and Vocational Education.* New York: Peter Lang, 1995.

Lazarsfeld, Paul F., ed. *Mathematical Thinking in the Social Sciences.* Glencoe, IL: Free Press, 1954.

Marx, Karl, and Friedrich Engels. *The Marx-Engels Reader*, edited by Robert C. Tucker. New York: Norton, 1978.

An Overview of Principles and Standards for School Mathematics. Reston, VA: NCTM, 2000.

Paulos, John Allen. *A Mathematician Reads the Newspaper.* New York: Basic Books, 1995.

Penn, Alexandra, and Dennis Williams. *Integrating Academic and Vocational Education: A Model for Secondary Schools.* Alexandria, VA: Association for Supervision and Curriculum Development, 1996.

Phillips, George McArtney. *Two Millennia of Mathematics: From Archimedes to Gauss.* New York: Springer, 2000.

Sarton, George. *The Study of the History of Mathematics and the Study of the History of Science.* New York: Dover, 1957.

Stasz, Cathleen, et al. *Classrooms That Work: Teaching Generic Skills in Academic and Vocational Settings.* Santa Monica, CA: RAND, 1993.

"Slates, Slide Rules, and Software." Smithsonian Institution [cited July 25, 2002]. <http://americanhistory.si.edu/teachingmath/>.

Stein, Mary Kay. *Implementation Standards-Based Mathematics Instruction: a Casebook for Professional Development.* New York: Teachers College Press, 2000.

Stillwell, John. *Mathematics and Its History.* New York: Springer, 2001.

Sylla, Edith Dudley. "The Emergence of Mathematical Probability from the Perspective of the Leibniz–Jacob Bernoulli Correspondence." *Perspectives on Science* 6, nos.1–2 (1998): 41–76.

Is free and unlimited file-sharing good for society?

Viewpoint: Yes, free and unlimited file-sharing through peer-to-peer systems such as Napster is good for society because it encourages choice, provides opportunities for underdogs, and spawns new and innovative technologies.

Viewpoint: No, free and unlimited file-sharing is not good for society. Rather, it is a form of piracy that threatens basic intellectual property rights that are rightly protected by law.

In 1999, a 19-year-old college student named Shawn Fanning developed a computer program and Web site that made it possible for users of the Internet to download and exchange musical recordings (and later, video recordings) for free. He called his system Napster, and in the next two years, Fanning would become a hero to many—and an arch-villain to many others. The controversy over Napster would open up larger questions concerning copyright, ownership, and the relationship of artists and media companies to the public.

The means of placing music or moving pictures on a computer file did not originate with Napster; that technology came in the form of the mp3, a compressed file format whose origins date back to the late 1980s, and which made possible the recording of large amounts of sound and video data on mp3 and mpeg files respectively. The innovation of Napster was in the file-sharing method applied by Fanning, who put into place a variety of what is known in the computer industry as peer-to-peer networking.

Up until the mid-1990s, most home computers or PCs in America existed on their own, as little islands on which information could be stored and retrieved. However, thanks to the spread of the Internet and the linkage of computers to the Net through Internet service providers, most PCs (not to mention larger computers used by business) soon were linked to networks, or systems of computers connected by communication lines. A network typically depends on a server, an extremely powerful computer that acts as a central governing mechanism, routing information and directing communication traffic. In a peer-to-peer network, however, there is no dedicated or full-time server; instead, each computer is at once independent and linked with the others in a non-hierarchical arrangement.

The particular genius of Fanning's system is in its file-storage mechanism. While Napster required a server to perform basic functions, the server did not have to function as a memory bank for its users' thousands upon thousands of mp3 files—which, though they are compressed, still take up far more disk space than most user files such as word-processing documents. Rather, "subscribers" of Napster (which charged no fee) downloaded Napster software, which made it possible to dedicate a portion of their hard drive to storing (and thus making available to other users) the songs and videos they downloaded.

That Fanning's software represented a stroke of genius is hardly a matter of dispute among people who understand enough about computers to appreciate the technology involved. The controversy surrounding Napster, however, does not revolve around questions of technological genius; rather, it is concerned with something much more visceral—property rights, and specifically, copyright. If Napster users could download songs without paying for the CD, did this not amount to a form of theft? Were they not taking something that did not belong to them, engaging in a form of mass shoplifting—and shoplifting of art, no less?

Record-company executives, along with a number of prominent musicians, answered these concerns strongly in the affirmative. The music, they said, belonged to those who had created it, and/or to those who owned the legal rights to the music. Many defenders of Napster responded to this obvious legal, moral, and ethical point with arguments that held little water. A number of these "arguments" took the approach that two wrongs could make a right. Thus it was claimed, for instance, that since record-company executives "steal" from musicians, there is nothing wrong with fans stealing. Others asserted that because some musicians make huge and in some cases presumably undeserved profits, they should be compelled, against their will, to "give something back" to the public. Some Napster partisans even offered up a watered-down form of Marxism, claiming that music "belongs to the people."

The emptiness of such claims unfortunately obscured the many arguments in favor of Napster as a form of technology. Setting aside the issues of copyright, a great many aspects of Napster did serve to recommend it. First of all, it was a brilliant technological creation, and quickly spawned a number of new technologies—among them encryption devices designed to protect a file from being "napped" and shared without the payment of royalties. Furthermore, Napster provided for a much-needed democratization of music, which had become a top-heavy industry dominated by millionaire rock stars and billionaire music-company moguls.

In this regard, the growth of Napster can be seen as part of a larger movement toward letting the "voice of the people" be heard in music. Also part of this movement was the adoption, in May 1991, of computerized Soundscan record sales data for use in calculating the *Billboard* charts. Up until that time, the magazine had relied on the reports of record-store owners and employees, a method that invited considerable error and even more corruption. Adoption of Soundscan helps to explain why radio became generally more gritty and edgy in the 1990s, after the bland commercialism of the 1980s. Yet radio, and along with it MTV, have remained the province of a few multi-million-selling acts, while the vast majority of artists must depend on near-constant touring to capture and maintain a following.

Clearly, record companies would benefit if they could figure out a way to harness the power of peer-to-peer networks, and provide fans with what they wanted, rather than what corporations say fans should want. This is exactly what the media conglomerate Bertelsmann AG set out to do when in 2002 it took control of Napster, shut down by court order in 2001. The new, "legal" Bertelsmann-controlled Napster would have to compete with Kazaa and numerous other file-sharing services that had sprung up in Napster's wake. Nevertheless, hopes remained high that Napster, once an outlaw technology, could become a part of the way that companies—in this case, record companies—do business over the Internet. —JUDSON KNIGHT

Viewpoint:

Yes, free and unlimited file-sharing through peer-to-peer systems such as Napster is good for society because it encourages choice, provides opportunities for underdogs, and spawns new and innovative technologies.

Many years ago, the great Scottish philosopher David Hume (1711–1776) wrote in his *Enquiry Concerning Human Understanding:* "There is no method of reasoning more common, and yet more blamable, than, in philosophical disputes, to endeavor the refutation of a hypothesis, by a pretense of its dangerous consequences to religion and morality." Hume's statement could as easily be applied to the public debate in the early twenty-first century over peer-to-peer file-sharing systems such as Napster. Because these systems make it possible for users to download music and other forms of entertainment from the Internet without paying for them, many dismiss the practice simply as immoral and illegal.

This opinion has the outspoken support, understandably enough, of the "big five" members of the Recording Industry Association of America (RIAA). These five companies—BMG, EMI, Sony, Universal Music, and Warner—control the vast majority of record labels, and exercise a near-oligopoly over the music industry. Equally unsurprising is the opposition to file-sharing services expressed by a wide array of recording artists, ranging from Neil Young to Dr. Dre to Lars Ulrich of Metallica. Ulrich may

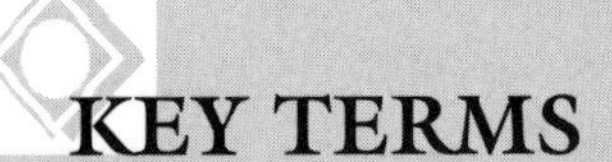

KEY TERMS

ENCRYPTION: The use of electronic codes to protect against theft of information stored on a computer.

NETWORK: A system of computers connected by communication lines.

OLIGOPOLY: An economic situation in which a few companies together exercise virtual control over the market.

PEER-TO-PEER NETWORK: A computer network in which there are no dedicated or full-time servers, nor any hierarchy among the computers; rather, each computer controls the security and administration functions for itself.

PUBLIC DOMAIN: When the copyright expires, intellectual property is considered part of the public domain, which means it can be used without charge or authorization.

SERVER: A computer that provides services such as routing, access to files, or the sharing of peripherals.

be a drummer for a loud heavy-metal band, but he has proven an articulate spokesman for his cause, as evidenced by an editorial for *Newsweek* in 2000. "[T]he critics of Metallica," Ulrich wrote, "keep asking, 'Who does Napster hurt?' Well, they're not really hurting us—yet—but I do know who they are hurting already: owners of small independent record stores."

Despite their differences in musical style, Ulrich, Young, Dre, and others have something in common when it comes to the debate over file-sharing: all of them sell millions and millions of albums, and thus, like the "big five," they perceive themselves as being economically affected by the free sharing of files over the Internet. Naturally, they do not speak of themselves as rich rock stars, but rather in terms of the moral issue (i.e., file-sharing as "stealing"), or like Ulrich, they portray themselves as advocates of the underdog.

In fact, the underdog—obscure and emerging artists, new labels, and most of all the fans—is likely to benefit, rather than suffer, because of file-sharing. Peer-to-peer systems such as Napster greatly encourage consumer choice, allowing listeners to circumvent the hierarchies that have controlled entertainment in general, and music in particular, for the better part of a century. Furthermore, the technology of file-sharing opens the door to all sorts of technological advances—including, ironically enough, ones in encryption. Thus, when the legal issues are worked out, the artists and companies that own the music will indeed get paid for it.

The "Big Five" of 1908 Once upon a time, Thomas Alva Edison (1847–1931) led his own effort to establish an oligopoly that would control an entire industry. Among the many great inventions to emerge from his Edison Manufacturing Company in New Jersey was an early version of the motion picture camera, which he patented as the Kinetoscope. By 1894, he had set out to make commercial use of his invention, but he soon recognized that increased competition in the nascent industry might cut into his profits. Therefore, in 1908 the "big five" film companies—Edison Manufacturing Company, Biograph, Vitagraph, Essanay, and Pathé—formed an alliance called the Motion Picture Patents Company (MPCC), through which they intended to dominate the market.

Yet as the "big five" leaders of the MPCC would discover, no one can bridle a technology whose time had come, and soon there emerged a legion of independent or "outlaw" filmmakers determined to make films regardless of what Thomas Edison wanted. One of them was Gilbert Anderson, who had starred in the Edison Company's 1903 motion picture *The Great Train Robbery,* generally regarded as the first true feature film in history. Determined to make films and gain a share of the profits himself, Anderson moved as far away from Edison's headquarters as he could, locating his production company outside Oakland, California. Edison responded to Anderson and others by dispatching a legion of patent attorneys to California, and exercised his influence to have renegade filmmakers arrested. The filmmakers responded by moving to a place in southern California from which they could easily escape to Mexico if need be—a Los Angeles suburb called Hollywood. But in 1912, these fears became moot when the U.S. government brought an anti-trust suit against the MPCC, which was declared illegal and dissolved in 1915.

Peer-to-Peer Computing Nearly a century later, much had changed, and yet nothing had. The former rogues and independents of Hollywood had become the entertainment establishment, and like Edison long before, they sought to establish oligopolistic control over the entertainment markets. Instead of grainy silent pictures that ran for less than 10 minutes, the technology in dispute now involved networks, or systems of computers connected by communication lines. And taking the place of Gilbert Anderson nearly a century after *The Great Train Robbery* was a 19-year-old college student named Shawn Fanning, who in 1999 developed a means for listeners to download songs from the Internet without having to pay for them. Fanning established a Web site at which music fans could do just that, and he called the site by

Shawn Fanning, the founder of Napster.
(AP/Wide World Photos. Reproduced by permission.)

a name that implied stealing or napping (as in *kidnap*): Napster.

Napster is an example of a peer-to-peer network. Most networks include at least one server, a computer that provides services such as routing, access to files, or the sharing of peripherals. By contrast, in a peer-to-peer network there are no dedicated or full-time servers, nor is there any hierarchy among the computers; rather, each computer becomes like a little server, controlling the security and administration functions for itself. This meant that Napster could operate on limited expenses, since instead of having its own dedicated server to store mp3s or other electronic music files, it relied on the users' hard drives. Each user would set aside a directory on his or her hard drive, and there would store music files for sharing on the network. Firewall or protective software would ensure that other users could not gain access to other parts of the users' hard drive, but as long as the user was connected to the Napster network, all other participants in the network would have access to all the files in the dedicated directory.

As Napster appeared on the scene, it caused a number of effects. Among the most obvious was the outcry by record companies and recording artists, who raised both legal and moral concerns. The result was a widespread public debate over the matter of ownership when it comes to works of art, and though the opponents of Napster were clearly correct in saying that it is wrong to take something without paying for it, the discussion soon involved terms more complicated than simple black and white or right and wrong. For example, as proponents of Napster pointed out, there is nothing illegal or immoral about a person downloading a piece of music that he or she already possesses on a legally purchased CD or cassette.

Another effect of the Napster brouhaha was the swift response on the part of the "big five" lawyers, who, like the patent attorneys under Edison's pay three or four generations earlier, sought to shut down Napster. Yet even as the "big five" sought to shut down Napster, the parent company of one of their members was involved in reconfiguring it. BMG, or Bertelsmann Music Group, is just one arm of Bertelsmann AG, a vast German-based media conglomerate that owns publishing companies (including Random House in the United States), TV and radio stations, newspapers, and so forth. Even as BMG was involved in legal action that shut down Napster in 2001, Bertelsmann acquired Napster's assets on May 17, 2002, and set about reshaping Napster as a fully legal enterprise that charges listeners a fee to download songs. The fate of the reconfigured Napster remains to be seen, but one thing is clear: for every peer-to-peer network, such as Napster, that corporate giants manage to shut down, another three or four—Kazaa, Gnutella, Morpheus, and so on—appears to take its place.

File-Sharing and the Future To some people—and not just those who stand to lose profits—Napster and its ilk seem to have posed a crisis. Others, however, see a range of opportunities in this "crisis." One of these opportunities is the vast array of new technologies that have appeared in Napster's wake, and even Kazaa and other second-generation peer-to-peer systems represent an improvement over Napster. For instance, Napster had a server that did the searching when users requested a particular song; Kazaa uses members' computers to search for the music. This led to the somewhat novel ruling by a Dutch court in 2002, to the effect that Kazaa *users* were violating copyright laws, but Kazaa itself was not.

This is just the beginning of the improvements in technology spawned by peer-to-peer computing, many of which have enormous potential applications in business. The linking of computers may make it possible for networks to avail themselves of the unused computing power of members, a fact exemplified by the SETI@home project: users all over the world devote the "downtime" of their computers to the search for extraterrestrial intelligence (SETI). While the computer is not in use, its

A poster advertising Edison's Kinetoscope. *(Corbis Corporation. Reproduced by permission.)*

resources are directed toward analysis of data collected from the skies, searching for radio signals indicating intelligent life. This makes it possible to replicate the processing abilities of a supercomputer for a fraction of the cost, and to log an amount of computer processing time that would be impossible on just one machine: in mid-2002, SETI@home announced that it had logged 1 million years' worth of computer processing time.

On a more down-to-earth level, businesses such as IBM, Hewlett-Packard, and Intel stand to benefit tremendously from systems that allow the sharing of processing power and the harnessing of resources that would otherwise be dormant. These companies, along with several smaller firms, formed the Peer-to-Peer Working Group in 2000 to exploit the potential of peer-to-peer networks. "If this was just about file-sharing," Grove Networks president Ray Ozzie told *InformationWeek,* "it would be the flavor of the month. This has big implications for supply-chain management. Partners in different companies will be able to work together very closely and define what they want each other to see." As Andrew Grimshaw of Meta-Computing LLC, a peer-to-peer software provider, explained in the same article, "The critical mass had been building. Now it's ready to explode."

Ironically, among the forms of technology influenced by peer-to-peer networking are more sophisticated encryption devices designed to protect copyrighted material from illegal file-sharing. Not only has such software been developed as a bulwark against Napster and other such services, but those companies who seek to cash in on the potential of peer-to-peer are developing such technology to protect the products on their own sites.

But the most important effect of peer-to-peer networking is societal rather than technological. Peer-to-peer democratizes music and other forms of entertainment, taking it out of the exclusive control of rapacious record companies that exploit artists and fans alike, and of the faceless executives who determine the extremely limited playlists of FM radio. Bands as diverse as the Grateful Dead and Kiss—or even, ironically, Metallica—have sold millions of albums despite receiving almost no airplay, or at least only airplay for selected radio-friendly hits. File-sharing offers the opportunity for fans to hear what they really want, instead of pop sensations manufactured by corporate America.

Will people pay for what they hear on peer-to-peer networks? Bertelsmann seems to have put its considerable influence behind the affirmative position. Furthermore, the fact that musicians today—in another outgrowth of the Napster phenomenon—often provide free song downloads as incentive for buying a new album, also suggests that record companies are catching on. One of the byproducts of the information age is an increasing democratization of society, with new technology and marketing platforms as diverse as Amazon and XM Radio making it possible for consumers to get what they want, instead of what a corporate executive tells them they should want. If sales of CDs are dropping, as many record companies claim they are, it is not because of Napster; it is because music fans are tired of the machinery that creates superstars based on the bottom line rather than on talent. Peer-to-peer networking thus offers a host of benefits to society, and to fear or oppose this is as futile—and ultimately as detrimental—as past opposition to new scientific discoveries such as the heliocentric universe or Darwin's theory of evolution. —JUDSON KNIGHT

Thomas Edison led his own effort to establish an oligopoly that would control an entire industry—the motion picture camera industry. *(The Bettmann Archive/ Corbis-Bettmann. Reproduced by permission.)*

Viewpoint:

No, free and unlimited file-sharing is not good for society. Rather, it is a form of piracy that threatens basic intellectual property rights that are rightly protected by law.

In the Good Ol' Days Anyone who has actually attempted to get permission to use copyrighted material in today's society will have at least one or two horror stories to tell. The process by nature is sometimes arduous, requiring several phone calls or e-mails and often a formal request in writing. It is no wonder that many people throw their hands in the air and give up in disgust. But nonetheless, you're supposed to try to do the right thing. You're supposed to credit your source. It's part of the professional honor system. And, in some cases, you may be required to pay a fee. For example, this is often true if you want to reprint someone's artwork or use a musician's song. It all depends on the requirements of the person or corporation that holds the copyright and therein lies the controversy surrounding free and unlimited file-sharing.

Just because something is on the Internet doesn't mean that it automatically falls into the public domain category. Or does it? With the advent of the computer age, the lines have gotten blurry. In fact, many would argue that they've gotten far too blurry. Plus, the anonymity of the Internet often allows information to be shared without accountability, and most of us have heard more than one account of how someone's work was used on the Internet without credit or permission. It wasn't that long ago when honesty actually counted for something, but for some people the end always justifies the means. Evidence of this can be found in the results of a Gartner Group survey, which was reported by Sylvia Dennis for Newsbytes. According to the survey, 28% of the people polled who listened to shared music files on the Web thought they were violating copyright law and 44% were unsure, but were still doing it anyway. So much for the honor system.

To complicate matters, the Framers of the Constitution certainly didn't anticipate a world revolutionized by computers, but they did build into the Constitution the ability to amend the law over time. However, change comes slowly and, in some cases, the law hasn't caught up with the technology. Whether we even need new laws is yet another aspect of the debate. After all, the existing copyright laws should be enough; they provide clear guidelines regarding intellectual property. However, people who favor the free and unlimited use of file-sharing say that material shared on the Internet falls into a different category and that the purpose of the Internet is to provide a vehicle for information-sharing without cost to the public.

On the surface, this argument seems logical and appealing. In fact, it is partially based in truth. Indeed, one of the most wonderful things about the Internet is the accessibility of information that can be found there. However, being able to disseminate information is one thing, but owning it is quite another. In the case of Napster, for example, the real controversy wasn't about sharing music files. It was about doing it without the express permission of the musicians or the record producers. It was, in essence, about professional courtesy, and about money—about big money. Napster argued that sharing music files actually increased product sales and promoted a musician's work. They claimed that the process of file-sharing was nothing more than what friends do when they lend each other a CD, but the music industry disagreed, claiming that thousands of dollars were lost in potential sales. Besides, musicians, unknown or otherwise, could create their own Web sites, if they wanted to; they didn't need Napster to do it for them. So, the swords were drawn on both sides and the legal battles that resulted made national headlines. Although Napster was ordered to change to a fee-based membership, the situation generated a lot of debate regarding what was fair and not fair.

A Society without Integrity On one level, the Napster controversy was about the unauthorized use of intellectual property with a clear market value. On a deeper level, however, there's much more at stake. Do we really want to live in a society without honor and personal integrity? Most of us learn about what it means to share when we are very young. We learn that the process embraces the idea that consent is requested by the borrower and granted from the lender. If one child grabs a toy from another's arms without asking, the child doing the grabbing gets scolded. This is because the behavior isn't nice; it isn't moral. This concept is no different when one considers the issues surrounding file-sharing. In the case of Napster, for example, allowing millions of people to listen to music for free without the consent of the owner is not morally right just because it is done on a grander scale. In fact, the activity is actually "piracy," not sharing.

At least that's the way United States District Court Judge Marilyn Patel saw it when she ordered Napster to shut down. She clearly saw the difference between a person buying a CD and lending it to a friend and the unauthorized distribution of music to millions of people for free—all of whom never paid a dime for the creative content. Napster representatives argued that no one was really being hurt by their activity, but that's simply not true. Damage was being done both monetarily and spiritually. When dishonesty is encouraged, society loses. It's that simple.

The Bargain Basement Mentality In an article she wrote for Internet.com entitled "The Future of File-sharing and Copyright," Alexis

Gutzman posed an interesting question: What if enough people vote for whatever you're selling to be distributed for free? Most of us would be outraged at the thought, wouldn't we? Then why do some people think that receiving free music through file-sharing is okay? Maybe it has something to do with inflation and the economy. Maybe some people are just so tired of paying outrageous prices for certain products that they'll leap at the chance to get them for free, whether it's ethical or not. Certainly, as Gutzman mentions, it doesn't help matters when you look at some of the price gouging that goes on in some of the record stores across the country. Combine that with the fact that many musicians aren't adequately compensated for their work and you don't always have a sympathetic public.

Nonetheless, Gutzman is quick to point out that sympathy is hardly the point. In a subsequent article for Internet.com, she identified two problems with everything being free. The first problem is philosophical in nature; it has to do with the mentality of the consumer. There are, in her definition, two types of consumers: fair traders and bottom feeders. Bottom feeders, she explains, always want things either for free or at prices so low that they are unbeatable. Fair traders, on the other hand, don't always demand to buy things below cost. Instead, they want to pay a fair price for a product, which ultimately helps fuel the economy. Sadly, too many people today embrace a bargain basement mentality. They don't care if the business they're dealing with succeeds or fails. All that matters to them is one thing: saving money. The second problem is practical in nature; it has to do with profit. When businesses start losing money, they are sometimes forced to make unpleasant changes. These

Lars Ulrich of Metallica (left) testified before the Senate Judiciary Committee, July 2000. Metallica sued Napster for copyright infringement.
(AP/Wide World Photos. Reproduced by permission.)

changes can result in the elimination of valuable services and can also mean that some people lose their jobs. And in the worst case scenario, the company can go out of business altogether. Then nobody wins.

Aside from the two problems mentioned above, there is another problem with wanting everything for free that is equally, if not more, disturbing. It has to do with the creative process. Let's use a musician as an example. Most true musicians will say they are artists first, business people second. However, that doesn't mean that they don't deserve financial compensation for their creative efforts. Indeed, many artists often complain that the creative process is thwarted enough by the constant concern over the financial bottom line. The last thing they need to face is the free and unlimited distribution of their work through file-sharing.

Taking a Stand Many corporations and educational institutions are taking a stand on file-sharing applications. Strict corporate policies are being developed that prohibit their use. Cornell University, for example, sent a memo to its faculty and staff warning them of the personal liability they face by violating copyright law. Along with the legal considerations involved, institutions are also concerned about the increased network traffic that file-sharing applications tend to generate. They can cause in-house servers to lag, and productivity suffers as a result.

Conclusion Proponents of free and unlimited file-sharing have brought up the argument that it gives people more choices and opportunities and encourages technological innovations. That may all be true, but does the end justify the means? Although we've moved into a technological age, it doesn't mean that we need to leave the ethics of the good ol' days behind. The honor system is as important today as it was before the Internet. Just because we have the ability to engage in free and unlimited file-sharing doesn't mean that we should or that it is good for society. —LEE ANN PARADISE

Further Reading

Clark, Don. "Napster Alliance Boosts Prospects for Encryption." *Wall Street Journal* (November 2, 2000): B-1.

Dennis, Sylvia. "Web Users Think Net File-sharing Violates Copyright—Gartner" [cited July 16, 2002]. <http://www.newsbytes.com/news/01/160251.html>.

Foege, Alec. "Bertelsmann's Quest to Harness the Napster Genie." *New York Times* (May 26, 2002): 4.

Gutzman, Alexis. "The Future of File-sharing and Copyright." Internet.com [cited July 16, 2002]. <http://ecommerce.internet.com/news/insights/trends/article/0,,3551_424781,00.html>.

———. "Something for Nothing; Nothing for Everyone." Internet.com [cited July 16, 2002]. <http://ecommerce.internet.com/news/insights/ectech/article/0,,9561_528571,00.html>.

Lanier, Jaron. "A Love Song for Napster." *Discover* 22, no. 2 (February 2001).

Lerman, Karen. "In Favor of Napster." *Writing* 23, no. 4 (January 2001): 15.

McClure, Polley. "Network Capacity and Policy Issues Arising from the Use of Napster and Other File-Sharing Programs." Cornell University Web site [cited July 16, 2002]. <http://www.cit.cornell.edu>.

McDougall, Paul. "The Power of Peer-to-Peer." *InformationWeek* 801 (August 28, 2000): 22–24.

Parkes, Christopher. "Peer-to-Peer Pressure." *Financial Times* (November 7, 2000): 4.

Richtel, Matt. "Music Services Aren't Napster, But the Industry Still Cries Foul." *New York Times* (April 17, 2002): C-1.

Snider, Mike. "No Copying, No Trading? No Kidding: Copyright Fight May Narrow Our Options." *USA Today* (March 6, 2001): D-1.

Ulrich, Lars. "It's Our Property." *Newsweek* (June 5, 2000): 54.

Is the increasing computerization of society healthy for the traditional arts?

Viewpoint: Yes, the increasing computerization of society is healthy for the traditional arts because new technology provides artists with better tools and greatly increases the exposure of their work within society at large.

Viewpoint: No, the increasing computerization of society is not healthy for the traditional arts, and in fact it threatens the creative process that drives artistic expression.

Every year, the pace of change in computer technology gains speed. Before the machine age, the computer was merely a concept, traceable through ideas dating to medieval times, or perhaps even to the ancient abacus. Then, at the end of World War II, with the development of ENIAC, the first truly electronic computer, the computer age was born. Thereafter computers loomed large in the popular imagination, but few people had direct experience of them. Most computers were large, expensive, cumbersome machines owned by businesses, universities, and government agencies.

The role of computers in daily life began to grow with the development of the personal computer (PC) and microchip during the 1970s, but even in 1982, when *Time* named the PC "Machine of the Year," only a fraction of Americans had PCs in their homes. The computer was then in a position like that of the television 40 years earlier: its time was coming, and just as televisions suddenly swept American homes in the 1950s, sales of computers boomed throughout the 1980s.

As more and more people began using PCs, the power of these computers increased apace. This phenomenon was due in part to Moore's law, a standard informally adopted by the computer industry after engineer and Intel Corporation cofounder Gordon E. Moore stated in 1965 that the power of computer chips tends to double every year. Moore later pulled back from his optimistic prediction, changing the figure to two years, but the growth of computing power has tended to occur at a rate halfway between his two estimates: for a period of at least four decades beginning in 1961, the power of chips doubled every 18 months.

With the growth of computing power, computers could do more and more tasks. Increasingly more complex programs, designed for ever more powerful systems, made it possible to create on a computer everything from spreadsheets to visual art to music. The increased linkage of home computers to the Internet, a movement that reached the take-off point in the early-to-mid 1990s, greatly accelerated this process. By the end of the twentieth century, the computer affected almost every aspect of human life—particularly the work of artists, architects, composers, and writers.

Today's architect relies on computer-aided design (CAD) programs that enable the visualization of structures in three dimensions. Musicians regularly use computers for composition and playback, acting as their own producers by changing the quality of sound and the arrangement of simulated instrumentation. Visual artists use computer graphics in creating and enhancing

their works, and the traditional darkroom of the professional photographer has given way to the digital camera and the PC. As for writers, their work has been affected immeasurably by a range of computer developments, from the simplest word-processing software to complex programs that assist in the writing of novels or screenplays.

Such is the situation, and the responses to it are manifold. On the one hand, these developments would seem to be purely, or almost purely, salutary for the arts. This is particularly so with regard to CAD and word processing, which provide the architect and the writer with tools that make their jobs much easier by eliminating such time-consuming tasks as drawing countless elevations or retyping a manuscript over and over. On the other hand, there are many who say that the use of computers as an aid in creating visual art or music or in writing novels encourages laziness on the part of the artist. In addition, there is an almost visceral concern with regard to the pace of technological progress, which seems to be undermining the human element in art as in much else.

Before the dawn of the industrial age in England during the latter part of the eighteenth century, people lived much as they had since ancient times. There had been varying levels of political order, and differing degrees of societal advancement in the arts and other endeavors, but the technology of daily life remained fairly much the same. Farmers plowed the ground using draft animals domesticated thousands of years earlier and took their crops to market using vehicles with wheels, an innovation that dates to approximately 3500 B.C. in Sumer. For cooking, heating, and lighting their homes, people used combustion technology that had developed in a few specific stages since the late Stone Age, but even the most advanced stoves and lights of the time still relied on the combustion of easily consumed materials such as wood.

The birth of industrialization changed everything and forced a complete rethinking of all ideas about progress. Since approximately 1800, people in the western world have become increasingly accustomed to the idea that the technology of the past no longer suffices for the present, but this awareness has not been won without a great deal of discomfort. For proof of the degree to which this uneasiness has permeated the human consciousness, one need only cite the vast number of books, movies, and other forms of art—Aldous Huxley's *Brave New World,* the *Terminator* films, or Radiohead's acclaimed *OK Computer* album—that depict technology as a sort of Frankenstein's monster that threatens to subjugate its creator.

Concerns about technology and the true meaning of progress are legitimate, and the argument that computers may serve ultimately to blunt the human imagination is unquestionably valid. Defenders of the role of computers in the arts, however, point to the advantages machines offer when manipulated by a human creator. Technological progress is almost inevitable, and as long as it is possible to use an easier and presumably better way to do something, people will do so.

It is in the nature of the restless human imagination, particularly the restless *western* imagination, to push boundaries, in the process surging beyond what most people find agreeable. In music, for example, the composers of the romantic era in the nineteenth century exhausted the limits of tonality, leaving the composers of the twentieth century to search for new musical meaning in atonality. Likewise, visual artists passed the frontiers of realism by the latter half of the nineteenth century, opening the way for movements that took art progressively through impressionism, expressionism, and abstraction. The same occurred in the literary arts with the end of the traditional novel and the development of experimental modes of narrative and nonnarrative.

Although the heat of these late nineteenth and early twentieth century battles in the arts has long subsided, one may with good reason ask whether the arts have truly benefited from the changes wrought by figures such as Arnold Schoenberg in music, Pablo Picasso in painting, and James Joyce in literature. There are strong positions on each side of the current controversy regarding the use of computers in the arts. Only a few aspects of this controversy are certain: that computer technology will keep changing, that people will keep finding new ways to apply the technology, and that many will continue to question the role of computers in the arts and other areas of life. —JUDSON KNIGHT

Viewpoint:

Yes, the increasing computerization of society is healthy for the traditional arts because new technology provides artists with better tools and greatly increases the exposure of their work within society at large.

Are computers, and the increasing computerization of society, good or bad for the arts? There are many who say *no*—that computers, by providing an aid or crutch to artists, take the creativity out of their work. Furthermore, these critics say, computers have had a deleterious effect on society, bringing with them dehumanization and a simple, gadget-driven world in which people, expecting "quick fixes" and instant gratification, are not willing to invest the time necessary to support and enjoy the arts.

The opposite is the case. The arts can be shown to have benefited extraordinarily from advances in technology. The term *arts,* incidentally, is used here in the broadest sense to encompass not only the visual and tactile arts, such as painting, sculpture, and architecture, and the performing arts, such as music, drama, dance, but also the literary arts. That most listings of the arts fail to include literature says a great deal about a deep underlying societal incomprehension concerning the arts—an incomprehension that can be improved with computers.

Information and the Discerning Consumer Computers help spread information. So do all means of communication, especially electronic communication. The more sophisticated the means, the more information can be disseminated. Because of the Internet, more data—more *experience* of the world—are more widely available to more people than has ever been the case. No longer is ballet, for example, the province only of wealthy patrons. A computer user with access to the Internet can find dozens of ballet Web sites and most likely can even download electronic files showing great performances.

That consumers have become more discerning through the use of computers has helped improve that most democratic of art media, television. In the 1970s, with the rare exception of programs such as *All in the Family,* situation comedies were forgettable at best and inane at worst. By contrast, television of the 1990s and 2000s was and is populated with exceptional programming, even in the arena of sitcoms—shows such as *Seinfeld, Friends,* and *Frazier,* all of them noted for witty repartee and clever insights, rather than the slapstick and sexual innuendo that passed for humor two decades earlier.

What has changed? The audience. Today's viewers, because of accelerated development of information technology, are much more discriminating, and much less easily amused, than were their counterparts 20 or 25 years ago. If the modern viewer does not like what is on television, there are many more outlets for expression, such as Internet message boards devoted to a particular show, which provide television executives with a ready and continuing opinion poll of viewers. Digital technology has enabled the spread of cable and satellite programming and with it networks such as Discovery and Bravo that further raise the stakes for entertainment in America. To compete, traditional networks must offer ever more innovative and interesting programs—quite an improvement over a time when electronic communication was limited to three networks and AM and FM radio.

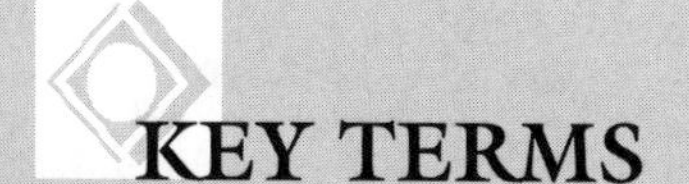

KEY TERMS

AUTOCAD: A computer-aided design (CAD) program for PCs developed by Autodesk

ENIAC: Acronym for *electronic numerical integrator and computer.* The first truly electronic computer.

E-ZINE: Short for *electronic magazine.* A periodical available on a Web site or via e-mail.

MOORE'S LAW: The amount of information stored on a microchip approximately doubles every year.

Critics of the role of computers in the arts may still be unimpressed and may dismiss television as a plebeian medium. Without addressing that viewpoint directly, it is easy enough to point out that the spread of information through the Internet has increased awareness of the "fine arts" as well. This is also true of literature, which gains more exposure through the World Wide Web than ever before. The Internet abounds with writers' sites and with "e-zines" that offer not only information about writing and interviews with writers but also opportunities for writers to publish their works on-line. Electronic publishing, whereby writers, for a fee, may arrange the distribution of their books in electronic form or in traditional book form on a print-on-demand basis, increases writers' exposure to a reading public.

Computers as a Tool for the Artist But what about the role of computer-based tools that aid the artist or writer? Are these not forms of "cheating"? According to this line of thought, Michelangelo would not have deigned to use AutoCAD for designing St. Peter's Basilica; Leonardo da Vinci would have shied away from three-dimensional imaging software; and Raphael would never have touched any sort of program that would have assisted him in creating visual art. There is no reason to believe that if those masters were alive today, they would not avail themselves of the technology available now, just as they did the technology to which they had access in their own time.

If the painters of the Renaissance had been antiprogress, they could well have opted not to paint their works using the chemical preservation techniques available at the time (for example, eggs as a setting agent in frescoes) and chosen instead to use the methods by which the

An interior view of St. Peter's Basilica at the Vatican, designed by Michelangelo.
(© G.E. Kidder Smith/Corbis. Reproduced by permission.)

Greeks applied color to their statues. Leonardo's *Last Supper* and Raphael's *School of Athens* might well be as faded and lost as the colors that once decorated the sculptures of Praxiteles.

It is absurd to imagine that Shakespeare would not have appreciated the value of software that assists writers. He might not use such software if he were alive today, simply because writing is an idiosyncratic vocation in which every practitioner has his or her own way of doing things. Shakespeare might, as many writers do today, opt to be alone with the empty page and a more basic form of software. But precisely because, as a writer, Shakespeare would know just how difficult writing is, it is difficult to imagine him turning up his nose at any form of technology that makes the task easier. Similarly, Tolstoy, who wrote a staggering eight drafts of *War and Peace,* would undoubtedly have benefited from word-processing software. Tolstoy's contemporary, Mark Twain (Samuel Langhorne Clemens), availed himself of what constituted state-of-the-art "word-processing" technology at the time. According to Clemens's autobiography, *The Adventures of Tom Sawyer,* published in 1876, was the first American novel submitted to the publisher as a typescript.

Is Technological Progress Bad? In effect, critics of the computer as a tool for the arts are saying that technology is ultimately bad for society. According to this line of reasoning, whatever is older and simpler is better. Such a view is reflected, for example, in the idea of "the good old days," which for Americans seems roughly applicable to the 80 years between the end of the Civil War and the end of World War II. It was an age, we grow up believing, when everything was simpler and better: when parents stayed married and only had children when they intended to raise them; when the government devoted its efforts to protecting the nation and its citizens instead of meddling in every aspect of human life; when students knew the Greek classics, Roman history, and the biblical stories and code of morality; and when talking and chewing gum—not shoot-outs and crack cocaine—were the biggest problems in school.

All of that sounds wonderful to most people, but lest we forget, it was also an age without penicillin, polio vaccine, magnetic resonance imaging, computed tomography, and an endless range of other medical benefits that came about in the mid twentieth century and afterward. It was a time when people smoked tobacco without any concept of the damage they were doing to their bodies, when people had little concept of dental hygiene as we know it today, and when most adults did not use deodorant. Without air conditioners and electric fans, washing machines and dryers, disposable diapers and canned baby formula, life was immeasurably more difficult than it is now.

Many defenders of antitechnological sentimentality claim, first, that society itself was more moral in an age when technology was simpler and, second, that technological progress and societal decline go hand in hand. But was society really more moral in the "good old days"? If we focus purely on the examples of the character of schools, marriage, and government then as opposed to now, a good case can be made for the past.

If we look beyond these carefully chosen examples, we see a past society that was morally coarse. It was a society in which, for example, most members of the white majority went along with a system in which blacks in the South were subject to Jim Crow laws and lynching and blacks in the North were treated as second- or third-class citizens. Tolerance of any difference from the perceived norm, in terms of religion, sexual orientation, or even physique, was much lower than it is now. The average marriage may have lasted longer, but there were many cases of spousal abuse, infidelity, frigidity, sexual dysfunction, and so on that simply went unknown or unreported. Although school shootings did not occur, it is instructive to compare the widespread horror and outpouring of emotion that followed the shootings at Columbine High School in Littleton, Colorado, in 1999 with the degree of apathy or even outright approval that regularly greeted the lynchings of a century earlier.

Does technological progress make people morally better? Of course not; it only enables them to gain a clearer picture of who they really are. It gives people greater exposure to ideas and by freeing people from drudgery and repetitive tasks increases their opportunities for self-reflection. These changes have affected artists and the arts as they have affected society as a whole. The changes can only be counted as positive. Because of computers, artists possess tools that make it possible to do things they could only dream of before, and they are surrounded by a much widely informed public that is vastly more likely to be "in touch with" their art—both figuratively and, through electronic communication, literally. The image of the artist laboring alone in a garret is a sentimental artifact of another century. Today's artist may still work alone, but the computer gives him or her access to millions of other people. —JUDSON KNIGHT

Viewpoint:

No, the increasing computerization of society is not healthy for the traditional arts, and in fact it threatens the creative process that drives artistic expression.

It's a beautiful day outside. The sunlight is streaming through your window, beckoning you to go outside and see the world. But you don't budge. Why should you? You're far too comfy for that. Plopped down in your new computer chair, complete with multiple swivel options, you needn't get up for hours. Feverishly you type away at the keyboard, attempting to save the universe from the latest computerized villain. Your eyes glaze over from the 3D accelerated graphics. Hours pass as food and sleep are forgotten. You're now officially a software junkie. Isn't the world of technology grand? At first glance, it might seem that way, but to people who are proponents of the traditional arts, the increasing computerization of society has some serious drawbacks.

To examine this issue carefully, it's important to discuss the definition of art. Webster's dictionary defines art as the activity of creating beautiful things. If this is so, is the activity of creating art meaningful when it is done on a computer? It certainly is not as personal as using traditional media. Children, for example, cannot create finger paintings in their truest sense on the computer. They cannot feel the paint squish between their fingers or enjoy the sensation that comes from blending colors on construction paper. For them, the "beauty" of the experience and the creative growth that comes with it are lost in a series of keystrokes and computer icons. The finished product is without texture; there is no real paint on a virtual canvas. Computer art is impersonal.

The Personal Nature of Art Anyone who attempts to define art in simplistic or immediate terms is doomed to fail. Even the

Mark Twain's *The Adventures of Tom Sawyer,* published in 1870, was the first American novel written on a typewriter. *(AP/Wide World Photos. Reproduced by permission.)*

world's greatest artists and philosophers have expressed various opinions on what constitutes art. What one person considers art, another may consider junk. Certainly there is a difference between refrigerator art and the renderings of an artist such as Michelangelo. However, there appears to be a thread that connects almost every valid definition of art. That is the concept or belief that the essence of creating art is personal in nature.

If art is the expression of human emotion, it is not difficult to see how an increasingly computerized society can thwart the creative process. Bean counters might argue that more works of art are being produced today than ever have been before, but what the counters fail to recognize is that quantity is not the same as quality. A depersonalized society damages the human psyche and ultimately the creative process. Computers do not replace human creativity and emotion.

The traditional arts, such as crafts, film, music, painting, storytelling, and theater, stem from the emotional connection between people and their environment and culture. The arts become an expression of life, reflecting the poignant elements of a person's world. As cultures change from region to region, so do the arts. Whether they present positive or negative impressions of that culture, the arts capture the human experience in a medium to be shared with others. It is impossible for a computer to possess the "humanness" required to convey such experiences, because computers cannot have any experiences of their own. They do not write, sculpt, or sing. They are incapable of presenting the infinite range of emotions that an actor can. The images, music, and text they display come from the creative endeavors of a human being, not from their own design. Although computers can enable us to learn about other cultures and their arts, they cannot help us truly understand the depth and soul of what is being expressed to us. They possess no more insight than a travel book or brochure. It is in the personal experience with a culture through its traditional arts that one can discover the diversity and customs of the culture, becoming engaged with other people through the experience of life.

The Virtual World versus the Real World Sitting in front of a computer screen may bring people together in a "virtual" sense, but not in a "real" sense. Part of what makes life worth living is the physical experience. A society that places too much emphasis on computerizing everything tends to lose touch with the human side of life. Let's take the AIDS quilt, for example. Looking at that quilt on a 17-inch computer screen is one thing, but actually seeing it stretch out for acres in person is quite another. Watching people embrace over a special hand-stitched square or actually hugging someone in pain as his or her tears hit your shoulder is far more personal than the coldness of a computer screen.

People who spend all their free time in front of a computer miss out on the richness and texture of life. Yet the more and more computerized our society becomes, the more difficult people find it to get away from computers. People begin to think like computer junkies; they find more pleasure in sitting in front of a computer screen than in actually engaging in life. They begin to think talking to an automated telephone operator is as good as talking to a real person. They send gifts they've never held in their hands to someone via the Internet. In essence, the computerization of society has a numbing effect on the soul, which is unhealthy for the traditional arts and is unhealthy altogether.

The "Want-It-Now" Mentality Computers have accelerated the pace of our daily lives. In some ways, that is a good thing. Numerous business activities take much less time to complete than they once did. However, the technology is a double-edged sword. Rather than using the extra time to engage in relaxing activities, such as the traditional arts, we are now trying to cram twice as much activity into a workday. The result is increased stress and a mentality that everything should be done in the interest of time. The quality of the experience is incidental. That approach is completely contrary to the one

needed to engage in the traditional arts. Artists understand that the process of creating a piece of artwork is as important as the artwork itself. Perception, contemplation, and experimentation are all part of the creative process. The traditional arts encourage a kind of creative thinking that is sometimes existential in nature. A person is not spoon-fed a subject with a couple of typed-in key words.

Computers have encouraged us to want and to expect things to happen instantly. This is an unhealthy environment for the traditional arts. Engaging in most of the traditional arts requires practice, patience, and time. In playing the violin, for example, one does not become proficient overnight. It takes discipline—and lots of it. The development of skill that embraces a combination of tenacity and contemplation is part of the creative process.

A Question of Values In a computerized society too much emphasis is placed on activities that have an immediate lucrative value. The negative effect on the traditional arts is obvious. One needs only to look at how the arts are underfunded in schools to see the message that is being sent to children. Although the arts foster self-expression and creative exploration, many people do not see the intrinsic value in supporting these programs. Yet such programs are extremely important on a variety of levels. They help to enrich children's lives with skills that the children can carry into adulthood.

Art therapists can attest to the importance of early emphasis on the arts. In the art therapy section of the *Gale Encyclopedia of Alternative Medicine,* Paula Ford-Martin points out that "art therapy encourages people to express and understand emotions through artistic expression and through the creative process." Ford-Martin lists self-discovery, personal fulfillment, and empowerment as some of the benefits gained from engaging in artistic expression. Personal goals such as these often are buried in an over-computerized world. Emotional laziness is encouraged as people anonymously hide behind pseudonyms in virtual chat rooms. The social skills required to engage in real-world relationships dwindle as people remain in the safety of their own homes. This is sometimes understandable given the complex world we live in; however, risk-taking is part of life, and even in failure we can sometimes benefit.

Conclusion The essence of artistic expression stems from the emotional connection between people and their environment. This connection fades with each day as our society becomes increasingly computerized. Our minds close as we allow our machines to think for us. Even as our world becomes smaller, the distance between humans continues to expand. With each day, we lose touch with our souls. Apathy and banality extinguish the creative spark that drives us to express ourselves through the traditional arts. As such, our culture and society suffer for this loss. We must push away from the computer desk and reimmerse ourselves in the living world, savoring both its beauty and its ugliness. Only then can we truly be human. —LEE A. PARADISE

Further Reading

Ford-Martin, Paula. "Art Therapy." *Gale Encyclopedia of Alternative Medicine* (July 26, 2002). <http://findarticles.com/cf_dls/g2603/0001/2603000173/print.jhtml>.

Hamber, Anthony, Jean Miles, and William Vaughan. *Computers and the History of Art.* New York: Mansell Publishing, 1989.

Hartman, Charles O. *Virtual Muse: Experiments in Computer Poetry.* Hanover, NH: University Press of New England, 1996.

Jody, Marilyn, and Marianne Saccardi. *Computer Conversations: Readers and Books Online.* Urbana, IL: National Council of Teachers of English, 1996.

Labuz, Ronald. *The Computer in Graphic Design: From Technology to Style.* New York: Van Nostrand Reinhold, 1993.

Lanham, Richard A. *The Electronic Word: Democracy, Technology, and the Arts.* Chicago: University of Chicago Press, 1993.

Robertson, Douglas S. *The New Renaissance: Computes and the Next Level of Civilization.* New York: Oxford University Press, 1998.

MEDICINE

Does the mercury used in dental fillings pose a significant threat to human health?

Viewpoint: Yes, the mercury used in dental fillings is dangerous to human health and can cause a variety of adverse effects.

Viewpoint: No, the mercury used in dental fillings does not pose a significant threat to human health. To the contrary, mercury-based fillings are safe, affordable, and durable.

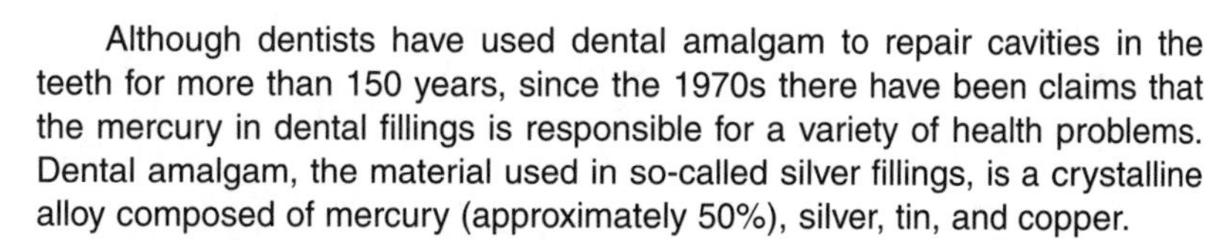

Although dentists have used dental amalgam to repair cavities in the teeth for more than 150 years, since the 1970s there have been claims that the mercury in dental fillings is responsible for a variety of health problems. Dental amalgam, the material used in so-called silver fillings, is a crystalline alloy composed of mercury (approximately 50%), silver, tin, and copper.

When made aware of the composition of dental amalgams, many people worry about the possibility that mercury might enter their body and have detrimental effects on their health. Chemists, however, note that when properly combined in the form of dental amalgam, the metals used to establish the proper hardness and durability of dental amalgams become tightly bound to each other in a form that is essentially safe, stable, and biologically inert. To make this point clear, Dr. Robert Baratz, president of the National Council against Health Fraud, explains that water is a chemical combination of hydrogen, a highly explosive gas, and oxygen, a gas that supports combustion. However, water does not explode or burst into flames.

A report on mercury fillings, titled "Poison in Your Mouth," that appeared on the television news magazine *60 Minutes* on December 23, 1990, played a major role in disseminating the claims of Hal A. Huggins, America's best known "anti-amalgamist." Narrated by Morley Safer, the program included patients who claimed to have recovered from arthritis and multiple sclerosis after their fillings were removed. One woman said her symptoms of multiple sclerosis disappeared overnight. Critics of the program noted that because mercury levels actually increase when amalgam is first removed, such an instantaneous recovery from "mercury toxicity" is impossible. Moreover, periods of remission and deterioration are common in multiple sclerosis and arthritis. The program and the media attention it attracted caused many people to demand the removal of their mercury fillings.

Anti-amalgamists claim that chronic mercury intoxication is a multisymptomatic illness. That is, it allegedly causes a variety of symptoms, including anxiety, irritability, fatigue, outbursts of temper, stress intolerance, loss of self-confidence, indecision, timidity, excessive shyness, memory loss, headache, depression, insomnia, unsteady gait, numbness and pain in the extremities, muscular weakness, drowsiness, edema, increased salivation, hair loss, nausea, constipation, diarrhea, and metallic taste.

Huggins, a Colorado dentist, is probably the most influential leader of the attack on dental amalgams and conventional dentistry. Huggins graduated from the University of Nebraska School of Dentistry in 1962. Before he became involved in anti-amalgam issues, Huggins promoted the premise that many diseases are caused by an imbalance of body chemistry and that he could prevent or cure such conditions by prescribing specific diets. During the 1970s, Huggins adopted the theory that amalgam fillings were the cause of many modern diseases, such as multiple sclerosis, arthritis, depression, cardiovascular disease, digestive disorders, epilepsy, chronic fatigue, Hodgkin's disease, and Alzheimer's disease. Huggins promoted his views in his 1985 book *It's All in Your Head: Diseases Caused by Mercury Amalgam Fillings,* pamphlets, seminars, lectures, tapes, private consultations, and his Toxic Element Research Foundation. Huggins later claimed that dental fillings, root canals, and crown materials caused autoimmune diseases, amyotrophic lateral sclerosis, birth defects, leukemia, and breast cancer.

According to Huggins, removing and replacing mercury fillings would reverse many illnesses and restore health. By the 1990s Huggins's clinic was attracting hundreds of patients, who paid thousands of dollars to undergo the Huggins protocol, a complex treatment regimen that involved much more than simple removal of mercury fillings. For patients, the first step in the Huggins protocol was to become educated about dental toxicity by reading books written by Huggins. Treatment involved a team of dentists and therapists, including physicians, intravenous sedation personnel, acupressurists, nutritionists, and detoxification doctors. After performing a dental examination and taking "electrical readings" on fillings and crowns, the dentist would determine the specific order in which removals and restorations should be performed. Blood samples were obtained so that an individual diet could be prescribed, the immune system could be assessed, and a "compatibility" test could be performed to determine which dental materials should be used. A sample of hair was used to analyze minerals, such as lead, mercury, cadmium, sodium, potassium, and calcium. For patients who could not come to his clinic, Huggins offered a consultation service that evaluated patients who sent in samples of hair and filled out questionnaires.

The Colorado State Board of Dental Examiners revoked Huggins's dental license in 1996. During the revocation proceedings, administrative law Judge Nancy Connick found that Huggins had consistently diagnosed "mercury toxicity" in all patients, even those without amalgam fillings, and had extracted all teeth that had had root canal therapy. The judge concluded that Huggins's treatments were "a sham, illusory and without scientific basis." Nevertheless, Huggins continued to write, publish, lecture, engage in consultations, and gain media attention for his attack on dental amalgam. Groups that support Huggins's conclusions include the American Academy of Biological Dentistry. In 1999, Huggins and Javier Morales, a Mexican dentist who followed the Huggins protocol, established the Center for Progressive Medicine in Puerto Vallarta, Mexico. By 2002, Huggins had published more than 50 articles and books on nutrition, child development, mercury toxicity, root canal, and cavities. His books and tapes include *Amino Acids, Dentistry and the Immune System, Mercury Issue Update, Further Studies on the Amalgam Issue, Why Raise Ugly Kids?, Mercury & Other Toxic Metals in Humans,* and *Uninformed Consent.*

Major national and international medical and dental organizations have investigated the safety and stability of dental amalgam and have concluded that the mercury in fillings does not pose a health hazard. A report issued by the U.S. Food and Drug Administration (FDA) in March 2002 concluded that "no valid scientific evidence has ever shown that amalgams cause harm to patients with dental restorations, except in the rare case of allergy." The U.S. Public Health Service concluded that there was no significant evidence that "avoiding amalgams or having existing amalgams replaced will have a beneficial effect on health." These conclusions have been supported by the American Dental Association (ADA), World Health Organization, European Commission, Swedish National Board of Health and Welfare, New Zealand Ministry of Health, and Health Canada.

Despite the evidence presented by the ADA, reputable medical groups, and public health organizations, anti-amalgamists have attempted to prohibit the use of dental amalgam. At the national level, legislators who have tried to ban dental amalgams include U.S. Representatives Diane Watson (D-Calif.) and Dan Burton (R-Ind.). Watson's bill, the *Mercury in Dental Filling Disclosure and Prohibition Act,* would amend the *Federal Food, Drug, and Cosmetic Act* to "prohibit the introduction of dental amalgam into interstate commerce" and would become effective January 1, 2007. A similar bill sent to the California legislature was defeated in committee, but in 1992 Watson succeeded in getting the California legislature to pass a law that requires the state dental board to issue a document listing the risks and efficacies of dental materials. The document issued by the dental board stated that "small amounts of free mercury may be released from amalgam filings over time" but that the amounts are "far below the established safe levels."

Watson's *Mercury Prohibition Act* would ban the use of amalgam in children younger than 18 years and in pregnant and lactating women and would require a warning that dental amalgam "contains mercury, which is an acute neurotoxin, and therefore poses health risks." Major scientific, medical, and dental associations oppose the bill and note that Watson's propositions are unfounded and that she apparently does not understand the fact that different forms of mercury have different properties and levels of toxicity.

Perhaps in the not too distant future the controversy over mercury in dental amalgam fillings will become moot because advances in science will fundamentally change the kinds of materials used to restore teeth and the way in which dentists prevent and treat cavities and gum disease. Dental problems eventually may be treated at the molecular level rather than with mechanical and surgical procedures. —LOIS N. MAGNER

Viewpoint:

Yes, the mercury used in dental fillings is dangerous to human health and can cause a variety of adverse effects.

In today's world, many people are concerned about the environment. This is not surprising when one considers the numerous industrial poisons that have become prevalent in the natural world. Worrying about toxins that might be in the air, our food, and the water we drink has become commonplace. News programs and public announcements seem to warn us daily about new health dangers. Many of us have heard about the things we need to avoid to protect our health. However, despite the new health consciousness that has permeated our culture, some of us might be shocked to realize that we already have one of the world's most dangerous poisons in our mouths: mercury.

Mercury, also known as quicksilver, is extremely toxic. In their book *Deadly Doses: A Writer's Guide to Poisons,* Serita Deborah Stevens and Anne Klarner indicate that a person weighing 150 pounds (68 kg) could experience a lethal dose by ingesting less than a teaspoon of the material. Although mercury can be ingested and absorbed through the skin, most poisonings come from the inhalation of mercury vapor that escapes from amalgam fillings when a person chews. In an article on mercury toxicity, Keith Sehnert, Gary Jacobson, and Kip Sullivan, a physician, dentist, and attorney, respectively, stated that the vapor can enter the bloodstream and be delivered to all parts of the body, including the brain. They emphasized that people with amalgam fillings have higher levels of mercury in their urine, blood, and brain than do people without amalgam fillings. In a study from the University of Calgary, M. J. Vimy, Y. Takahashi, and F. L. Lorscheider, faculty in the department of medicine, found that pregnant sheep with new amalgam fillings had elevated levels of mercury in their fetuses within two weeks of the placement of the fillings. The investigators concluded that mercury travels to the gastrointestinal tract, kidney, liver, and brain.

Mercury vapor is produced when the chemical boils at 40°F (4°C). That is why, for example, the chemical is so useful in thermometers. According to Stevens and Klarner, with high concentrations of vapor, the victim usually dies of ventricular fibrillation (abnormal heart rhythm) or pulmonary edema (excessive fluid in the lungs). Mercury builds up in the body, so smaller doses could eventually accumulate into dangerous levels. These lower-level poisonings can lead to numerous neurological diseases, including muscle tremors, hallucinations, migraines, and even psychosis.

The ways in which people suffer from having amalgam fillings vary. Some people may feel acute symptoms; others may appear to be completely unaffected. Nonetheless, anyone with amalgam fillings has at least some mercury built up in the body. Vasken Aposhian, professor of cellular and molecular biology at the University of Arizona, was quoted in an article by Francesca Lyman for MSNBC as saying that the number of amalgam fillings one has is highly important in terms of how much mercury the body absorbs. Of course, exposure to mercury can come from different sources; however, a large number of researchers agree that the greatest exposure comes from amalgam dental fillings. At a congressional hearing, Aposhian emphasized the seriousness of this exposure and stated that pregnant women should be especially careful, because mercury can cross the placenta and harm the developing nervous system of the fetus.

The United Kingdom Department of Health has issued warnings that pregnant women should avoid amalgam fillings, and the New Zealand Ministry of Health is reviewing its policy on the use of mercury amalgam for tooth fillings. In a letter to the editor of the *New Zealand Medical Journal,* Dr. Michael Godfrey, a leading environmental physician, brings attention to the fact that several major amalgam manufacturers issue material safety data sheets and directions for use that warn dentists of the many dangers associated with amalgam use. Warnings against using amalgams next to fillings containing other metals, in patients with kidney disease, and in pregnant women and children younger than six years are only some of the restrictions mentioned. The manufacturers warn that even in low concentrations, amalgam fillings can induce psychiatric symptoms, such as depression and mental deterioration. Given the seriousness of these warnings, it is not surprising to learn that many of Godfrey's patients who have chronic fatigue have an average of 15 amalgam

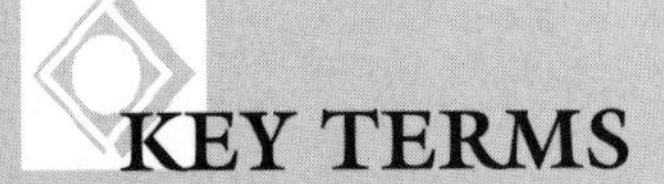

KEY TERMS

AMALGAM: The material used in silver dental fillings. It is composed of approximately equal parts mercury and an alloy containing silver, tin, and copper. It is solid or liquid at room temperature according to the proportion of mercury present.

AMALGAM ILLNESS: The set of symptoms sometimes attributed to dental amalgam.

COMPOSITE RESINS: Tooth-colored, plastic materials made of glass and resin and used in some dental fillings.

DOSE-RESPONSE CURVE: A graph of the relation between a toxin and its symptoms.

PLACEBO EFFECT: An effect in medicine in which improvement after treatment is caused by the patient's positive expectations not by the treatment itself.

TOXICITY: The ability to cause injury to the body by chemical means.

fillings and display many of the symptoms the manufacturers warn against.

Considering the health risks mercury poses to human physiology, it is difficult to understand why anyone would want to put even the tiniest amount in his or her mouth. Yet amalgam dental fillings have been used for quite some time. An April 2001 news release presented by Bio-Probe Inc. discussed how the ADA at first denied that leaks of mercury vapor could occur from what are commonly called "silver" fillings. However, the ADA later modified its stand when numerous studies pointed to the contrary. The vapor leaks from these fillings, which typically consist of 48%–55% mercury, are now being linked to several health problems. In fact, some dentists refuse to fill their patients' teeth with mercury amalgam, choosing to use a safer composite material instead.

The Bio-Probe news release also presented findings based on data from the National Health and Nutrition Examination Survey (NHANES III). This study, according to the mission statement of the National Center for Health Sciences, is critical because the results can ultimately influence public health policy in the United States. With the help of a company in California that specializes in statistical analysis, links between dental fillings and adverse health conditions were sought. The study showed that 95% of those surveyed who were classified with disorders of the central nervous system, such as epilepsy and multiple sclerosis, had dental fillings. It also was found that although 18% of Americans are filling free, 33% of the survivors of circulatory disorders, such as heart and arterial disease, are filling free. Ernie Mezei, an activist with Citizens for Mercury Relief, who is also a chemist and electrical engineer, stated, "clearly the survivors of heart disease have a much higher rate of being dental filling free, and we know heart disease is the number one killer in the United States." In a study led by Jukka T. Salonen, M.D., at the University of Kuopio in Finland, researchers found that mercury exposure and heart disease were clearly linked. The researchers also reported that a clear correlation exits between amalgam fillings and the risk of heart attack. Although the circulatory and central nervous systems are known to be the major areas damaged by mercury exposure, the risks to human health do not stop there.

The ADA is well aware of the dangers attributable to the improper handling of dental amalgam. The spokeswoman for the International Academy of Oral Medicine & Toxicology, Pam Floener, was quoted in the *Bio-Probe News* as follows: "The metallic mercury used by dentists to manufacture dental amalgam is shipped as a hazardous material to the dental office. When amalgams are removed, for whatever reason, they are treated as hazardous waste and are required to be disposed of in accordance with OSHA regulations. It is inconceivable that the mouth could be considered a safe storage container for toxic material." Even with this knowledge, the dental profession itself has not been without its tragedies. In *Deadly Doses,* Stevens, a registered nurse, and Klarner present a case history involving a middle-aged dental assistant who died of mercury intoxication attributed to her regular handling of amalgams with 40% mercury compounds. In her article for MSNBC on the safety of dental fillings, Francesca Lyman pointed to a study from scientists at the Battelle Centers for Public Health Research and Evaluation in Seattle that linked exposure to mercury vapor from dental amalgam fillings to central nervous system toxicity among dental personnel.

Continuing studies have begun to uncover even more health problems related to mercury exposure. Some connections are being drawn between mercury exposure and Alzheimer's disease. The Web site <http://www.holisticmed.com> highlights a significant study led by the renowned researcher, Dr. Boyd Haley. In the study, rats were exposed to mercury vapor in a way that accounted for the size difference between humans and rats. The tissue damage that developed in the rats was described as being "indistinguishable" from Alzheimer's disease. Haley was quoted as saying, "I'm getting the rest of my fillings taken out right now, and I've asked my wife to have hers replaced, too." This

MERCURY AND THE MAD HATTER

It was no "tea party" for the makers of the very popular felt top hats in Victorian England, whose occupational disease gave rise to the popular expression of the day, "mad as a hatter," the inspiration for Lewis Carroll's famous character in *Alice in Wonderland.* The demented host of "A Mad Tea-Party" exhibited symptoms all too commonly found among hat makers in the nineteenth century, symptoms of mercury poisoning.

Hat makers worked in poorly ventilated rooms. They dipped the furs used for the felt hats in hot mercuric nitrate solution as they shaped the felt to make the hats. The workers readily absorbed mercury through their skin in the process. Their shaking and slurred speech were so common that the symptoms were described as the hatter's disease. In the United States, Danbury, Connecticut, was a center for the hat industry. There the symptoms of mercury poisoning were so common they were called the "Danbury shakes." Mercury is no longer used in the hat business.

Mercury poisoning was not unknown at the time of Alice's famous tea party. Pliny, the Roman philosopher and author of *Natural History,* in approximately A.D. 50 described the poisonous nature of mercury. In his work Pliny described a method for cleaning mercury by squeezing it through leather. He also noted that laborers in mercury mines were dying at work.

Although it is speculation, there are indications Isaac Newton was a victim of mercury poisoning. For a time Newton turned from his interest in physics to alchemy. Around 1690 he wrote in his notebook about heating mercury. For approximately three years Newton was reclusive, suffering from the typical mental and physical symptoms of mercury poisoning.

As ancient as the knowledge is, mercury continues to take victims. In 1997 a Dartmouth College chemistry professor died of acute exposure to probably the most toxic compound of mercury, dimethylmercury, when less then half a milliliter was spilled and went through the disposable latex gloves the scientist wore for protection. By the time the symptoms were diagnosed, the poisoning had progressed too far to save the victim.

—M. C. Nagel

is especially troubling because not everyone agrees on the level of risk, which complicates the issue even further. Although not everyone agrees that it is necessary to have a patient's amalgam fillings removed, interesting studies have been conducted on the subject. One such study involving 118 subjects was conducted by H. Lichtenberg, a dentist in Denmark, and the results were published in the *Journal of Orthomolecular Medicine.* After the amalgam fillings were removed, 79% of the symptoms experienced by the subjects were reduced or eliminated. The results also indicated that patients with mild reactions to metals were more likely to have fewer or no symptoms than were patients with strong reactions to metals.

Despite the mounting epidemiological data, many organizations such as the ADA continue to ignore the risks posed by the use of mercury amalgam dental fillings. Furthermore, dentists have not been prevented from placing approximately 80 million of these fillings in American mouths each year. This practice seems almost ludicrous when one considers that mercury poisoning commonly causes gingivitis, damage to the gum line, and loosening of the teeth. An interesting analogy posed by Sehnert and his coauthors helps to illustrate the gravity of the situation. They point out that in general an amalgam filling weighs 1 gram, which is half mercury, and that if half a gram of mercury were to find its way into a 10-acre lake, a fish advisory for the lake would be warranted. Therefore it is no surprise that many people are calling for a ban on the use of mercury amalgam. Sweden has already banned it. Health authorities in Austria, Canada, Denmark, and Germany are trying to restrict the use of amalgam. Legislation requiring informed consent has been passed in California, and Representative Watson has proposed legislation to ban mercury amalgam altogether. Opponents to the legislation might argue that there are not many other options, and sadly, they would be correct. The materials available for dental fillings are composites, which come in a variety of brands, and mercury amalgam. To complicate matters, some dentists believe that composite material is not as durable as mercury amalgam, especially when used to fill molars. This often leaves patients and dentists with difficult and limited choices. Research is needed to develop a wider range of safe and durable products that can be used to fill teeth. In the meantime, this does not mean that we should quiet the controversy or turn a blind eye to the problem.

In an editorial in the *Bio-Probe News,* Sam Ziff states that many Americans are simply unaware of the multiple health risks associated with amalgam fillings. In Ziff's opinion, not enough is being done, and it must be made very clear to the public that amalgam fillings not only leak mercury vapor but also do so continually. This vapor is then directly inhaled into the lungs, where 80%–100% of the toxin is absorbed. As of yet, there are no scientific data for determining the least amount of mercury that will not cause cellular damage. It is known, however, that the longer a person is exposed to mercury vapor, the higher the dose will be. Therefore the problem must be dealt with immediately, not some time in the future. After all, the controversy over amalgam dental fillings is not new; it has been around for more than 150 years. However, the fact that dentists have been filling patients' teeth with amalgam for so long has been one of the arguments the ADA has embraced to justify its use. As each year passes, more Americans will have pieces of deadly toxin placed in their mouths. "We've always done it this way" no longer seems an appropriate justification. Perhaps Lord Baldwin, joint chairman of the British Parliamentary Group for Alternative and Complementary Medicine, said it best: "To point to the fact that amalgam fillings have been used for a hundred years is not a proof of safety anymore than it is a proof of safety to claim that tobacco smoking must be safe because people have been smoking for a long time." —LEE ANN PARADISE

Viewpoint:

No, the mercury used in dental fillings does not pose a significant threat to human health. To the contrary, mercury-based fillings are safe, affordable, and durable.

Dental amalgam, the material used in silver fillings, is approximately one-half (43%–54%) mercury. The other half is composed of an alloy containing silver, tin, and copper. Each of these elements by itself can be poisonous if taken into the body in sufficient amounts. Mercury, in particular, is known for its toxicity—the ability to cause injury to the body by chemical means. On the surface, it seems quite plausible that the mercury in fillings might slowly poison the body and cause a host of health problems. Plausible, but wrong. This argument is disproved by a basic fact of chemistry: The mercury in dental amalgam chemically binds to the other elements; the result is a substance that is hard, stable, and safe.

Dental amalgam has been used for more than 150 years. During that time, particularly over the last decade, its safety and effectiveness have been studied extensively. In March 2002, the FDA issued a consumer update on the findings of these studies. The FDA concluded that "no valid scientific evidence has ever shown that amalgams cause harm to patients with dental restorations, except in the rare case of allergy." Other major national and international organizations reached similar conclusions after reviewing the research. They include the ADA, U.S. Public Health Service, World Health Organization, European Commission, Swedish National Board of Health and Welfare, New Zealand Ministry of Health, and Health Canada.

Birth of a Health Scare If amalgam fillings are safe, why are people so worried about them? In the United States, much of the public concern can be traced to Hal Huggins. In the 1970s, Huggins began promoting a theory that amalgam fillings were at the root of a wide range of modern ills, including multiple sclerosis, depression, heart and blood vessel problems, arthritis, some cancers, chronic fatigue, digestive disorders, and Alzheimer's disease. Huggins went on to publish books and pamphlets and make videotapes and media appearances in which he advocated having amalgam fillings removed and replaced. By the early 1990s, Huggins's center was treating more than 30 patients per month, many of them desperately ill. The patients were charged up to $8,500 each for two weeks of intensive therapy. Huggins's dental license was revoked in 1996, but he continues to publicize his controversial views.

Meanwhile, other voices have joined the anti-amalgam chorus. Much of the outcry has been fueled by the media, including the report that aired on *60 Minutes* in 1990. Among the more outlandish claims made in that program was the story of a young woman with multiple sclerosis who said that on the very day her fillings were removed, she was able to throw away her cane and go dancing. Considering that the act of removing amalgam fillings actually temporarily increases mercury levels in the body, this sounds more like the power of positive thinking than a medical miracle. Yet just months later, nearly half of 1,000 Americans surveyed by the ADA said they believed dental amalgam could cause health problems. The level of public fear and misinformation is still high. In fact, the situation is so bad that the American Council on Science and Health, a nonprofit consumer education organization, has branded it one of the "greatest unfounded health scares of recent times."

Mercury is known to be toxic to human health. Here, a sign warns against eating too many bass from a mercury-poisoned lake.
(© Charles Philip/Corbis. Reproduced by permission.)

A Question of Quantity The unfortunate truth is that most consumers and even many dentists have heard only one side of the story. They are not aware of the large body of research backing up the safety and benefits of dental amalgam. One key question is how much mercury exposure is too much. It is true that some mercury vapor may be released from amalgam fillings by chewing or tooth grinding. However, the exposure lasts only a few seconds, and the amount taken into the body seems to be extremely small. It is worth noting that mercury is already everywhere in our environment, including our water, air, and food, especially fish. Although high levels of mercury are toxic, the lower levels that most people encounter every day, whether or not they have amalgam fillings, do not seem to pose a health risk. It is a classic example of the scientific principle that "only the dose makes a poison." This is the same principle that explains why drugs and even vitamin supplements can be helpful in small doses but harmful in excess.

One difficulty lies in measuring precisely how much mercury is released by and absorbed from amalgam fillings. The process can be affected by everything from how many teeth a person has to how he brushes them, and from how someone eats to whether she breathes through her mouth. Currently, the amount of mercury in the body is measured through analysis of breath, blood, and urine, but these methods need improvement. It still is difficult to discern whether mercury found in the body comes from fillings or other sources. Nevertheless, the best evidence indicates that the amount of mercury released by fillings is inconsequential.

Some of the most compelling data come from studies of dentists. Because dentists breathe in mercury vapors every time they insert or remove an amalgam filling, it stands to reason that their risk should be considerably greater than that of the average person. Studies have shown that dentists do tend to have higher levels of mercury in the urine than does the general population. Yet dentists still have no greater risk of illness or death. In other words, even dentists' above-average mercury levels do not seem to be high enough to cause harm.

A number of studies have compared the health of people with amalgam fillings and the health of those without amalgam fillings. These studies failed to support the claims of any related health problems other than rare allergic reactions. Even when researchers compared people who believed they had filling-related health problems with those who had no such complaints, they found the level of mercury was similarly low in both groups. This finding is revealing, because a true toxic response should show up on a dose-response curve—a graph of the relation between the toxin and the symptoms. The more severe the symptoms, the higher the exposure to the toxin should be. Yet researchers found no link between the severity of symptoms blamed on amalgam and the number of fillings people had or the level of mercury in their blood or urine.

Drilling Out the Filling There are still plenty of people who truly believe they have problems caused by dental amalgam or who swear by the benefits of having old fillings replaced. A new term—amalgam illness—has even been coined for the catchall set of vague symptoms that have been attributed amalgam fillings. For some of these people, the problem may lie more in the mind than in the head. Several studies have shown that people who claim to have amalgam illness are more likely than others to have signs of various psychological disorders. For other people, the physical symptoms may be rooted in a very real medical condition, such as multiple sclerosis or Alzheimer's disease. Sadly, it is not uncommon to hear of seriously ill people having their fillings replaced in a vain attempt to find a cure for an incurable ailment. At the very least, these people are risking disappointment. A bigger danger is that some may give up standard medical care in the false belief that all they need to do is get their teeth fixed.

After having their amalgam fillings removed and replaced with another material, some people say they feel better, at least for a time. However, this is likely the result of the placebo effect—a well-known effect in medicine in which improvement after a treatment is caused by the person's positive expectations, not by the treatment itself.

There is no scientific evidence that removing perfectly good fillings has any beneficial effect. To the contrary, having fillings removed is expensive and often painful. In addition, it damages healthy tooth structure, which must be drilled out to accommodate the new filling. Ironically, replacing a filling also may expose people to more mercury than if the tooth had been left alone, because drilling into the filling releases mercury into the air. Of course, old fillings may eventually have to be replaced anyway, when they start to wear away, chip, or crack. Most dentists stop short of encouraging patients to replace sound fillings.

As long as there are people willing to pay to have fillings removed unnecessarily, there will be dentists willing to do it. A few have resorted to rather dubious tests to justify the procedure. For example, some use a symptom questionnaire that asks about a wide range of medical problems and a long list of symptoms, such as tiredness, boredom, and restlessness. The list is so inclusive that almost everyone would have some of the signs. Other dentists use an industrial mercury detector, which seems reassuringly objective. However, the amount of mercury released varies throughout the day, depending on what a person is doing. The test typically is done by having the person chew gum vigorously for several minutes, then checking the breath for mercury. This reading is used to estimate the total daily dose. Because chewing briefly releases mercury, the estimate is likely to be much too high. Even blood testing is of limited value, because blood concentrations are so low—typically less than 5 parts per billion—that they are difficult to detect and too small to identify the source of the mercury.

Hard Numbers versus Soft Science Then there is urine testing. It may sound frightening to detect mercury in a person's urine. Remember, however, that mercury is all around us. As a result, most people, with or without fillings, have up to 5 micrograms or so of mercury per liter of urine. Practicing dentists typically have urine mercury levels of approximately 10 micrograms per liter or less. To put these numbers in perspective, the legal limit on mercury exposure for industrial workers is 50 micrograms per cubic meter of air, eight hours per workday, 50 weeks per year. Exposure to this extent produces urine mercury levels of approximately 135 micrograms per liter—more than 13 times the usual level found in dentists. Yet even this amount of mercury exposure is considered safe.

The fact is, amalgam fillings are needed in dentistry. They are safe, relatively inexpensive, and long-lasting. Although there has been a push in recent years to find new and better alternatives, amalgam still can hold its own against other filling materials. No other material is as versatile or easy to put in place. And unlike some materials, amalgam can be used to fill large cavities on stress-bearing areas of the teeth. There are some disadvantages, of course. Perhaps the biggest is appearance. Because many people do not want a silver filling to show, composite resins—tooth-colored, plastic filling materials made of glass and resin—often are used on front teeth or wherever a natural look is important. As far as health goes, however, the main risk of amalgam seems to be an allergic reaction. Yet such reactions appear to be quite rare. Fewer than 100 cases have been reported.

Despite the solid track record of amalgam, some people would still like to see it banned. The latest tactic is to attack amalgam through lawsuits and laws. The bill introduced by Representatives Watson and Burton in April 2002 if passed would ban the use of dental amalgam in the United States by 2007. A statement by Watson outlining her reasons for the bill shows the kind of muddled science that has given amalgam a bad reputation. Watson said that "mercury is an acute neurotoxin." In truth, although mercury can be toxic in some forms, the mercury in amalgam is chemically bound, making it stable and therefore safe. Watson goes on to say that the mercury in fillings constantly emits "poisonous vapors." The truth is that the minuscule amount of mercury that the body might absorb from fillings is far below the level known to cause health problems.

It is impossible to prove anything 100% safe. Yet dental amalgam is the most thoroughly tested material in use for dental fillings, and its safety remains unchallenged by mainstream scientific groups. The risk posed by the mercury in fillings seems to be minimal to none. On the other hand, the danger of losing a valuable option because of nothing more than unfounded fears and emotional arguments is quite real. —LINDA WASMER ANDREWS

Further Reading

"Amalgam/Mercury Dental Filling Toxicity." *Holistic Healing Web Page* [cited July 15, 2002]. <http://www.holisticmed.com/dental/amalgam/>.

American Council on Science and Health [cited July 15, 2002]. <http://www.acsh.org>.

American Dental Association Council on Scientific Affairs. "Dental Amalgam: Update on Safety Concerns." *Journal of the American Dental Association* 129, no. 4 (April 1998): 494–503.

American Dental Association [cited July 15, 2002]. <http://www.ada.org>.

Baldwin, E. A. A. "Controlled Trials of Dental Amalgam Are Needed." *British Medical Journal* 309 (1994): 1161.

Dodes, John E. "The Amalgam Controversy: An Evidence-Based Analysis." *Journal of the American Dental Association* 132, no. 3 (March 2001): 348–56.

Godfrey, Michael E., and Colin Feek. "Dental Amalgam" [letter to the editor]. *New Zealand Medical Journal* 111 (1998): 326.

Huggins, Hal. Home Page [cited July 15, 2002]. <http://www.hugnet.com>.

Lichtenberg, H. Home Page. "Symptoms Before and After Proper Amalgam Removal in Relation to Serum-Globulin Reaction to Metals" [cited July 15, 2002]. <http://www.lichtenberg.dk/>.

———. "Elimination of Symptoms by Removal of Dental Amalgam from Mercury Poisoned Patients, as Compared with a Control Group of Average Patients." *Journal of Orthomolecular Medicine* 8 (1993): 145–48.

Lyman, Francesca. MSNBC.com. "Are 'Silver' Dental Fillings Safe?" July 11, 2002 [cited July 15, 2002]. <http://www.msnbc.com/news/599087.asp>.

Mercury in Dental Filling Disclosure and Prohibition Act.. 107th Congress, 2nd sess., H.R. 4163.

Quackwatch [cited July 15, 2002]. <http://www.quackwatch.com>.

Salonen, J. T., et al. "Intake of Mercury from Fish, Lipid Peroxidation, and the Risk of Myocardial Infarction and Coronary, Cardiovascular, and any Death in Eastern Finnish Men." *Circulation* 91 (1995): 645–55.

Sehnert, Keith W., Gary Jacobson, and Kip Sullivan. "Is Mercury Toxicity an Autoimmune Disorder?" [cited July 15, 2002]. <http://www.thorne.com/townsend/oct/mercury.html>.

Stevens, Serita D., and Anne Klarner. *Deadly Doses: A Writer's Guide to Poisons.* Cincinnati: Writer's Digest Books, 1990.

U.S. Food and Drug Administration [cited July 15, 2002]. <http://www.fda.gov>.

"U.S. Govt. Data Connects Human Disease and Dental Fillings." *Bio-Probe News.* April 18, 2001 [cited July 15, 2002]. <http://bio-probe.com/ReadNews.asp?article=34>.

U.S. Public Health Service. *Dental Amalgam: A Scientific Review and Recommended Public Health Strategy for Research, Education and Regulation.* Washington, DC: U.S. Department of Health and Human Services, Public Health Service, 1993.

Vimy, M. J., Y. Takashasi, and F. L. Lorscheider. "Maternal-fetal Distribution of Mercury Released from Dental Amalgam Fillings." *American Journal of Physiology* 258 (1990): 939–45.

Ziff, Sam. "Editorial on the Alzheimer's/Mercury Connection." *Bio-Probe News.* March 2001 [cited July 15, 2002]. <http://bio-probe.com/ReadNews.asp?article=32>.

Do HMOs or insurance providers have the right to require the use of generic drugs as substitutes for name-brand drugs in order to save money?

Viewpoint: Yes, HMOs and insurance providers have the right to require the substitution of generic drugs for name-brand drugs, as generic drugs have been scientifically proven as safe and effective.

Viewpoint: No, HMOs and insurance providers do not have the right to require the substitution of generic drugs for name-brand drugs. Even if the active ingredients are the same, differences in the inactive ingredients might have a significant effect on some patients.

According to the Food and Drug Administration (FDA), and many pharmacists, low-cost generic drugs are as safe and effective as name-brand drugs. As part of its Center for Drug Evaluation and Research the FDA established the Office of Generic Drugs. The FDA guidelines for generic drugs require testing to prove that the generic drug is equivalent to the original name-brand product in terms of its therapeutic effects and its bioavailability. In other words, generic drugs must contain the same amounts of the active ingredients that are found in the name-brand product. Moreover, the manufacturer must demonstrate that the generic drug delivers the same amount of the active ingredient into the bloodstream as the name-brand drug.

The trend to generic drugs has markedly accelerated since the 1980s when generics comprised less than 20% of all prescription drugs. The Generic Pharmaceutical Association, a trade group that represents generic drug manufacturers, claims that 44% of all prescriptions filled in the United States today are generic drugs. The Hatch-Waxman Act of 1984 was an important factor in stimulating the growth of the generic drug industry. The act was sponsored by Senator Orrin Hatch (R-UT) and Representative Henry Waxman (D-CA).

The objective of the Hatch-Waxman Drug Price Competition and Patent Term Restoration Act was to allow generic drug manufacturers to market copies of name-brand drugs after their exclusive patents had expired. From 1984 through 1993, the Office of Generic Drugs received over 7,000 applications to manufacture and market generic drugs. About half of the applications were finalized and approved. A series of scandals shook the generic drug industry in 1989, prompting the FDA to reevaluate hundreds of generic drug applications and expand testing procedures.

Concerned with the competition from generic drugs, pharmaceutical companies have been fighting to extend the period of patent protection. Patents for "innovator" or "pioneer" drugs are issued for a 17-year period from the point at which a new drug application is made. Although manufacturers of generic drugs must satisfy FDA guidelines, generic or "copycat" drugs do not have to undergo the same costly array of animal and clinical trials of safety and efficacy that are required for new drug applications. The FDA allows manufacturers of generic drugs to apply for an abbreviated new drug application, in order to demonstrate the bioequivalence of the generic drug.

Faced with ever-increasing drug costs, many health maintenance organizations (HMOs) and insurance providers are attempting to force their clients to substitute generic drugs for higher priced name-brand drugs whenever possible. The savings involved in using generic drugs rather than their name-brand counterparts are substantial. Indeed, the Congressional Budget Office estimates that the use of generic drugs results in savings of 8–10 billion dollars a year. Nevertheless, these savings seem small in comparison to the $121.8 billion spent on prescription drugs in 2000.

Although patients are concerned about affordable drugs, many worry that generic drugs may be of lower quality than name-brand products. Critics of generic drugs argue that even if the active ingredients are the same, differences in the inactive ingredients, such as coatings, colorings, binders, and fillers might have a significant effect on some patients. In the case of a small percentage of drugs, designated narrow therapeutic index (NTI) drugs, very small changes in dosage level appear to have a substantive effect on efficacy. Some tests of narrow-therapeutic-range drugs, however, suggested that the failure rate for generic drugs was similar to that of name-brand drugs.

Critics of generics contend that even though the FDA requires proof of bioequivalence, the testing, which is usually done on a small number of healthy patients, may fail to reveal differences that can occur when the drug is taken by a wide variety of seriously ill patients. Thus, physicians and patients are particularly concerned when HMOs force patients to use the lowest-priced drug available. They argue that despite the importance of containing health care costs, accountants and administrators should not be making decisions about which drugs a patient should be taking. Many physicians object to having HMO administrators interfere with the doctor/patient relationship and overruling decisions made by the prescribing physician. Substituting a generic drug for the drug selected by the physician, therefore, seems to demonstrate more concern for profits than for patients. Patients may also feel that the generic drug must be different from a drug that enjoys a very distinctive, familiar, and even comforting brand name. For example, consumers may feel more secure with a brand name like Valium than a generic equivalent known as diazepam.

Arguments about drug safety and efficacy are not limited to the battle between manufacturers of generic drugs and companies marketing name-brand drugs. Questions arise about whether heavily promoted "innovator" drugs are actually more effective than older prescription drugs, or even over-the-counter medicines. For example, some researchers suggest that ibuprofen and other anti-inflammatory drugs are just as effective at relieving arthritis pain as Celebrex and Vioxx, the newest high-priced prescription arthritis drugs. Moreover, there is growing evidence that Celebrex and Vioxx do not produce fewer side effects than less costly anti-inflammatories. Celebrex (Pharmacia) and Vioxx (Merck) have provided enormous profits since they were approved by the FDA in 1998 and 1999, respectively. Sales of Celebrex were $3.1 billion in 2001; sales of Vioxx reached $2.6 billion.

Critics and medical insurers, however, insist that these very expensive drugs are not significantly better than less expensive alternatives. Indeed, critical reviews of the studies used to promote Celebrex and Vioxx found evidence of "serious irregularities" and "misleading results," especially those dealing with gastrointestinal complications and cardiovascular problems. Thus, critics charge that Celebrex and Vioxx became best sellers, not because they were really safer and more effective than over-the-counter drugs, but because of massive advertising campaigns, especially those that appealed directly to consumers.

Rather than assuming that the newest, most expensive drugs are always the best, patients and doctors might consider adopting a more skeptical attitude and turn to the most costly drugs only if they find that the lower-priced alternatives are unsatisfactory. Of course consumers cannot be certain that the ingredients in high-priced, name-brand products are always superior. For example, in May 2002 Schering-Plough Corporation, manufacturer of the best-selling allergy medications Claritin and Clarinex, was facing allegation that it had been using imported, inexpensive chemical ingredients not approved for use in the United States in its products. Although pharmaceutical companies often buy chemical ingredients made by foreign suppliers, they must obtain FDA approval if those ingredients are used in prescription drugs sold in the United States. —LOIS N. MAGNER

Viewpoint:

Yes, HMOs and insurance providers have the right to require the substitution of generic drugs for name-brand drugs, as generic drugs have been scientifically proven as safe and effective.

You can buy them at your local pharmacy or supermarket for a fraction of the cost: substitutes for name-brand drugs called generic drugs that the Food and Drug Administration (FDA) affirms are as effective as their name-brand counterparts.

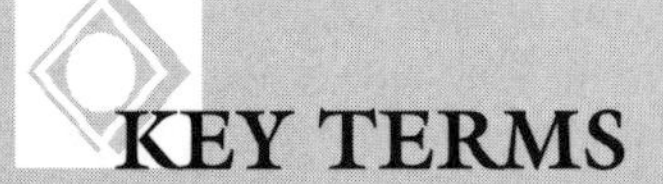

KEY TERMS

ANDA (ABBREVIATED NEW DRUG APPLICATION): Shortened version of an NDA that may be submitted to the FDA for approval of a new formulation of an existing drug or a generic equivalent of an existing drug.

BIOAVAILABILITY: The extent and rate at which the body absorbs a drug.

BIOEQUIVALENCE: Term used to describe a drug that acts in the body in the same manner and to the same degree as the drug to which it is being compared. For example, a generic drug must be bioequivalent to the original brand-name drug of which it is a copy.

BRAND-NAME DRUG: An "innovator drug" that is the first of its kind to appear on the market. Its manufacturer, usually a research-based pharmaceutical company, must demonstrate to the FDA that it can produce a drug that is safe and effective for the disease or condition for which it is intended.

GENERIC DRUG: A "copycat" version of a brand-name drug that the FDA has declared to have the identical amounts of active ingredients as its brand-name counterpart.

HATCH-WAXMAN ACT OF 1984: Also known as the Drug Price Competition and Patent Term Restoration Act of 1984, this act made it possible for the first time for generic drug makers to manufacture and market copies of name-brand drugs whose patents had expired.

"INNOVATOR" OR "PIONEER" DRUG PATENTS: Patents issued for a 17-year period from the point at which a new drug application is made.

NARROW THERAPEUTIC INDEX (NTI) DRUGS: A class of drugs in which very small changes in dosage level appear to have a substantive effect on efficacy.

NDA (NEW DRUG APPLICATION): Document submitted by a research-based pharmaceutical company to the FDA as a request for approval to market a given drug. The NDA is submitted after the company has significant proof of the safety and efficacy of the drug.

Forty-four percent of all prescriptions filled in this country today (as measured in total countable units, such as tablets and capsules) are generic drugs, explained Clay O'Dell of the Generic Pharmaceutical Association, a trade group that represents generic drug manufacturers. Eighteen years earlier that figure, a mere 18.6%, jumped to 36.6% in 1998 before it surged to today's high of 44%. The most acknowledged contributing factor to that dramatic rise in sale of generic drugs is the Hatch-Waxman Act of 1984 (also known as the Drug Price Competition and Patent Term Restoration Act of 1984). The act, sponsored by Utah conservative Orrin Hatch in the Senate and California liberal Henry Waxman in the House of Representatives, made it possible for the first time for generic drug makers to manufacture and market copies of name-brand drugs whose patents had expired.

With these increases in generic prescriptions in mind, the Congressional Budget Office recently estimated that consumers save 8–10 billion dollars a year at retail pharmacies by using generic drugs. That saving is seen by many as quite substantial in light of the staggering $121.8 billion spent on prescription drugs in the United States in 2000, a startling increase in spending from the previous year of 17.3%. Increases such as this have been associated with the 11% increase in insurance premiums reported in this country in 2001. Many health maintenance organizations (HMOs) and insurance providers have been, as a result, requiring that generic drugs be substituted whenever possible to save money. Answers to the questions, " Are generic drugs as effective as name-brand drugs?" " Why is it possible to offer consumers generic drugs at costs lower than the costs of name-brand drugs?" and " Who benefits from the savings when a generic drug is substituted for a name-brand drug?" spotlight reasons why HMOs and insurance providers have the right to substitute generic drugs for name-brand drugs in order to save money.

Are Generic Drugs the Same as Name-brand Drugs? Generic drugs are as safe and effective as name-brand drugs, according to Vincent Earl Pearson, Pharm. D., Clinical Coordinator for Drug Information for the Department of Pharmacy at Johns Hopkins Hospital. Dr. Pearson explained that the FDA regulates all generic drugs using strict guidelines, requiring the same testing as name-brand drugs to ensure that the generic is interchangeable with its name-brand counterpart or that the two drugs are therapeutically equivalent.

Since generic drugs usually sell for less than name-brand drugs, some consumers deduce (as they do when they evaluate generic food and household products) that they are paying less for a lower standard of quality and, therefore, that generic drugs must be inferior to name-brand drugs. That is simply not the case, according to Doug Sporn, Director of FDA's Office of Generic Drugs, who argued that generic drugs are not inferior to name-brand drugs. Even though they are less expensive (sometimes half the price of their name-brand equivalent) than name-brand drugs, generic drugs contain exactly the same active ingredients as name-brand drugs, and they are just as safe and effective. Ronni Sandroff, editor of *Consumer Reports on Health* concurred with Sporn when he stated that consumers often "get the same for less" when they opt for generic drugs.

The FDA, the agency that ensures that drugs are safe and effective for their intended uses, dictates that generic drugs must contain the same, identical amount of active ingredients as their name-brand counterparts, and in the identical dosage. Furthermore, the generic drug must deliver the same amount of those active ingredients into a patient's bloodstream within the same time frame as the name-brand drug. And, finally, the generic drug must also fall into "acceptable parameters" established by the FDA for bioavailability, which is the extent and rate at which the body absorbs the drug.

So, besides price, generic drugs are essentially the same as name-brand drugs with the only real discrepancy being the inactive ingredients, such as coatings, fillers, binders, colorings, and flavoring. These inactive ingredients, which have been associated with differences in the size, color, and shape of generic and name-brand pills and capsules, are, in general, harmless substances that have little or no effect on the consumer's body. However, in some sensitive consumers, inactive ingredients may cause unusual and sometimes severe adverse reactions. For example, preservatives, such as sodium metabisulfite, have been associated with asthmatic allergic reactions in a number of consumers. As another example, consumers who are allergic to wheat may experience a reaction to drugs that use wheat fiber as a bulking agent (to facilitate quick passage through the gastrointestinal tract), while some other consumers might be allergic to tartrazine, a coloring agent used in certain medications. Two points regarding reaction to inactive ingredients in drugs are worth mentioning. First, in most situations where consumers encounter problems with allergic or adverse reactions, the reactions are treated before they are able to affect the health of the consumer. Second, allergic reactions to inactive ingredients in drugs are just as likely to occur with name-brand drugs as they do with generic drugs.

Besides being associated with allergic or adverse reactions, inactive ingredients in drugs have also been associated with the extent and rate at which the body absorbs active ingredients (i.e., the drug's bioavailability). Some researchers and physicians express concern that a decrease in absorption in certain drugs can make the drugs ineffective or that an increase in absorption can render them dangerous, but most agree that for most people taking most drugs, the slight difference in absorption of active ingredients is insignificant.

Why Is It Possible to Offer Consumers Generic Drugs at Costs Lower Than the Costs of Name-brand Drugs? The reason consumers are able to purchase generic drugs at costs lower than the costs of name-brand drugs is because the procedure required for FDA approval of generic drugs does not require the costly, lengthy clinical trials to prove safety and effectiveness that FDA approval for name-brand drugs requires.

FDA Approval of Name-brand Drugs. Since name-brand drugs, also referred to as "innovator" drugs, are the first to appear on the market horizon, their manufacturers must obtain FDA approval via a new drug application (NDA). The NDA contains data and information from time-consuming and expensive animal and clinical trials that demonstrate to the FDA that the manufacturer can produce a drug that is safe and effective for the disease or condition for which the manufacturer proposes to market it. According to the Pharmaceutical Research and Manufacturers of America (PhRMA), the lobby group for name-brand drug companies, the intricacy and time that extensive animal and clinical testing command are the prime reasons research costs often aggregate into hundreds of millions of dollars. Once the NDA is initiated, the manufacturer is issued a patent that lasts 17 years. During that time, no other drug manufacturer is allowed to create and market a generic equivalent of that drug. It is important to point out that since a sizable portion of the 17 years is sometimes spent in the FDA approval process, the name-brand manufacturer has less than 17 years to market the new drug. One case in point is Claritin (one of the first non-drowsy allergy products), whose parent company, Schering-Plough, spent over 10 years in the FDA approval process, leaving the pharmaceutical giant only seven years to market and sell the drug.

FDA Approval of Generic Drugs. The manufacturers of generic drugs, also known as "copy cat" drugs, are not required to replicate the extensive testing and clinical trials that were required of pioneer or name-brand drug manufacturers, since the generic drug is a copy of the name-brand drug and since the research will have been recently completed by the name-brand manufacturer. Instead, once the name-brand drug's patent expires, the generic drug manufacturer begins the procedure for FDA approval via an abbreviated new drug application (ANDA). The approval process usually takes about two years, in contrast to the approval process for name-brand drugs, which as mentioned earlier, can take up to 10 years. The ANDA contains data and information that show that the generic drug manufacturer can produce what is essentially a copy of the name-brand drug along with information that shows that the copy is bioequivalent to the name-brand drug, meaning that the generic drug must deliver the same amount of active ingredients into the patient's bloodstream in the same amount of time as the name-brand drug.

The affordability of prescription drugs is a key issue in the debate over generic drugs. Here, former President Bill Clinton urges government action to make drugs more affordable for rural seniors. *(© Reuters News Media, Inc./Corbis. Reproduced by permission.)*

Who Benefits When Generic Drugs Are Substituted for Name-brand Drugs? Generic drugs continue to represent one of the greatest values in U.S. health care, and as such are benefiting a number of different individuals and organizations.

Financial Benefits. The first and most obvious beneficiary of the money saved by substituting generic drugs for name-brand drugs is the patient who has been prescribed a given drug. If the patient is paying for a generic drug out-of-pocket, chances are that the cost would be one-third to two-thirds the cost of the name-brand drug. For the patient who is covered by health care insurance (e.g., HMOs, PPOs), the benefit would likely be lower premiums and lower co-payments. And finally, the benefit for the patient who insists on purchasing the name-brand drug would be a reduced price because of competition from the generic drug.

HMOs and insurance providers would benefit by making more profit on lower premiums. Even though the cost of reimbursement would be lower, the benefits would mount because the lower premiums would attract more patients, resulting in more incoming premiums.

Pharmacies and pharmacy benefit manager (PBM) programs would benefit from a higher margin of sales because the lower prices of generics would spur more sales.

Finally, society as a whole would benefit because substitution of generic drugs for name-brand drugs would threaten the profit margin of the name-brand pharmaceutical companies, motivating them to develop more new and novel drugs that they could patent and market for monopolistic prices.

Health Benefits. The substitution of generic drugs for name-brand drugs would help to alleviate some of the health care problems initiated by the high cost of prescription drugs. Generic drugs would make affordable to millions of Americans many of the medications their physicians have been prescribing. As a result, there would be less unnecessary suffering, and in some cases, less deaths. Generic drugs would also minimize practices being used to extend prescriptions, such as "self-prescribing," where patients take medicines less frequently than prescribed, or "half-dosages," where patients take half the prescribed dose. Needless to say, these practices are dangerous because they exacerbate existing health problems or create new ones.

Future Issues Several new issues have arisen in the debate on generic versus name-brand drugs. Pharmaceutical companies have found loopholes in the Hatch-Waxman Act that have allowed them to inappropriately extend their patents. The Greater Access to Affordable Pharmaceuticals Act (GAAP Act) has been designed to combat this. Sponsors of the act, Senators John McCain and Charles Schumer, recently predicted that the act could save consumers up to $71 billion in 10 years.

Another issue involves the narrow therapeutic index (NTI) drugs, which make up less than 5% of all prescription drugs. NTI drugs have been described as drugs in which even very small changes in dosage level can produce serious side effects or decreased efficacy. In spite of the FDA's specific statement in 1998 that generic NTIs are equivalent to name-brand NTIs, and that there are no safety or effectiveness concerns about switching from one to the other, many physicians and pharmacists are unwilling to switch patients from a name-brand to generic NTI or visa versa. The drugs known to have an NTI include Premarin (hormone replacement therapy), Dilatin and Tegretol (anticonvulsants), Theophylline (asthma and lung diseases), Coumadin (blood thinner), Lithium carbonate (antidepressant), and Cyclosporine (antirejection drug for organ transplant patients). —ELAINE H. WACHOLTZ

Viewpoint:

No, HMOs and insurance providers do not have the right to require the substitution of generic drugs for name-brand drugs. Even if the active ingredients are the same, differences in the inactive ingredients might have a significant effect on some patients.

Measuring Value: The Controversy over Generic Drugs There aren't very many people in the United States today who are unaffected by the rising cost of health insurance. Managed care companies often play a big role in the equation, claiming that they help to keep costs down by eliminating unnecessary services and grouping together physicians who agree to provide treatment at set prices. In exchange, the physicians or providers, as they are sometimes called, are placed on the insurance company's referral list. A prescription card is often a separate part of the plan and can be offered by the primary insurance provider or an adjunct provider.

With pressure from all sides, insurance companies have been forced to examine their premiums and try to keep costs down. The managed care departments of large companies often negotiate contracts with various insurance companies in an effort to obtain the best possible group rates. This, of course, is in the best interest of the employer, since many of them pay a portion of their employees' insurance premiums as a company benefit. So, it appears that everyone has something to gain by cutting costs. But what is the price to the patient? How does this emphasis on cost cutting affect the quality of health care?

Many HMOs and insurance companies claim that patients still receive what they need, but at a lower cost. However, as we all know, there are good insurance companies and there are bad insurance companies. There are good HMOs and bad HMOs. And the truth is that these companies are businesses; they operate for profit. As such, they apply corporate values to the management of health care and that doesn't always mean that they have the patient's best interest at heart. After all, there's a reason why the word "managed" comes before the word "care."

Many insurance companies are driven by the "bottom line." The leaders of these corporations must answer to their shareholders and many times are driven by their own need to maintain personal wealth. To complicate matters, they also face pressure from various consumer groups who demand lower cost insurance, especially with regard to prescription drugs. So, ultimately a dollar amount is placed on health.

But here's the problem: how do you place a value on human life? What do you think your health is worth? Shouldn't your doctor be the one to choose what drug is best for you, not your insurance provider? What happens if you need to take a drug that is classified as a narrow therapeutic index drug (a drug that requires careful dosing)? Are all generic drugs truly the same as name-brand drugs? These are just a few of the questions people are asking as the controversy over generic drugs continues.

Close Is Not Enough Those who favor the use of generic drugs claim that they are the same as their name-brand counterpart. They argue that generic drugs cost less because they don't require the same level of marketing and development expenses. Even the Unites States Food and Drug Administration (FDA) has told us that generic drugs are just as good as name-brand drugs. Why then do some people question the substitution of generic drugs for name-brand drugs? Issues of bioequivalency are at the heart of the debate. In an article on generic drugs from the Crohn's & Colitis Foundation of America (CCFA), it is explained that a generic drug may contain the same active ingredients as the original drug, but it may not be the same. So, the FDA requires that manufacturers submit proof of bioequivalency. This means that if two drugs are to be considered bioequivalent, they are supposed to produce the same therapeutic effect and have the same level of safety. However, not everyone is convinced that every generic drug on the market is truly bioequivalent. In fact, all you have to do is analyze the way generic drugs are produced and tested to have some doubts regarding their bioequivalency.

In an article for the American Council on Science and Health, Jean Eric, a physician, explained that unlike name-brand drugs, generic drugs are rarely tested for safety and efficacy on patients. Instead, Eric explains, the innovator drug and its generic counterpart are administered in a highly structured setting to a small number of healthy patients—almost always fasting, young male adults—and blood concentrations of the drug are measured over time after just one dose. Eric goes on to say that if the blood levels for each product are similar (usually within 20% of each other), the FDA will declare the generic product equivalent to the innovator product. There is an obvious problem, however. Under real life circumstances, the drugs will be taken in multiple doses and by patients who fall into a variety of demographic and medical categories. Factors such as age, sex, diet, and physical activity can all affect drug absorption. In addition, generic drugs can contain fillers that are not present in the name-brand drug. These fillers can sometimes affect the drug's absorption into the bloodstream and can cause unexpected side effects. And the HMOs and insurance companies are aware that side effects from generic drugs can occur. Some of them even have policies in place to allow patients to appeal the requirement of having to take a generic drug when one is available. Unfortunately, most companies require that the appeal be made after the adverse reaction has occurred, which may seem logical from a cost-savings point of view, but it's the patient who suffers the consequences. Naturally, patients who can afford name-brand drugs might decide to simply pay the out-of-pocket expenses rather than go through all the red tape, but what about those patients who can't afford it? Does this make them any less deserving of quality health care?

A significant study conducted by Dr. Susan Horn, a senior scientist at the Institute for Clinical Outcomes Research at the University of Utah School of Medicine, was published in the *American Journal of Managed Care* and raised serious questions regarding generic drug equivalency. The study analyzed over 12,000 patients in six different HMOs. It was found that formulary limitation, which is the practice of forcing physicians to prescribe cheaper generic drugs instead of more expensive name-brand drugs, resulted in many more emergency-room visits and hospitalizations among those patients who received the generic substitutes. The irony was that it cost the HMOs more money in the long run.

The Integrity of the Doctor/Patient Relationship The integrity of the doctor/patient relationship should be maintained. The HMO should not be able to override decisions made by the examining physician who knows his or her patient and assumes the legal liability for the patient's health. In fact, unreasonable restrictions placed on physicians by some HMOs and insurance companies have caused some physicians to opt out of certain insurance plans, leaving patients stranded or faced with difficult financial choices.

From an ethical point of view, the real decision-making power should be balanced between the patient's wishes and the physician's advice. Many patients are learning the importance of being "informed" and often make decisions in tandem with their physicians. In fact, studies have shown that when a patient feels empowered, healing is facilitated. Insurance companies and HMOs should promote this process with flexible plans. This increasing trend by HMOs and insurance companies to limit a physician's decision-making ability with regard to his or her patient's treatment is a scary one and ultimately ill-advised, especially when you consider that certain drugs require careful dosing and must be monitored closely. These drugs are usually referred to as "narrow therapeutic index" drugs. Blood thinners and anticonvulsants fall into this category. For patients who need to take these types of drugs, a generic drug may not be the best choice. In fact, in an article for the Center for Proper Medication Use, David V. Mihalic cautions that it is not advisable to switch between name-brand and generic drugs in this category. He points out that generic drugs are rated with a two-letter code and that although AA-rated generic drugs have essentially the same absorption rate as name-brand drugs, most generic drugs are rated AB. For a drug to be

THE REDUX DEBACLE

Although pharmaceutical companies and the FDA can be aware of potential health risks associated with the use of prescription drugs, that in no way means they will acknowledge these risks until it is too late. The FDA approved the short-term use of Pondimin (fenfluramine) and Redux (dexfenfluramine) for treating obesity, despite being aware that the drugs could cause primary pulmonary hypertension (PPH), a potentially fatal heart disease. To make matters worse, some people allege that the manufacturers even went so far as to downplay the drugs' link to PPH, ignoring the potential health risks. No approval to use the drugs together had ever been given by the FDA. Nor was approval given for these drugs to be used in conjunction with Phentermine, the most commonly prescribed appetite suppressant. Because of the effectiveness these combinations, or "cocktails," provided in weight management, Fen-phen (Pondimin and Phentermine) and Dexfen-phen (Redux and Phentermine) became extremely popular and widely prescribed. By the summer of 1997, the Mayo Clinic began reporting numerous cases of heart valve troubles and PPH directly associated with the use of Pondimin, Redux, and their cocktails. Ironically, obese patients hoping to lose weight and avoid the risk of heart disease now found themselves with a serious threat to their health. As thousands of heart disease cases poured in, the FDA pulled Pondimin and Redux from the market in September of 1997. Class action lawsuits continue to this day, and many patients remain under close medical supervision for their life-threatening heart damage.

—Lee A. Paradise

rated AB, it means that any bioequivalency problems relating to absorption have been resolved to the satisfaction of the FDA. However, B-rated generics do not have the same bioequivalency as their name-brand counterparts. Especially interesting is the fact that there is usually only one co-payment listed on a patient's prescription card with regard to generic purchases; there aren't several co-payments to reflect the two-letter coding system associated with generic drugs. Some might say it's easier to just have one co-payment for all generic purchases and that so much money is saved on the use of generic drugs that there is no need to have different co-payments. However, a more skeptical person might suggest that only one co-payment is used so that the differences between various generic drugs are obscured.

The Politics of It All Maybe a little skepticism is a good thing. After all, the patient often gets lost in a political tug of war that makes the truth sometimes hard to find. One needs only to read the March 22, 2000, letter addressed to the Department of Health and Human Services from Sidney M. Wolfe, M.D., director of the Public Citizen's Health Research Group, to see how pressure from powerful sources can possibly affect FDA drug approval. In his letter, Dr. Wolfe asked that the FDA be held to the Code of Ethics for Government and under no circumstances should the agency be allowed to succumb to outside pressures. When FDA medical officers Robert Misbin and Leo Lutwak opposed the approval of Rezulin, Dr. Wolfe states that an internal affairs committee harassed them and ultimately their concerns about the drug's safety went unheeded. The result was disastrous. Although the drug was approved in January 1997, it was withdrawn from the market in March 2000, because it was linked to liver toxicity and the FDA had to admit that their medical officers had been right all along. So why did Rezulin, which later had a generic counterpart, get FDA approval, even when the FDA's own medical officers expressed serious reservations? The answer is simple: politics. The FDA does a lot of good work, but the agency isn't infallible.

Add to the equation the pressure for new drugs and instant cures and it isn't hard to see why some drugs may be approved, even if they shouldn't be. If a name-brand drug can be approved in the face of strong opposition from FDA scientists and physicians, so can a generic drug. In fact, the chances might even be greater, because a generic is not a "new" drug; it is often viewed merely as a lower-priced copy of the original. Of course, generic drugs aren't approved by the FDA automatically, but the process for getting them approved is often less stringent than obtaining approval for name-brand drugs. That is why only the attending physician—not the HMO or the insurance company—should have the right to decide which drug a patient should receive and whether a generic substitute is acceptable. After all, it is the physician who is responsible for the well-being and health of the patient, not an administrator at an insurance company.

Furthermore, patients, along with their employers, *pay* for medical benefits. A government employee in the state of Texas, for example, might pay as much as $280 a month to cover a spouse and three children. This doesn't include the equal contribution that is made by the employer. For over $400 a month (which doesn't include all the $20 co-payments here and there for prescriptions and doctor's visits), doesn't the patient have a right to obtain good medical care? It isn't as though patients expect charity from the HMOs and the insurance companies. People pay premiums and in exchange they expect quality health care. For most people, that means they should be able to choose their physician and get the best treatment for their specific condition.

Conclusion When it comes to "rights," a patient's well-being should be more important than the financial interests of an HMO or insurance company. However, many HMOs and insurance companies continue to regard the "bottom line" as their number one priority. This shortsightedness blinds them to the long-term results of their decision-making, especially when it comes to forcing generic drugs on patients. Health care is not based on short-term solutions. Cutting costs in the area of patient care will only lead to more spending tomorrow.
—LEE A. PARADISE

Further Reading

Beers, Donald O. *Generic and Innovator Drugs: A Guide to FDA Approval Requirements.* New York: Aspen Publishers, 1999.

Cook, Anna. *How Increased Competition from Generic Drugs Has Affected Prices & Returns in the Pharmaceutical Industry.* Collingdale, PA: DIANE Publishing Co., 1998.

Eric, Jean. "Are Generic Drugs Appropriate Substitutes for Name-brand Drug?—No." American Council on Science and Health [cited July 16, 2002]. <http:www.acsh.org/publications/priorities/1001/gendrugno.html>.

Jackson, Andre J., ed. *Generics and Bioequivalence.* Boca Raton, FL: CRC Press, 1994.

"Letter to the Department of Health and Human Services Urging That They Implement and Enforce the Code of Ethics for Government (HRG Publication #1516)." The Health Research Group. The Public Citizen Web site [cited July 16, 2002]. <http://www.citizen.org/publications/release.cfm?ID=6717>.

Lieberman, M. Laurence. *The Essential Guide to Generic Drugs.* New York: HarperCollins, 1986.

"Medications: Generic Drugs." The Crohn's and Colitis Foundation of America Web site [cited July 16, 2002]. <http://www.ccfa.org/medcentral/library/meds/generic.htm>.

Mihalic, David V. "Generic Versus Name-brand Prescription Drugs." The Center for Proper Medication Use Web site [cited July 16, 2002]. <http://www.cpmu.org/Generics.html>.

Mohler M. and S. Nolan. "What Every Physician Should Know About Generic Drugs: Are Generic Drugs Second-class Medicine or Prudent Prescribing?" *Family Practice Management* 9, no. 3 (2002).

Sandruff R. "Can You Trust Generics?" *Consumer Reports on Health* 13, no. 7 (2001): 2.

Sporn Doug. "FDA Ensures Equivalence of Generic Drugs." Food and Drug Administration website [cited June 15, 2002]. <http://www.fda.gov/fdac/special/newdrug/generic.html>

Is universal assessment of students a realistic solution to the prevention of school violence?

Viewpoint: Yes, to prevent school shootings, which have become a threat to the public health, school psychologists or social workers should evaluate all students for signs of mental instability or a potential for violence.

Viewpoint: No, universal assessment of students is not a realistic solution to the prevention of school violence.

Although juvenile violence declined during the last decade of the twentieth century, a series of school shootings shook the nation. Between 1995 and 2000, students at 12 schools planned and carried out shootings that resulted in the deaths of several students and teachers at each school. In other incidents, shootings resulted in at least one death per school. School shootings occurred in Washington, Alaska, Mississippi, Kentucky, Arkansas, Pennsylvania, Tennessee, Oregon, Virginia, Colorado, Georgia, New Mexico, Oklahoma, Michigan, Florida, and California. In July 1998, President Bill Clinton said that this series of school shootings had "seared the heart of America."

In response to these shocking acts of violence, school authorities, psychologists, government officials, law enforcement agencies, and criminologists began to explore and debate possible responses to what some news reports called an epidemic of violence threatening all American schools. Critics of the news media argued that school shootings, though a tragedy, did not represent a real public health emergency. Reporters were accused of creating a crisis of fear by labeling a small series of incidents a trend and increasing the potential for copycat crimes. Educators, psychologists, law enforcement agents, and others debated whether the threat of school violence was pervasive enough to justify massive interventions that would affect the climate of American schools and the rights of individual students.

Most studies of the problem of school violence focused on strategies for creating a safer environment and preventing violence at school. Suggestions included stricter gun control policy, increased security, assigning police officers or security guards to patrol schools and their entrances, metal detectors, television monitors, searches of school lockers, telephones in all classrooms, anger-management courses, enriched after-school activities, mentoring, recreational programs, safety drills, suspensions and expulsions, and a zero tolerance policy for students who made threats, carried out any act of violence, or brought weapons to school.

The National Alliance for Safe Schools suggested that if schools were effective in dealing with minor problems, such as disruptive behavior, fighting, bullying, class cutting, and truancy, they might decrease the likelihood of catastrophic events. Some observers argued that external security measures create a false sense of security. Establishing a safe environment at school had to include efforts to monitor and modify the behavior of violence-prone students. One way to focus on students who might present a threat was to have psychologists or social workers evaluate all students for signs of mental instability or a potential for violence.

Forensic psychologists argued that universal assessment would lead to early identification of students who might commit acts of violence at school. Having identified at-risk students, psychologists could send troubled students to appropriate counseling and intervention programs. Such programs would be dedicated to anger management and to teaching alternatives to violent behavior. Denouncing this approach as student profiling, some authorities argued that all students, beginning in kindergarten, should participate in age-appropriate programs that would promote social and cognitive skills, teach problem solving and conflict resolution, enhance self-esteem, and develop positive behavioral attitudes. School programs could be linked to similar programs at the family and community levels.

Mandatory assessment programs raise important civil rights issues of privacy and freedom to choose. Those who support universal assessment of students counter that when the public health is threatened, mandatory interventions, such as vaccinations, can be required, even though certain vaccines may pose some risk to the individual. With methods developed for psychological profiling, which law enforcement agents use to find criminals, psychologists have developed checklists of warning signs that allegedly identify children who may express violent behavior. These programs have been denounced for profiling and stereotyping students on the basis of vague and unproven criteria. Assessment might well lead to a large number of false-positive and miss students who eventually engage in violent acts. Whether or not the tests have any predictive value, the risk of mislabeling and stigmatizing students who fit particular profiles is very real. Critics of universal assessment also argue that universal screening would divert scarce resources from more important school programs.

The most effective approach to school shootings remains a matter of debate, but researchers hope that further studies will provide a better perspective for analyzing school safety issues and focusing the debate on the best way to identify and alleviate the true sources of violence in schools and communities. Although many solutions have been recommended, most observers agree that there are no quick or simple solutions. —LOIS N. MAGNER

Viewpoint:

Yes, to prevent school shootings, which have become a threat to the public health, school psychologists or social workers should evaluate all students for signs of mental instability or a potential for violence.

Making It Mandatory Whenever regulations are proposed that impinge on an individual's freedom to choose, controversy arises, even when those regulations have a positive outcome. For example, mandatory vaccination against certain childhood diseases before a child can enter the public education system prevents, and in some instances has totally eradicated, serious and fatal infectious diseases (smallpox, for example). Yet some parents believe, because of a small risk of serious adverse reactions to certain vaccines, they should have the right to decide not to have their children vaccinated. Allowing individual decisions in certain cases poses a serious risk to public health. Many children are not vaccinated unless it is mandatory, and unvaccinated children can carry a disease without displaying symptoms and can infect others, spreading the disease and placing the entire community at risk. In an ideal world, mandatory regulation would have no negative effects. In reality, negative effects almost invariably exist. Although we must, as a society, respect each individual's right to choose, we also must respect the greater good of society. This perspective must be applied to the concept of mandatory evaluation of all students to help determine the existence of mental instability and potential for violent behavior that could result in the murder of innocent people.

Why School Shootings Occur and How to Prevent Them In the wake of students' indiscriminately and brutally shooting teachers and fellow students, officials are scrambling to find ways to prevent—as far as is possible—future incidents. Although the final decade of the twentieth century saw a reduction in juvenile violence, it also saw an increase in school-based violent episodes in which more than one person was killed. According to an article in the journal of the American Association of School Administrators, between 1992 and 1995 only two school shootings resulted in multiple deaths. From late 1995 until February 2000, there were 12 separate incidents. Society reels at the horror and heartache of children murdering their classmates and teachers, and the question "Why?" hangs in the air. In the meantime, law enforcement agencies, policy makers, and school officials struggle with the question "How?" How can further tragedies be prevented? Prevention methods such as metal detectors, searches, police guards, and television monitors in hallways; zero tolerance of students who perpetrate violent acts, make violent threats, or carry weapons to school; anger-management courses; hostage drills; and the like are all part of the effort. Criminologist William Reisman commented in *The Memphis Conference: Suggestions for Preventing and Dealing with Student Initiated Violence* that metal detectors, videos, and other similar preventive measures only create a false sense of security. According to Reisman, "The problem is the heart and mind of the kid."

Identification and Early Intervention Many behavioral specialists agree that external approaches fail to reach the root of the problem. These experts feel strongly that children are much less prone to violence if they have strong, nurturing relationships with caring, involved adults from a very early age. A 1993 report by the American Psychological Association Commission on Youth Violence concluded that programs aimed at strengthening family relationships, reducing "detrimental life circumstances" early in a child's life, and incorporating similar programs into the "climate and culture of schools" have the greatest positive effect on violence prevention. The report advocates methods such as promoting social and cognitive skills, teaching alternatives to violent behaviors, self-esteem enhancement programs, problem solving and anger management training, and peer negotiation guidance from kindergarten through high school. According to social worker Joann R. Klein writing for the National Association of Social Workers, children reap great benefits from individual and group programs aimed at teaching and enhancing social and cognitive skills. Klein states that the most effective programs involve children young enough to develop positive social behavioral attitudes before they adopt violent methods of dealing with conflict. In accordance with these philosophies, many schools encourage teachers and students to identify children who are aggressive or who are at risk of failing school. These children are connected with social workers or psychologists. Other programs attempt to identify individuals and families at risk and integrate them with school-based programs consisting of community members, parents, families, and peers.

An article titled "Profiling Students: Schools Try a Controversial New Violence Prevention Tool" on <abcnews.com> noted that in light of the increased number of school shooting deaths, school psychologists are designing checklists of warning signs to help teachers identify children who may be predisposed to violence. Several organizations have developed character checklists to help teachers, social workers, and school psychologists evaluate a student's potential for dangerous behavior. These organizations include the National School Safety Center in Westlake Village, California, The American Psychological Association in conjunction with MTV, and criminologist Reisman, who lists 50 indicators in *The Memphis Conference.*

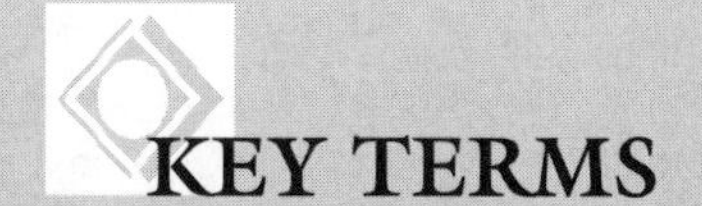

KEY TERMS

INCIDENT PROFILING: Recording details of particular types of incidents to determine commonalities.

PROSPECTIVE PROFILING: A method of creating a profile of the next perpetrator of a particular crime on the basis of analysis of previous perpetrators of the same type of crime.

STUDENT PROFILING: Similar to criminal profiling and FBI behavioral assessment.

TARGETED VIOLENCE: A violent attack in which both the attacker and the target are identifiable before the attack. The target may be a person, such as a specific teacher, or a building, such as the school itself. A term coined by the Secret Service.

WARNING SIGNS: Checklists of traits and behaviors commonly displayed by violent offenders.

Psychological profiling is widely used in law enforcement to help find certain types of criminals. Although historically this method has been avoided as taboo in many public schools, there are signs it is becoming more widely accepted. For example

- In September 2000, the Federal Bureau of Investigation (FBI) National Center for the Analyses of Violent Crime in Virginia issued a report listing dozens of risk factors usually found among school shooters. More than 150 educators, health officials, and police officers, many involved in school shootings, studied 18 cases. (The report emphasizes the list is not a profiling tool and that a student displaying more than one trait is not necessarily a shooter.)
- *Early Warning, Timely Response: A Guide to Safe Schools,* a collaborative effort commissioned by President Clinton between the National Association of School Psychologists, the U.S. Department of Education, and other agencies, was distributed to schools in 1998. It cautions against labeling a child if early warning signs are identified.
- In Granite City, Illinois, school staff are required to identify students who fit an at-risk profile. Factors include the student's preference for violence in media and entertainment and his or her writing of essays expressing "anger, frustration and the dark side of life."

A yearbook photo of Eric Harris, one of two students who carried out the April 1999 massacre at Columbine High School in Littleton, Colorado. *(AP/Wide World Photos. Reproduced by permission.)*

- Certain school districts in Connecticut and Massachusetts have developed districtwide profiles to identify potentially violent students.

The word *profiling* carries negative connotation because of past racial bias and abuse by law enforcement agencies. (The FBI, a recognized authority on the use of profiling, prefers to use the term *behavioral assessment.*) Advocates and opponents alike caution that extreme care is necessary for assessing students. For example, a checklist of early warning signs may be misinterpreted to wrongfully identify some students and cause stereotyping, discrimination, and alienation from teachers and other students. However, well-designed programs used by well-trained authorities can be valid and valuable tools to help identify troubled students and place them in remedial programs.

Stepping Out to Intervene Peter Blauvelt, president of the National Alliance for Safe Schools in College Park, Maryland, is a strong proponent of behavioral assessment. He agrees these programs have potential for abuse but believes there is potential for abuse in many areas in the education system. In an article in *School Administrator,* Scott LaFee quotes Blauvelt as follows: "People just have to be careful not to cross the wrong line." Blauvelt, also the former director of school security in Prince George's County, Maryland, which has a student population of 125,000, prefers to call the method he has implemented "incident profiling." By closely monitoring aggressive and disruptive behavior and reporting details such as where and when an incident occurs and who is involved, school officials can more accurately anticipate a serious problem. Blauvelt calls it a preventive management tool. By analyzing the reports, school authorities can determine which children are sent to the office for which types of violations, can compare incidents to determine commonalities among offenders, and can take steps to avoid more serious situations. Blauvelt remarked that most school districts already collect information about school disruptions but feels that few review or use the information to full benefit. "Reasonable educators just need to develop better internal abilities to monitor and assess what's going on," he said.

Once a student is identified as being potentially dangerous, the question remains what happens next. Steve Balen, superintendent of the Granite City, Illinois, Community Unit School District—probably the first to institute a written policy for psychological assessment of potentially violent students—is quoted by LaFee as saying that although checklists are good, creating a vehicle to respond to troubled students is essential. In Balen's program, the profile of a troubled student includes a record of behavior that may lead to armed violence, grade history, past behavior reports, medical records, and pertinent material from outside sources. The report is evaluated by a team composed of a school resource officer or counselor, a social worker, a psychologist, and an administrator. To maintain distance, three members of the team must not be part of the school district. If the team identifies a problem, the child's parents are approached and asked for permission for psychological testing of the child. Test findings can result in the parents' being ordered to pursue remedial treatment of the child. If the parents refuse, the child can be placed in an alternative education program chosen by the school board. Either way, removing the troubled child from school is not the objective—getting the appropriate help is.

Different forms of profiling have been used for decades, not only in law enforcement but also in industry. Many large companies require potential employees to undergo evaluations specifically designed to help determine emotional intelligence, potential for success in business and in life, personality traits, management potential, and the like. Sophisticated technology allows mental health experts to design software programs to evaluate individuals for self-esteem, integrity, achievement and potential for life success, arousal and sensation seeking, sociability, submissiveness, alcohol and drug use, parental attitudes, emotional empathy, and many other traits and tendencies, including violence and aggression. One such program is the Mosaic-2000, which can provide schools with law enforcement assessment capabilities similar to those used to evaluate threats on Supreme Court justices and Hollywood stars. Mosaic-2000 is being field tested in 25 public schools, primarily in the Los Angeles area, with the participation of the Federal Bureau of Alcohol, Tobacco, and Firearms. In *The Christian Science Monitor,* Gail Chaddock quotes promotional material from the firm that designed and is testing the program: "School is the workplace of children . . . [and] the strategies learned by industries and government should be available to school administrators."

This test program—as do many other evaluation programs already in place—profiles situations rather than individuals. Most individual evaluations are geared toward students who already show significant signs of serious aggression or other antisocial behavior. Mental health and education professionals believe early intervention is the most effective means of curbing violent behavior. Mandatory student evaluation could begin as early as first grade and be conducted once a year, adjusted for age appropriateness, to track each student's emotional and social developmental progress (just as their educational developmental progress is evaluated with tests and examinations). Such a program would facilitate early identification of children with a predisposition to violence or who live in detrimental life situations. Well before violence becomes an integral part of their behavioral patterns, children could be referred to programs that provide the counseling and support necessary to help them become well-adjusted members of society.

Scientifically Unproven but Socially Imperative The Center for the Study and Prevention of Violence (CPSV), at the University of Colorado, Boulder, has published a position summary pertaining to the issue of behavioral assessment. The document notes that no scientific research has been done to determine the effectiveness of such measures. It states that profiling efforts by law enforcement agencies have not been encouraging but goes on to say, "Despite this, profiling seems to remain popular with law enforcement authorities and educators alike."

Mary Leiker, a school administrator for 25 years and superintendent of Kentwood, Michigan, Public Schools, a district that educates 8,600 students, introduced an assessment program similar to that implemented by Balen. Leiker believes that school administrators in other districts will eventually establish similar programs, regardless of their hesitation, because such programs will be a "social imperative." Scott LaFee quotes Leiker: "Profiling isn't something most of us think we're going to do. But when you think about the top issues of education, creating safety in schools is one of the biggest. Profiling doesn't guarantee absolute safety. There is no 100% accurate predictor, but it can be an effective tool. And the fact is, I have to live with myself. If I, as a superintendent and educator, left one stone unturned in trying to keep children safe, if I lost one child because of it, I don't know how I would cope."
—MARIE L. THOMPSON

Viewpoint:
No, universal assessment of students is not a realistic solution to the prevention of school violence.

It is important to remember that although shocking and tragic, school shootings are not the public health epidemic the media would have us believe. In fact, school violence declined in the 1990s, and fewer students carried weapons on school grounds, according to the U.S. Departments of Education and Justice. Nor are school shootings a new phenomenon. The earliest documented case occurred in 1974, and a total of 40 incidents have been recorded since then.

This is not to say that we should not try to prevent school violence, and not just mass killings such as those in Colorado. We must, however, resist the temptation to apply a "quick fix," and instead look for proven alternatives that improve the school experience for everyone involved. There is no universal solution to problems that stem from human behavior, simply because human behavior, especially among teenagers, is unpredictable.

In "Evaluating Risk for Targeted Violence in Schools," Marisa Reddy and colleagues point out that teens who commit murders are *less* likely to have a history of mental illness, arrests, or school adjustment problems than are teens who have committed nonviolent crimes. In addition, teens who commit murders are less likely to have a history of violent behavior than are teens who have been convicted of assault. To further complicate matters, teens who commit targeted school violence (a school attack in which the attacker and the target are identifiable before the attack) may be different even from teens who commit other forms of homicide or delinquency. In reality, it is impossible to predict which student may become violent. Even if the purpose of the assessment is to identify immediate risk (rather than to predict future risk), a universal screening program will not be effective and may do more harm than good. Instead of focusing on the entire student population, resources should be available to attend to individuals who are brought to the attention of school professionals.

The Question of Available Resources Those who call for implementation of a universal assessment program may not realize what resources are needed to follow through and make such a program effective. Finding an assessment tool that is reliable is only the beginning. And if the right tool for assessing risk is found, a system has to be built to handle the assessment and the results.

The mental health professionals who administer assessments have to be trained in correct administration and interpretation. For assessment of an entire student population, we have to find people dedicated only to school assessments and interpretations. School psychologists and counselors are already stretched to the limit and will not be able to shoulder this additional burden. And when the results are obtained, we have to deal with the students who are identified as at risk.

Identifying a student as at risk is only the first step, Joanne McDaniel, director for the Center for Prevention of School Violence in North Carolina, points out. A system is needed for referring students to the appropriate mental health facilities for treatment. The system should ensure students are receiving the help they need, even if their families are unable to foot the bill. Although this is true whenever a student is identified as at risk, a universal assessment program would have some false-positive results. Some students therefore would unnecessarily go through at least the initial steps of the mental health and legal systems. A system would be needed to address the consequences of false-positive results, from purging student records to assisting students in handling the possible emotional and social consequences of false accusation. The resources needed to make universal assessment effective cannot be justified by what we know today about our ability to assess the next perpetrator of school violence.

The Problem of "Prospective Profiling" Assessing students to find potential perpetrators of violent school crimes can be done, according to some school professionals, through a technique known as prospective profiling. With this technique analysts look at previous perpetrators of a crime, such as school shooting, find commonalities, and attempt to predict what the next perpetrator will look like and how he or she might behave.

Ironically, one of the problems posed by prospective profiling is that students who may pose a danger in the school will not set off any alarms because they do not "fit the profile." In an evaluation of school shootings, the U.S. Secret Service studied 37 incidents in which "the attacker(s) chose the school for a particular purpose (not simply as a site of opportunity)." The researchers used primary sources, including interviews with 10 of the shooters. The report of the results, *Safe School Initiative: An Interim Report on the Prevention of Targeted Violence in Schools,* issued in October 2000, states: "There is no accurate or useful profile of 'the school shooter'." An FBI report titled *The School Shooter: A Threat Assessment Perspective* reached the same conclusion.

Because the profile is unreliable, and because the incidence of targeted school violence is so low, prospective profiling carries with it a considerable risk of false-positive results. This means a student may fit a profile the school has in mind, such as a loner who wears black and dislikes athletes, but this student will probably never pose a risk to the school. The risk then becomes stigmatization of students, and possible legal action against them, when these students are never a danger in the first place.

Reddy and coworkers point out results from a survey conducted by *Time* magazine and the Discovery Channel that "indicate that the majority of students polled (60%) disapprove of the use of profiling in schools. . . . Their concerns and fears are based on the potential for unfair use of profiling against students who are not likely to be violent." Students and parents alike worried that prospective profiling would be used to target students different from most of the other student body, as in race or sexual orientation.

Other Assessment Techniques To date, no proven tool will give an accurate assessment of the "next school shooter." Checklists often are used by mental health professionals and school officials and faculty to evaluate students brought to their attention for one reason or another. However, no reliable research has established criteria for evaluating a potential perpetrator of school violence. The checklists used today are based on characteristics of violent teens and psychiatric patients. Perpetrators of targeted school violence differ from these groups in several ways.

The psychological tests used to look for mental disorders are of no use in looking for the next school shooter. According to Reddy and coauthors, there is currently no information about the "prevalence, nature, or role" of mental disorders in targeted school violence. No research has proved the current psychological tests for mental disorders useful in identifying a student who may perpetrate targeted school violence.

Dylan Klebold (right) and Eric Harris (left) murdered fourteen classmates and one teacher at Columbine High School on April 20, 1999.
(AP/Wide World Photos. Reproduced by permission.)

McDaniel points out that assessment tools may work well in a research environment, but implementing them in the schools changes the conditions for the assessments. The staff is overworked and may not be trained properly, for example. This change in conditions harms the integrity of the assessment tool and may lead to an increased incidence of false-positive findings.

Potential for Abuse in Psychological Assessments In 1995, elementary schoolchildren in Pittsburgh were subjected to psychological testing without many parents' consent or knowledge. Children five to ten years of age were evaluated with a tool called the Pittsburgh School-Wide Intervention Model (P-SWIM). Funded partially by a grant from the National Institute of Mental Health, the battery of psychological tests required children to answer questions such as "Have you ever forced someone to have sex?" and "Have you ever tortured an animal?" Even five-year-olds had to answer these questions. Parents were not allowed to see the results of the tests or the interpretations. The incident came to light when parents started to notice marked changes in their children's behavior. Children were displaying stress-related symptoms such as nightmares, bed-wetting, and headaches. The parents filed suit in 1997. The suit was finally settled four years later with a monetary award and a promise to destroy the records pertaining to the test. Supposedly, the testing was conducted to diagnose hyperactivity and other behavior problems in the students.

According to McDaniel, there is no federal policy regulating informed consent in psychological assessment of students. In some areas, simply enrolling in a school implies consent to various assessments that may be performed on the students, and the parents may not be aware of the kind of questions in the assessment. If psychological assessments are to become routine for preventing violence in schools, informed consent of parents and full sharing of the results must become federally mandated. It is too easy to abuse a system in which implied consent for psychological testing is the rule. Assessment records stay in students' files and may follow the students into adulthood, possibly interfering with the pursuit of higher education and a career.

Alternatives to Universal Screening

Although the threat of school violence is not of epidemic proportions, it is important to take steps to minimize this threat as much as possible. Universal screening is clearly not the answer, but what alternatives do we have?

The first step should be a shift in attitude, as Reddy and colleagues point out, from *prediction* to *prevention*. The goal should be to prevent school violence, not to predict who may carry out the next attack. To that end, schools should strive to create an environment of trust and support, where communication between students and adults is fostered and encouraged.

The Secret Service noted that an overwhelming majority of school attackers had told at least one peer about their plans. These conversations were rarely reported to adults. One of the implications, according to the Secret Service report, is that schools should strive to decrease the barriers between students and faculty and staff.

In an overwhelming number of incidents, the person who carried out the attack was already on the "radar screen" of a peer or even a school official. Many times, the would-be attacker displayed suicidal tendencies before the attack or had difficulty coping with a recent major event in life. Schools with systems for handling these students in a timely manner are still the exception rather than the rule. Why go through the trouble of assessing the entire student population if we still cannot help those we know to be at risk?

The CSPV launched a national violence prevention program in 1996. Blueprints for Violence Prevention identifies programs around the country that have been scientifically proved to reduce adolescent violence, substance abuse, and delinquency. The programs range from school-based curricula to community-based therapeutic approaches. These programs can be adapted to fit various communities in the United States at relatively low cost and have a proven record of success.

The most reliable tool for identifying students at risk is the classroom teacher. In a school environment that cares and nurtures students, the teachers are sensitive enough to know whether a student's change of behavior is reason for concern. Because "normal" behavior for children is such a wide continuum, it is teachers, not psychologists, who can determine best which student needs help. They should be able to do so not with checklists but through their knowledge of their students as individuals.

The U.S. Department of Education treats violence prevention efforts as a pyramid. At the base of the pyramid is a schoolwide foundation. This foundation is built, in part, on the following principles:

- Compassionate, caring, respectful staff who model appropriate behavior, create a climate of emotional support, and are committed to working with all students.
- Developmentally appropriate programs for all children to teach and reinforce social and problem-solving skills.
- Teachers and staff who are trained to support positive school and classroom behaviors.

This foundation facilitates identification of at-risk students, according to the Department of Education, because it fits the needs (in terms of improving behavior and academic performance) of all but the students who need extra intervention. For those who do need extra intervention, the next step in the pyramid is early intervention. This step, for which mental health professionals are needed, may be anything from brief counseling to long-term therapy. Students who exhibit very severe academic or behavioral problems go to the tip of the pyramid—intense intervention. The system relies on knowledgeable, caring staff—not assessment tools—to identify students with early warning signs. The signs that a student needs help are fairly obvious to a caring teacher. But once a student is identified, the teacher should be able to refer him or her to a facility for helping children. This is where our resources would be much better spent.

"For every problem, there is a solution which is simple, neat, and wrong." H. L. Mencken was not talking about response to school violence. The FBI, however, in *The School Shooter,* aptly chose this quote in warning about what it characterized as "a knee-jerk reaction, [in which] communities may resort to inflexible, one-size-fits-all policies on preventing or reacting to violence." The call for universal screening of students is exactly such a reaction. It is risky and not useful. Resources to combat targeted school violence are much better spent in establishing community-wide programs equipped to handle and support troubled students. According to McDaniel, research has shown that when they keep education as their mission and priority with safety and security seamlessly integrated into the learning environment, schools are more successful overall, even when operating in a challenging environment. Available resources, therefore, would be much better spent on creating schools that teach rather than schools that police their students. What we need is a system ready to help, not a system that is busy predicting what a student may do. —ADI R. FERRARA

Further Reading

"Profiling Students: Schools Try a Controversial New Violence Prevention Tool," *abcnews.com* September 7, 1999 [cited July 15, 2002]. <http://www.abcnews.go.com/sections/us/DailyNews/profiling990907.html>.

Blueprints for Violence Prevention. Center for the Study and Prevention of Violence [cited July 15, 2002]. <http://www.colorado.edu/cspv/blueprints/>.

Center for the Study and Prevention of Violence. *Safe Communities—Safe Schools Planning Guide.* Boulder, CO.: Center for the Study and Prevention of Violence, 2000.

"Student Profiling." *CSPV Position Summary.* Center for the Study and Prevention of Violence [cited July 15, 2002]. <http://www.colorado.edu/>.

Chaddock, Gail Russell. "A Radical Step for School Safety." *The Christian Science Monitor.* January 13, 2000 [cited July 15, 2002]. <http://www.csmonitor.com/>.

Dwyer, K., and Osher, D. *Safeguarding Our Children: An Action Guide.* Washington, DC: U.S. Departments of Education and Justice, American Institutes for Research. 2000 [cited July 15, 2002]. <http://cecp.air.org/guide/actionguide/Action_Guide.htm>.

Dwyer, K. P., D. Osher, and C. Warger. *Early Warning Timely Response: A Guide to Safe Schools.* Washington, DC: U.S. Department of Education, 1998.

Fey, Gil-Patricia, J. Ron Nelson, and Maura L. Roberts. "The Perils of Profiling." *School Administrator* 57, no. 2 (February 2000): 12–16. American Association of School Administrators. February 2000 [cited July 15, 2002]. <http://www.aasa.org.publications/>.

"Home Snoops." *The Washington Times.* October 25, 1999.

Klein, Joann R. "Violence in our Schools: The School Social Work Response." *The Section Connection* 4, no. 2 (August 1998): 8–9. National Association of Social Workers.

LaFee, Scott. "Profiling Bad Apples." *School Administrator* 57 no. 2 (February 2000): 6–11.

Reddy, Marisa, et al. "Evaluating Risk for Targeted Violence in Schools: Comparing Risk Assessment, Threat Assessment, and Other Approaches." *Psychology in the Schools* 38, no. 2 (2001): 157–72.

Reisman, William. *The Memphis Conference: Suggestions for Preventing and Dealing with Student Initiated Violence.* Indianola, IA: Reisman Books, 1998.

The School Shooter: A Threat Assessment Perspective. Federal Bureau of Investigation. 2000 [cited July 15, 2002]. <http://www.fbi.gov/publications/school/school2.pdf>.

Vossekuil, Bryan, et al. *U.S.S.S. Safe School Initiative: An Interim Report on the Prevention of Targeted Violence in Schools.* Washington DC: U.S. Secret Service, National Threat Assessment Center. October 2000 [cited July 15, 2002]. <http://www.secretservice. gov/ntac/ntac_ssi_report.pdf>.

Are OSHA regulations aimed at preventing repetitive-motion syndrome an unnecessary burden for business?

Viewpoint: Yes, OSHA regulations are an unnecessary burden for business: not only are the costs prohibitive, but the benefits are highly questionable.

Viewpoint: No, OSHA regulations aimed at preventing repetitive-motion injuries are straightforward and save businesses time and money in the long run.

Evidence of illnesses caused by occupational conditions and hazards is very old. Indeed, *On the Diseases of Trades*, a work often lauded as the first account of the diseases of workers, such as stone mason's consumption, painter's palsy, and potter's sciatica, was published by Bernardino Ramazzini in 1700. In twentieth-century America, Alice Hamilton served as pioneer of occupational medicine and industrial toxicology and worked tirelessly for the prevention of work-related injuries and illnesses. By 1916 Hamilton was internationally known as America's leading authority on industrial toxicology and occupational medicine. Having established the dangers of lead dust, she went on to investigate the hazards of arsenic, mercury, organic solvents, radium, and many other toxic materials. When Hamilton began her exploration of the "dangerous trades," about the only method used to deal with occupational poisoning was to encourage rapid turnover of workers.

Despite considerable progress in occupational medicine since Hamilton's era, employers and governments have generally been reluctant to accept and enforce laws that guard worker safety, health, and well-being. In 1970, as the result of widespread worker complaints and union activity, the United States passed the Occupational Health and Safety Act, which established the Occupational Safety and Health Administration (OSHA). Congress passed this act in order to "assure safe and healthful working conditions for working men and women; by authorizing enforcement of the standards developed under the Act; by assisting and encouraging the States in their efforts to assure safe and healthful working conditions; by providing for research, information, education, and training in the field of occupational safety and health."

To combat repetitive-motion injuries and other musculoskeletal disorders (MSDs), OSHA developed regulations based on the science of ergonomics and the study of workplace injuries. Ergonomics is a branch of ecology that deals with human factors in the design and operation of machines and the physical environment. Generally considered an applied science, ergonomics makes it possible to design or arrange the workstation of an individual in order to promote safety and efficiency.

The task of drafting many of the OSHA regulations for repetitive-motion disorders began during the administration of George Herbert Walker Bush (1988–1992). In 1990 Secretary of Labor Elizabeth Dole announced that the department was working on rules to protect workers from the painful and crippling disorders associated with the work environment. The actual rules were not published until the end of the decade. In 2001, however, the administration of George W. Bush moved to roll back OSHA's ergonomic regulations and substitute voluntary guidelines established by industry.

Critics of OSHA argue that evidence of workplace-related MSDs is ambiguous and misleading, but the National Institute of Occupational Safety and Health points to hundreds of articles on carpal tunnel syndrome and other MSDs published in respected peer-reviewed journals. MSDs affect millions of workers and result in lost productivity as well as pain and suffering. In a report published in 2001, the prestigious National Academy of Sciences also concluded that the evidence linking workplace risk factors and MSDs was compelling. Even the General Accounting Office and the American Conference of Governmental Industrial Hygienists agreed that there was rigorous evidence to support the proposition that ergonomically designed workstations can reduce MSDs. The Bureau of Labor Statistics reported that ergonomic disorders, especially carpal tunnel syndrome associated with personal computers, were the most rapidly growing category of work-related injuries reported to OSHA. But carpal tunnel syndrome and other MSDs are also found among factory workers, meatpackers, and sewing machine operators.

Opponents of OSHA argue that the cost of implementing the regulations designed to reduce work-related repetitive-motion injuries constitute an enormous and unnecessary burden for industry. OSHA regulations are often described as overly broad, intrusive, confusing, and expensive. In particular, OSHA regulations are criticized as "unfunded mandates" that amount to a form of taxation that reduces productivity and increases the costs for American businesses.

According to OSHA officials, the amount that would be saved in worker compensation, retraining programs, and so forth, by preventing repetitive-motion injuries would be much greater than the costs of implementing the regulations. However, representatives of industry dispute these claims and argue that OSHA consistently underestimates the true costs and problems associated with the regulations. Moreover, opponents of OSHA regulations argue that psychosocial factors such as stress, anxiety, and depression are more likely causes of MSDs than repetitive work activities. Thus, although there is little doubt that many workers suffer from work-related repetitive-motion injuries, disputes about the impact of OSHA regulations often resemble a political polemic rather than a debate about medical issues. —LOIS N. MAGNER

Viewpoint:

Yes, OSHA regulations are an unnecessary burden for business: not only are the costs prohibitive, but the benefits are highly questionable.

When Americans hear the term "federal government" in association with the performance of almost any task other than law enforcement or national defense, the response is likely to be a groan or even a shudder. Washington is notorious for its wastefulness and inefficiency, combined with the excruciating burdens it places on the productive sectors of the national economy—not only middle-class taxpayers, but to an even greater extent, businesses.

The politicians and bureaucrats that make up the federal government depend on business to generate income for them, in the form of taxes, so that they can secure their own power by doling out those resources to others. Given this situation, one would think these politicians and bureaucrats would do everything in their power to enhance the ability of business to create wealth, the fuel that runs that exceedingly inefficient engine called government. Yet, contrary to Washington's ultimate best interests, its legislators and executive decision-makers continue to fight a war on American business, seemingly with the intention of hamstringing productivity and increasing the costs of operation. Among the most nefarious tactics in this war is the use of "unfunded mandates": directives, issued by the government, that require costly compliance procedures on the part of businesses, thus amounting to a *de facto* form of taxation.

Nowhere is the sickness inherent in government and its practice of unfunded mandates more apparent than in the case of OSHA, the Occupational Safety and Health Administration (OSHA). Established by the Occupational Safety and Health Act of 1970, OSHA, which opened its doors in 1971, came at the end of a decade of prosperity in which the federal government hurled buckets of cash down the twin sinkholes of the Vietnam War and President Lyndon B. Johnson's "War on Poverty." The latter succeeded only in creating more poverty. The unrestrained spending of the 1960s helped bring about the recessionary 1970s. With American business already in great trouble—the 1970s saw the shift of leadership in the automotive and electronic industries from the United States to Japan—OSHA placed additional fetters on business that all but crushed American productivity.

Ostensibly, OSHA is charged with enforcing safety and health regulations, and with educating employers and employees alike regarding hazards on the job. The reality is rather different, as illustrated by prominent New York attorney Philip K. Howard in *The Death of Common*

KEY TERMS

CARPAL TUNNEL SYNDROME: A condition that occurs when the median nerve in the hand's carpal tunnel is compressed, causing weakness, pain, and disturbances of sensation in the hand.

ERGONOMICS: An applied science that deals with the design and arrangement of the spaces people occupy so as to ensure the safest and most efficient interaction between the human and his or her environment.

MUSCULOSKELETAL DISORDERS (MSDS): Health problems relating to or involving both musculature and skeleton.

OCCUPATIONAL SAFETY AND HEALTH ADMINISTRATION (OSHA): A branch of the U.S. Department of Labor whose mission is to prevent work-related injuries, illnesses, and deaths.

Sense. "For 25 years," Howard wrote in 1996, "OSHA has been hard at work, producing over 4,000 detailed rules that dictate everything from the ideal height of railings to how much a plank can stick out from a temporary scaffold. American industry has spent billions of dollars to comply with OSHA's rules." After speculating that "All this must have done some good," Howard went on to answer his own implied question: "It hasn't. The rate of workdays missed due to injury is about the same as in 1973."

OSHA Introduces Yet Another Directive In a 2000 poll conducted by *Safety and Health*, OSHA tied with the Internal Revenue Service for last place in a rating of "customer satisfaction" among government agencies. Given the huge popular dissatisfaction with OSHA, as well as its track record for failure and abuse, the business response to new regulations in 2000—these designed supposedly to address injuries caused by repetitive tasks—were unsurprisingly negative.

According to OSHA, workers who perform repetitive tasks such as those involved with manufacturing or even administrative work may be subject to increased risk for disorders of the muscles, skeleton, and nervous system. Statistics furnished by the Labor Department, which oversees OSHA, indicated that nearly 600,000 workplace absences in 1999 were caused by so-called repetitive-stress disorders, including carpal-tunnel syndrome, tendinitis, and lower-back pain. Therefore the administration of President Bill Clinton, represented by Labor Secretary Alexis M. Herman and OSHA director Charles Jeffress, put forth new directives designed to address the problem. These directives paid special attention to ergonomics, an applied science that deals with the design and arrangement of the spaces people occupy so as to ensure the safest and most efficient interaction between the human and his or her environment.

Yochi J. Dreazen and Phil Kuntz in the *Wall Street Journal* described the effects of the repetitive-stress directives: "Businesses with a worker who experiences a single injury covered by the rule would be required to set up a comprehensive program to address the problem in that job. That would mean assigning someone to oversee the program, training workers and supervisors to recognize and deal with ergonomics-related injuries, figuring out how to eliminate or sharply reduce hazards, and providing injured workers with medical care and time off at reduced pay." The repetitive-stress rules were the result of 10 years' worth of work on the part of OSHA, and where bureaucracies are concerned, more time simply leads to ever more imperfect laws: "...in rewriting the rules," Dreazen and Kuntz explained, "the agency made the document substantially longer and seemingly more confusing."

Staggering Costs and Questionable Benefits What about the cost of this unfunded mandate? According to OSHA, it would be a mere $4.5 billion a year—obviously, a sizable chunk of money, even by the standards of the federal government, but considering that it would be spread out over millions of U.S. businesses, this would not pose an extraordinary expense. However, OSHA's estimate was far to the low end compared with other figures.

Even the figure determined by the National Association of Manufacturers, which estimated the cost of compliance with the ergonomics rule at $18 billion a year, was still modest compared to most other appraisals. According to the National Coalition on Ergonomics—which would presumably have had every reason to support the rule, since it would virtually ensure full employment for its members—the rule would cost business $90 billion a year. At the high end, the Employment Policy Foundation placed the figure at $125.6 billion a year, an amount more than 25 times greater than that of OSHA. Assuming about 25 million American workers in the sectors of the economy affected by the ergonomics rule, including administrative and office workers, employees in manufacturing and labor, as well as truck drivers and others subject to repetitive tasks, then a single OSHA ruling would result in businesses paying out about $5,000 extra per worker per year.

Compared to these staggering figures, the one presented by United Parcel Service (UPS) seems fairly small: just $20 billion initially, and $5 billion a year to comply with OSHA's ergonomics directive. However, that figure is for UPS *alone,* and the yearly cost would be twice the company's annual net income!

And for all that expense, would workers actually be helped by the new directives? Not according to physician Alf Nachemson, testifying before a panel assembled by the then-recently installed administration of President George W. Bush in July 2001. According to Nachemson, "complex individual and psychosocial work-related factors," including anxiety, stress, and depression, had much more to do with back pains than did repetitive activities. Nor did Nachemson suggest that workers with such problems take time off: "It is better to stay at work with a little pain," he explained, "than to adopt a lifestyle of disability."

Even if one should judge Dr. Nachemson's prognosis too harsh, and even if one fully accepts OSHA's claims regarding the threat of repetitive-stress injury, experience offers little to suggest that OSHA regulations will improve the problem. In fact, quite the opposite is likely to be the case. Wrote Howard in *The Death of Common Sense,* "A number of years ago, two workers were asphyxiated in a Kansas meat-packing plant while checking on a giant vat of animal blood. OSHA did virtually nothing. Stretched thin giving out citations for improper railing height, OSHA reinspected a plant that had admittedly 'deplorable' conditions only once in eight years. Then three more workers died—at the same plant. The government response? A nationwide rule requiring atmospheric testing devices in confined work spaces, though many of them have had no previous problems."

Who Wins and Loses? To anyone who has observed the federal government in action over the past few decades, it hardly takes a crystal ball to foresee the outcome of OSHA's repetitive-stress rule. One need only look at the "War on Poverty," which began in the mid-1960s and never really ended as much as it turned into a protracted battle of attrition reminiscent of the western front in World War I. It is estimated that, for the amount of money spent on "fighting" poverty, the federal government could have simply bought the 500 largest corporations in America and turned them over to the nation's poor. And for all that expenditure, today there are more poor people in America than ever, and their lives are vastly more hopeless than in 1965, before the government set out to, in effect, create a permanent underclass.

The OSHA ergonomics mandate is, like the "War on Poverty" (or the equally absurd "War on Drugs"), a typical example of government meddling and mismanagement. For this reason, the Bush administration opted to drop the original OSHA plan, and in April 2002 announced that it would call for voluntarily established industrial guidelines rather than unfunded government mandates. Yet this is not to say that the OSHA directive was or is without supporters; quite the contrary is true.

Former U.S. Secretary of Labor Alexis Herman announces new OSHA standards in November 1999.
(Photograph by Joe Marquette. AP/Wide World Photos. Reproduced by permission.)

In its ergonomics plan, OSHA not only had the backing of the Clinton administration, but even after Clinton left office in January 2001, it continued to enjoy the support of key Democrats in Congress. Among these Democratic leaders is Senator Edward Kennedy, long noted as an advocate of a bigger, more intrusive, and allegedly more benevolent government. Regarding the Bush administration's ruling on the OSHA plan, Kennedy said that it "shows that when it comes to protecting America's workers, this administration's goal is to look the other way and help big business get away with it."

Equally predictable was the response of the labor unions, including the umbrella organization of the AFL-CIO, whose director of safety and health, Pat Seminario, called the Bush plan "a sham." Then there were the sort of protesters who had made themselves an ever more ubiquitous fixture of national and international economic discussions, starting with the 1999 World Trade Organization summit in Seattle, Washington: three of these, one wearing a George W. Bush mask, interrupted testimony at the final Bush administration hearings on the ergonomics rule in July 2001, shouting "Big business loves the ergonomics scam."

The protesters do not so much support OSHA as they oppose its opponents, since for them protest is apparently an end in itself. As for the labor unions and politicians such as Kennedy, it is easy to see why they would be natural constituents of OSHA. The AFL-CIO and the unions it represents are in the business of getting more pay for less work, and they seem to be of the impression that they somehow benefit by cutting into companies' profits. These are the same unions that bleed companies dry, then decry the loss of jobs; the same unions whose leaders seem befuddled as to why more and more of their jobs are going to laborers in other countries. As for Kennedy and his ilk, their support of OSHA is even more natural. Even in the case of the distinguished senator from Massachusetts, with his wealth and family name, nothing compares to the power and influence that comes from having one's hand on the government feeding trough. More power for OSHA means more influence for politicians and bureaucrats, more jobs that can be handed out, and less power for business—the golden goose that Washington seems willing to kill for its eggs.

These are the beneficiaries of OSHA rules, but who are the losers? It may be hard to shed a tear for wealthy, powerful corporations such as UPS, but in the final analysis, it is not the corporation that suffers—it is the worker. For several generations now, Washington has devoted considerable energy to convincing workers that it is taking money from their bosses and giving it to them, but in fact the only party benefiting is the government itself, and workers would be far better off taking home more of the money that goes to Washington. The Bush administration changes to the ergonomics laws are promising, but this should be just the first step in rolling back OSHA's strangling influence on business. If the federal government is truly interested in dealing with repetitive-stress injuries, and not simply enamored with increasing its power, this can be done for a fraction of the cost. As with most things in public life, treatment of musculoskeletal injuries, if it is to be done efficiently, should be performed by the private sector and not by Washington. —JUDSON KNIGHT

Viewpoint:

No, OSHA regulations aimed at preventing repetitive-motion injuries are straightforward and save businesses time and money in the long run.

A common misconception is that ostriches hide their head in the sand when they feel threatened, hoping that the danger will go away. Nothing could be further from the truth. Imagine this large bird with its head buried and its body protruding, feathers waving in the wind. Any species that responds to a threat in this manner is fast on its way to becoming extinct. Fighting against standards for ergonomics and repetitive-motion injuries is a "hide-your-head-in-the-sand" reaction to a safety issue that is too big to disappear. Further, the perceived threat itself is misconstrued.

Business and political opposition to regulations targeting worker safety and health is nothing new. In fact, it is to be expected. Although occupational cancer was identified in English chimney sweeps as far back as 1775, bills were not passed to help ensure any type of workers' health and safety until the nineteenth century. When mechanization and changes in mass-production practices came about at the beginning of the twentieth century, they had a direct and adverse effect on workers' health, primarily through accidents and work-related diseases. Nevertheless, businesses routinely balked or sidestepped regulations to improve working conditions. It was not until 1970, after years of worker complaints and unionized efforts, that the first comprehensive health and safety regulations for workers in the United States were passed in the form of the Occupational Health and Safety Act. Much of the business community is so used to crying wolf when regulations are imposed on them that they fought tenaciously against the implementation of the Fair Labor Standards Act in 1938, which included the first effective national child labor regulations. Opponents proclaimed that industry could not withstand the burden if child labor was eliminated.

Opponents to the Occupational Safety and Health Administration (OSHA) standards have argued that the science of ergonomics is not sufficiently advanced to implement regulations on how businesses should handle repetitive-motion injuries and other musculoskeletal disorders (MSDs). Although research has not provided all the answers and more studies are needed, more evidence exists for these types of injuries than for any other injury or illness related to the workplace. In 1999, more than 200 peer-reviewed, scientific articles were published on carpal tunnel syndrome alone. In a review of more than 600 peer-reviewed, epidemiological studies, the National Institute for Occupational Safety and Health stated that sound evidence supports the relationship between workplace risk factors and a large number of musculoskeletal disorders. In

January 2001, the National Academy of Sciences also published study results concluding that there is abundant scientific evidence to demonstrate that repetitive workplace motions can cause injuries. Furthermore, the academy stated that evidence clearly shows that these injuries can be prevented through worker safety interventions. Further support based on scientific studies that ergonomic interventions can help reduce repetitive-motion injuries and other MSDs has come from the U.S. General Accounting Office and the American Conference of Governmental Industrial Hygienists, an internationally recognized occupational health organization.

Studies have also delineated the impact that repetitive-motion injuries and other MSDs have on the workplace. Each year, more than 600,000 workers miss work while recuperating from MSDs. The median number of missed workdays because of each incident is seven days. Incidents of carpal tunnel syndrome typically result in 25 missed days of work; many miss six months or more of work. According to Bureau of Labor Statistics data as reported by employers, an estimated 24–813 of every 1,000 general industry workers, depending on their industry, are at risk of missing a workday due to MSD over the lifetime of their employment career.

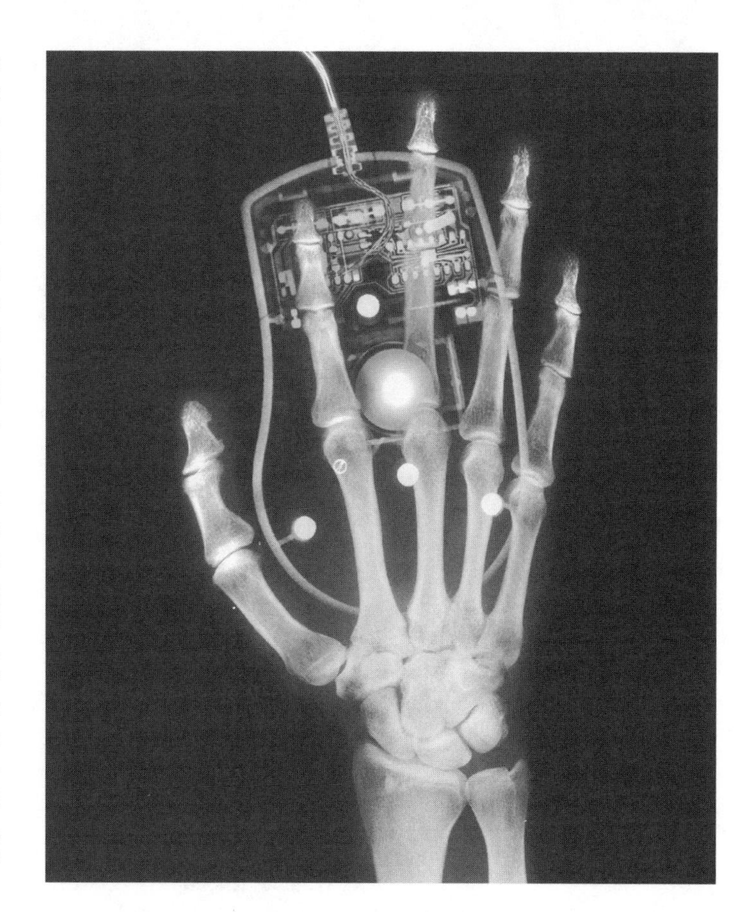

An x ray of a hand on a computer mouse. Repetitive motions such as those required in computer use can lead to carpal tunnel syndrome and other injuries.

(© Lester Lefkowitz/Corbis. Reproduced by permission)

A Look at the Costs When it became apparent that scientific data was abundant, opponents' next salvo against OSHA focused on the cost of implementing regulations targeting repetitive-motion and other injuries. However, economic feasibility is also not a viable argument against establishing such standards. According to OSHA estimates, the annual costs of implementing its regulations would be around $4.5 billion dollars while the overall savings would be in the vicinity of $9 billion per year by preventing injury to about 300,000 workers. For example OSHA estimates that employers who would need to correct problems would spend an average of $150 a year per workstation fixed.

According to an article in the United Auto Workers magazine *Solidarity,* nearly 2 million workers suffer from "strains, sprains, carpal tunnel syndrome, and other ergonomic injuries" each year. Calculated over the year and at 24 hours a day, 18 workers are injured each second every day of the year. According to the National Academy of Sciences, "reported" injuries in terms of lost productivity and medial expenses alone are estimated to cost $45–$54 billion per year. Broken down these costs include $15–$20 billion annually for workers' compensation and $30–$40 billion annually in various other expenses, including medical care.

According to the Bureau of Labor Statistics, ergonomic disorders like repetitive-motion injures due to typing at computers are the most rapidly growing category of work-related illnesses reported to OSHA. When personal computers were first sold to the public in 1981, only 18% of all occupational illnesses reported to OSHA were repetitive-motion injuries. Over the ensuing years, that figure continued to grow to 28% in 1984, 52% in 1992, and an estimated 70% in 2000. As a result, it is estimated that the 70 million PCs used in businesses in the United States cost companies approximately $20 billion a year due to repetitive-motion injuries.

Considering the high costs of ignoring ergonomic-related disorders, it is clear that preventing these injuries and addressing the workplace causes of them will save both businesses and society as a whole far more than it would cost. For example, small and large businesses that already have high standards in worker safety in various areas consistently save money in workers compensation and retraining costs, while improving efficiency and productivity. According to benefit-cost analysis estimates by the Washington state Department of Labor and Industries concerning its own ergonomic work rules, implementation of the rules will yield annual social benefits worth $340.7 million in their state alone while incurring only $80.4 million in annual compliance costs on employers, resulting in a benefit-cost ration of 4.24. Overall the department estimated that their ergonomic rules will save employers more than $2 for every $1 they spend to meet the rules.

Perhaps the most hotly debated provision in the OSHA regulations is that it requires employers to provide full pay and benefits to workers who either are removed from their jobs or reassigned to less strenuous work because of their injuries. Employers would have to maintain such pay and benefits for at least six months. While employers view this as an unnecessary burden, it is essential so workers who are injured will report their problems without fear of losing their job or having their pay reduced.

Complications Are in the Eyes of the Beholder As the British novelist Graham Greene once wrote, "Behind the complicated details of the world stand the simplicities. . . ." The same can be said about the proposed regulations by OSHA. Contrary to opponents' anguished cries that the rules are too broad, complicated, intrusive, overreaching, and confusing to be followed in a fiscally or rationally sound manner, the OSHA rules are simple and straightforward. Although there are hundreds of pages in the OSHA guide concerning these rules, the vast majority are related to scientific data and measurements. The rules do cover a broad range of workers, from nursing home aides who lift heavy patients to people who work at computers or on assembly lines. But the rules themselves are only 9–12 pages long and are written in plain and simple language. They are based on two primary principles: the standard should focus on jobs where ergonomic problems are significant and feasible solutions are available.

OSHA's rules merely ask employers who have employee safety problems to implement a program that is similar to successful programs already implemented by thousands of employers throughout the United States. The guidelines target only those jobs where the risk of injuries and illnesses is highest, such as manufacturing jobs and other occupations that have a demonstrated association with ergonomic injuries.

In simple terms, OSHA regulations require employers to establish a system that is basic to all effective safety and health programs, including management leadership/employee participation, hazard identification and reporting, job hazard analysis and control, training, and program evaluation. However, it does not direct the employers to develop a specific kind of reporting system. It does not tell them how many hours of training effected workers must receive, or what the content of training must be. It does not even tell business how workers should be involved in the planning, implementation, and evaluation of program elements. The plan's flexibility also includes a "grandfather" clause that permits employers who have already implemented an ergonomics program to continue its operation as long as it meets the minimum OSHA requirements. There is even a "Quick Fix" provision that enables many employers to eliminate their hazards without having to implement a full ergonomics program and an "Exit" provision that allows employers whose programs have successfully addressed their ergonomic hazards to return to a basic program. The OSHA regulations also permit companies to approach their problems incrementally, trying one method and then another if it does not work.

Profits and Productivity The development of OSHA guidelines for workplace health concerning repetitive motion and other similar injuries did not happen overnight. Work on ergonomics standards started under President George W. Bush's 1988–1992 administration. Then–secretary of labor Elizabeth Dole was already convinced of the evidence demonstrating the need for workplace economic standards. In a 1990 address announcing the beginning of work on such a rule, Dole said, "These painful and sometimes crippling illnesses now make up 48% of all recordable industrial workplace illnesses. We must do our utmost to protect workers from these hazards, not only in the red meat industry but all of the U.S. industries." Over the rest of the decade, OSHA reviewed hundreds of studies and worked on the rules for eight years before announcing them in November 1999.

To expect companies to universally control ergonomic hazards voluntarily without government control is wishful thinking. For example, the landmark Federal Coal Mine Health and Safety Act of 1969, which changed mining practices to protect miners' safety and provide compensation for those suffering from black lung disease, was implemented only after years of work by the United Mine Workers Association trying to convince Congress to enact such measures. In Washington, the state's Department of Labor and Industries noted that for 10 years it encouraged voluntary control of ergonomic hazards by providing information and technical assistance to businesses. Nevertheless, a department survey of some 5,000 employers revealed that 60% of the businesses in Washington did nothing to control these hazards. It is also interesting to note that occupations most affected by MSDs tend to be lower-wage jobs that employ high numbers of minorities. These workers are typically non-unionized and have little or no say in how they perform their jobs.

In addition to the cost advantages outlined earlier, instituting federal guidelines for ergonomic safety in the workplace would also save government, business, and employees from becoming snarled in expensive and time-consuming lawsuits. In the late 1980s, the United Food and Commercial Workers (UFCW) union filed a number of complaints about repetitive-motion injuries in meatpacking plants. The result was numerous settlements, often in the millions of dollars and a plethora of congressional hearings that led to OSHA developing special guidelines for that industry. One of the results of the repeal of the OSHA ergonomic regulations by Congress in early 2001 is that businesses and its employees are once again going to court. For example, the UFCW filed a federal complaint with OSHA alleging hazardous working conditions at a chicken plant in Texas. According to union officials, this is only the first salvo in many more efforts to force government to address debilitating problems associated with ergonomic factors in production work, particularly in industries dominated by low-skilled immigrants.

It is time for businesses and government to stop hiding their heads in the sand. Yes, there is danger out there, but it is not to the ability of big or small businesses to run a tight ship and make a profit. The real burden is on the nearly 5,000 workers who are injured by ergonomic hazards at work each day while no standards are in place to protect them. —DAVID PETECHUK

Further Reading

Brooks, Nancy Rivera, and Marla Dickerson. "OSHA Drops Home Office Safety Order: Agency Reacts to Outcry." *Los Angeles Times* (January 6, 2000): C-1.

Chen, Kathy. "Bush Proposal on Repetitive-Stress Injuries Relies on Voluntary Industry Guidelines." *Wall Street Journal* (April 8, 2002): A-28.

Cleeland, Nancy. "Union Decries Conditions at Pilgrim's Pride Chicken Plant: Federal Complaint Is Filed to Spotlight Injury for Workers Who Make Repetitive Motions." *Los Angeles Times* (February 27, 2002): C-2.

"Did Regulation Fears Fuel the Fight Against the OSHA Rule?" *IIE Solutions* 33, no. 5 (May 2001): 14.

Dreazen, Yochi J., and Phil Kuntz. "New OSHA Proposal Enrages Businesses—Ergonomics Plan Is Tougher Than One That Caused Big Budget Stalemate." *Wall Street Journal* (November 8, 2000): A-2.

Dul, J., and B. A. Weerdmeester. *Ergonomics for Beginners: A Quick Reference Guide*. London, England; Washington, DC: Taylor and Francis, 1993.

Ergonomics, A Question of Feasibility: Hearing before the Subcommittee on Oversight and Investigations of the Committee on Education and the Workforce, House of Representatives, One Hundred Fifth Congress. Washington, DC: U.S. Government Printing Office (G.P.O.), 1997.

Ergonomics Desk Reference. 2nd ed. Neenah, WI: J.J. Keller and Associates, Inc., 2000.

Gupta, Anuj. "Forum Reveals Divisiveness of Debate over Ergonomics: Business, Labor, and Health-Care Interests Are at Odds over How to Protect Workers from Repetitive Stress." *Los Angeles Times* (July 17, 2001): C-3.

Hoover, Kent. "Study: Ergonomics Rule Costs $91 Billion a Year." *Denver Business Journal* 52, no. 3 (September 1, 2000): 29-A.

Howard, Philip K. *The Death of Common Sense: How Law Is Suffocating America.* New York: Warner Books, 1996.

Karr, Al. "OSHA Ties for Last Place in 'Customer Satisfaction'." *Safety and Health* 161, no. 3 (March 2000): 16.

Kroemer, Karl, K. H. Kroemer, and Anne Kroemer. *Office Ergonomics.* New York: Taylor and Francis, Inc., 2001.

Mayer, Caroline E. "Guidelines, Not Rules, on Ergonomics: Labor Dept. Rejects Pay for Repetitive-Stress Injury." *Washington Post* (April 6, 2002): E-1.

OSHA's Industrial Hygiene and Ergonomics [cited July 16, 2002]. <http://www.osh.net/hygiene_02.htm>.

"'A Slap in the Face': Bush Ergonomics Plan Offers No Action for Injured Workers." *Solidarity* (May 2002): 9–10.

Straker, L., K. J. Jones, and J. Miller. "A Comparison of the Postures Assumed When Using Laptop Computers and Desktop Computers." *Ergonomics* 28, no. 4 (1997): 266–68.

Should parents have the right to refuse standard childhood vaccinations?

Viewpoint: Yes, a properly run voluntary system could produce higher vaccination rates while also protecting parents' rights.

Viewpoint: No, mandatory vaccinations have greatly reduced the incidence of many diseases and should be maintained.

Controversies over vaccination are not a new phenomenon. Learned debates about the safety, efficacy, morality, and even theological status of vaccination have raged since eighteenth-century physicians began to investigate ancient folk practices meant to minimize the danger of smallpox. Analyzing the risks and benefits of immunization, however, requires sophisticated statistical approaches to general mortality rates, case fatality rates for specific diseases, and studies of the safety and efficacy of various vaccines.

Until the nineteenth century, reported mortality rates were generally very crude estimates, but attempts to provide more accurate measurements of vital statistics through the analysis of the weekly "Bills of Mortality" began in the seventeenth century. During this time period, smallpox was so common that distinguished physicians considered it part of the normal human maturation process. Many eighteenth-century physicians were influenced by Enlightenment philosophy, which inspired the search for rational systems of medicine, practical means of preventing disease, improving the human condition, and disseminating the new learning to the greatest number of people possible. Johann Peter Frank's *System of Complete Medical Police* provides a classic example of the goals and ideals, as well as the sometimes authoritarian methods, employed in public health medicine. Of course harsh, even draconian public health measures can be traced back to the Middle Ages, as the authorities struggled to contain bubonic plague.

One of the most significant medical achievements of the eighteenth century was recognition of the possibility of preventing epidemic smallpox by inoculation (the use of material from smallpox postules) and vaccination (the use of material originally derived from cowpox postules). Smallpox probably accounted for about 10% of all deaths in seventeenth-century London. About 30% of all children in England died of smallpox before reaching their third birthday.

Smallpox left most of its victims with ugly pitted scars, but blindness, deafness, and death might also occur. Peasants in many parts of Europe attempted to protect their children by deliberately exposing them to a person with a mild case. Some folk healers took pus from smallpox pustules and inserted it into a cut or scratch on the skin of a healthy individual. In China, some practitioners had patients inhale a powder made from the crusts of smallpox scabs. Although such practices were dangerous, patients who survived enjoyed immunity to the naturally acquired infection. When inquisitive individuals like Mary Wortley Montagu and Cotton Mather brought smallpox inoculation to the attention of physicians, clergymen and physicians denounced this "unnatural" practice. Some theologians argued that inoculation was a dangerous, sinful affront to God, who used disease as a test of faith and a punishment for immorality.

When smallpox struck Boston in 1721, Mather and Dr. Zabdiel Boylston initiated a test of inoculation, but their efforts generated great fear and hostility. One critic threw a bomb through Mather's window along with a note that read, "COTTON MATHER, You Dog, Dam you, I'll inoculate you with this, with a Pox to you." After performing more than 200 inoculations, Boylston concluded it was "the most beneficial and successful" medical innovation ever discovered. In contrast, Dr. William Douglass, Boston's best-educated physician, argued that it was a sin to transmit a disease by artificial means in order to infect people who might have escaped the naturally acquired disease. Boylston and Mather later presented statistical evidence to show that the mortality rate for naturally acquired smallpox during the epidemic had been significantly less than that for the inoculated group.

Inoculation had important ramifications for medical practitioners and public health officials, ramifications that are still part of the modern debate about compulsory vaccinations. As Benjamin Franklin explained, weighing the risks and benefits of inoculation became an awesome responsibility for parents. In 1736 Franklin printed a notice in the *Pennsylvania Gazette* denying rumors that his son Francis had died of inoculated smallpox. The child had acquired natural smallpox before he could be inoculated. Knowing that some parents refused to inoculate their children because of fear that "they should never forgive themselves if a child died under it," he urged them to consider that "the regret may be the same either way, and that, therefore, the safer should be chosen."

By the second half of the eighteenth century, inoculation was a generally accepted medical practice. Inoculation paved the way for the rapid acceptance of vaccination, the method introduced by Edward Jenner in the 1790s. While practicing medicine in the English countryside, Jenner became intrigued by local folk beliefs about smallpox and cowpox. Although cowpox usually caused only minor discomfort, it provided immunity against smallpox. Critics warned that transmitting disease from a "brute creature" to human beings was a loathsome, dangerous, and immoral act. However, despite considerable skepticism and debate, Jennerian vaccination was rapidly adopted throughout Europe and the Americas. As early as 1806, Thomas Jefferson predicted that thanks to Jenner's discovery smallpox would eventually be eradicated.

Some opponents of vaccination insisted that any interference with "nature" or the "will of God" was immoral and that deliberately introducing disease matter into a healthy person was an abomination. Others warned that even if inoculations appeared to be beneficial, the risks outweighed the benefits. Other critics objected to the enactment of laws that infringed on personal liberty. On the other hand, Johann Peter Frank said that vaccination was "the greatest and most important discovery ever made for Medical Police." Frank predicted that if all states adopted compulsory vaccination, smallpox would soon disappear. Vaccination was made compulsory in the United Kingdom in the 1850s, but the vaccination laws were not generally enforced until the 1870s. Despite a dramatic reduction in the death rate from smallpox, critics continued to denounce the statutes as "a crime against liberty."

In the 1910s epidemiologists were still complaining that the United States was "the least vaccinated of any civilized country." Many states passed laws prohibiting compulsory vaccination. After World War II compulsory vaccination laws were extended and more vigorously enforced. The risk of contracting smallpox within the United States became so small that in 1971 the Public Health Service recommended ending routine vaccination. Medical statistics demonstrated the risks of the vaccine itself: 6 to 8 children died each year from vaccination-related complications. Unfortunately, with the global eradication of smallpox and the end of smallpox vaccination, the threat that the disease could be used as a weapon by bioterrorists escalated. According to officials at the Centers for Disease Control and Prevention, the appearance of smallpox anywhere in the world would have to be regarded as a global public health threat.

In describing the 10-year campaign for the eradication of smallpox led by the World Health Organization, Donald A. Henderson proposed the next logical step: what had been learned in the smallpox campaign should form the basis of global immunization programs for controlling diphtheria, whooping cough, tetanus, measles, poliomyelitis, and tuberculosis. However, as demonstrated by the 1976 campaign for mass immunization against an expected swine flu pandemic, reports of any adverse effects from a vaccine can quickly lead to widespread fears, criticism of public health authorities, and skepticism about the relative risks and benefits of all vaccines. Compulsory childhood vaccination programs remain particularly controversial, as shown by the following essays, but in an age of extensive and rapid movement of people, the existence of infectious epidemic diseases anywhere in the world is a threat to people everywhere. —LOIS N. MAGNER

Viewpoint:

Yes, a properly run voluntary system could produce higher vaccination rates while also protecting parents' rights.

Public concern about vaccination has reached an alarming stage. A series of congressional hearings have brought into public focus not only the potential undesirable side effects of vaccines, but also raised issues of our century-old policy of compulsory vaccination. Blue ribbon panels of the National Academy of Sciences' Institute of Medicine have for years been examining alleged links of vaccinations to a rising incidence of serious adverse events and potential links to disease. Many citizen and health care provider groups have called for an end to requiring infant vaccinations for entry into schools, infant day care, and participation in childhood activity programs. Some groups advocate an antivaccine position not only on religious or conscientious philosophical objection, but because they attribute to vaccines the cause of many growing childhood and adult maladies.

How has the trust in immunization programs, one of the marvels of modern medical advances, come to be so diminished? Why should parents be forced by law to have their infants vaccinated? Why have the potential harms of vaccines not been made better known? Why cannot parents make the decision about vaccinations for their children, as is common in other western societies?

Vaccines are an effective method of stopping epidemics and preventing disease. But they are not entirely safe—nor, unfortunately, are they always 100% effective. They are not for every infant. Nor are they consistent with deeply held convictions of many in our society. Our laws have consistently recognized an exception on religious and philosophical grounds as well as on medical grounds, when allergic reaction or poor health compromises the immature immune system.

However, those who would decline to have their children vaccinated are often intimidated by their physicians and humiliated or ostracized by school authorities and administrators of childhood and youth programs. Sometimes the threats may include legal sanctions and penalties. Many parents go so far as to isolate themselves from mainstream medical care and community involvement to avoid being coerced to immunize their children. The approach of compulsory vaccination is fraying. Confidence in the protection against disease provided by vaccination is ebbing. Many families not only decide to forego immunization; they are choosing to distance themselves from the well established benefits of modern medicine altogether.

Reasons for the Current Debate A major reason for the current public debate over the value of childhood vaccination lies in the type of decision-making behind it. When common childhood epidemics ravaged the population, there was little time or opportunity to take into account the dissident views of the portions of society that balked at vaccination. In 1905 and again in 1922, the U.S. Supreme Court decided that the government could require vaccination to prevent the spread of contagious diseases that threatened death, disability, and serious illness not only to individuals contracting the disease but to the population as a whole. Compulsory vaccination was the only way to control outbreaks of lethal disease.

Surely, fresh in the minds of all at the time of the 1922 Supreme Court decision was the experience of the deadly influenza epidemics after World War I, which took over 40 million lives worldwide. Although no influenza vaccine was dreamed of at the time, the clear model for vaccinating the population was the way we had countered the dreaded smallpox virus that had so often devastated populations around the world. As Daniel A. Salmon and Andrew W. Siegel point out in their article in July-August 2001 number of *Public Health Reports,* "Many of the [state] laws were written with specific reference to smallpox and later amended to include other VPDs [vaccine preventable diseases]." Drastic measures to prevent enormous numbers of deaths were imperative.

Even today, after smallpox has been eradicated from the world, the threat of unleashing a bioterrorist attack of this disease looms as a dreaded potential in our nation, which, ironically, ceased requiring the vaccination by the 1980s.

What made sense in the past when so-called invisible and unknown "killers" were unleashed in a community, seems now to be unwarranted. As regards smallpox, for example, now that we know that vaccination against smallpox may, in fact, kill, disable, and disfigure some small percentage of those who receive it, we hesitate to encourage or recommend it except for those whose duties would place them at immediate peril of the disease. But other diseases for which we now have vaccines do not pose such fearsome threats to the population.

In part, this is owing to the success of vaccines. "Before the vaccines were introduced," reads a 2001 report in the respected *Consumer Reports* magazine, "the toll of 10 of these vaccine-preventable diseases—diphtheria, measles, mumps, pertussis, polio, rubella (German measles), tetanus, hepatitis B, pneumococcus, and HiB [childhood influenza]—was nearly 2

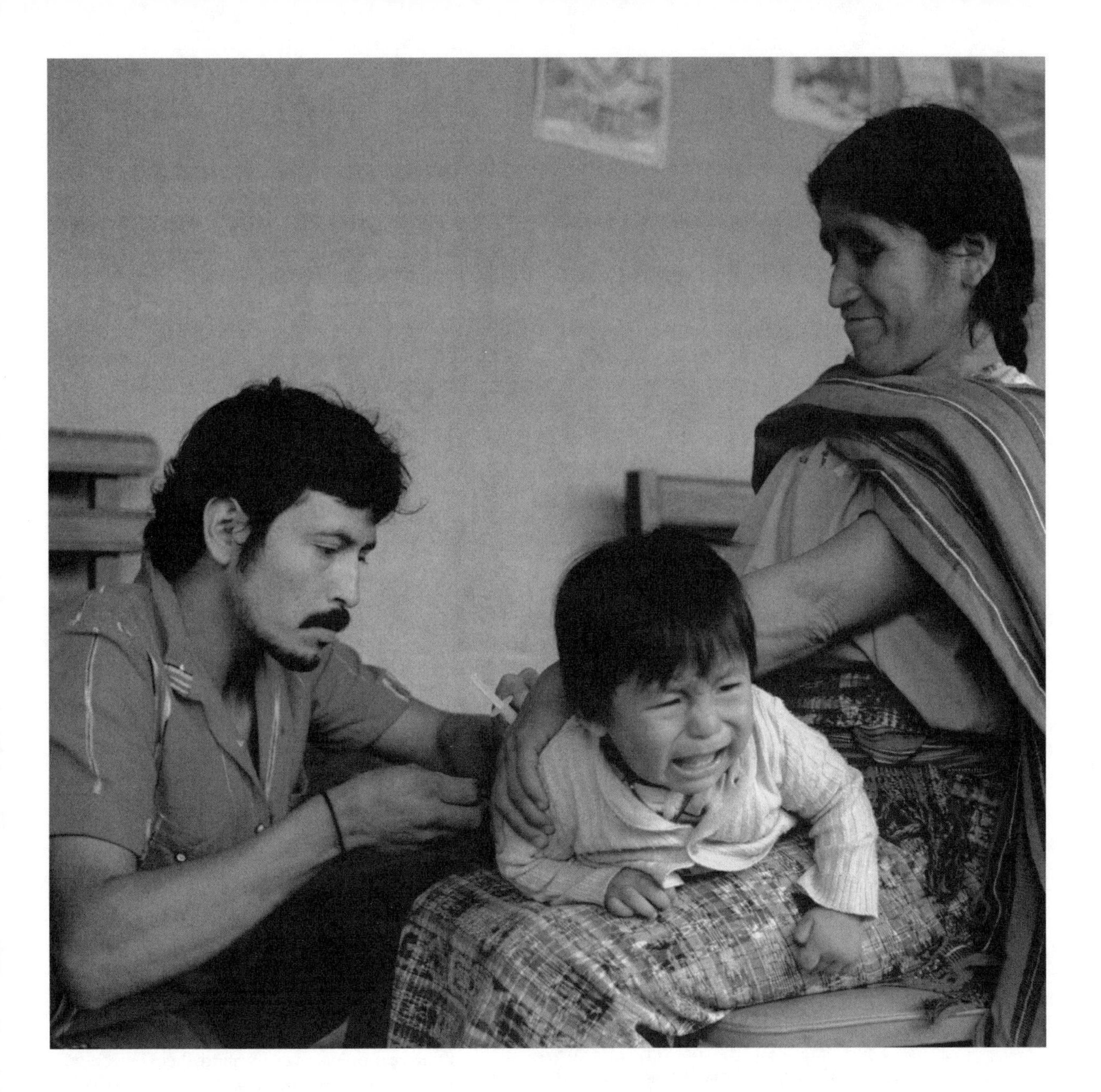

A Guatemalan woman holds her child as a doctor gives the child a vaccination.
(© Bill Gentile/Corbis. Reproduced by permission.)

million reported cases per year. Even the 'mildest' vaccine preventable disease, chicken pox, claimed 100 lives each year."

So our standard of evaluating costs vs. benefits is different. We were ready and willing to accept compulsory vaccination when it could save thousands of lives and prevent deafness, blindness, and brain disorders. But now, as some of the unavoidable adverse effects of vaccines emerge, we do not have the background of immediate danger against which to measure the response of society. Preventing 1,000 deaths by vaccination is clearly desirable. But what if there are 10 vaccine-related injuries in the process? Is the vaccine worth the risk? For those who must protect the public health, the answer is unclear.

Parents' Rights If we try to incorporate the values our society holds important into the equation, cost-effectiveness is no longer the sole yardstick. In addition to the now-well-established risk to the individual child, what of the parents' rights to decide about the well being of the child? At present, parents are confronted with little or no choice in whether their children are subjected to mass vaccination. A program of no-fault compensation has been established by the federal government to cover injuries related to vaccine administration, but no amount of money can compensate for loss of the child or permanent physical and mental disability.

The risks a parent of a very sick child may be willing to accept to treat or cure the child are judged in the context of the desperate hope to save the child's life. But vaccines are administered to healthy children, to infants who are only a few months old. These children are not ill, though they face a possibility of illness from a vaccine-preventable disease. Consequently, some parents would rather take their chances

KEY TERMS

ALLERGIC REACTION: Heightened reaction to a substance (allergen) that triggers abnormal responses in the body's protections system (immune system).

ANTIBODY: Protein produced in the blood as part of an immune response to a foreign substance in the body.

ANTIGEN: Foreign substance that stimulates the body's immune response.

DIPHTHERIA: Disease in which the bacteria *Corynebacterium diphtheriae* produce a deadly toxin that produces a thick gray coating in the back of the throat, which in turn inhibits breathing and swallowing.

HEPATITIS B: Contagious, viral liver disease leading to liver failure and liver cancer, causing approximately 5000 deaths a year and remaining dormant but contagious in symptomless, unsuspecting "carrier" individuals capable of spreading the disease.

***HAEMOPHILUS INFLUENZAE* TYPE B (HIB):** Bacteria responsible for one type of meningitis.

MEASLES: Respiratory disease caused by virus that grows in the cells that line the back of the throat and lungs.

MUMPS: Disease in which bacteria of the genus *Paramyxovirus* causes inflammation of various body parts, particularly the parotid glands.

PERTUSSIS (WHOOPING COUGH): Disease caused by *Bordetella pertussis* that causes severe coughing spells that can last for several weeks.

POLIO: Acute viral disease causing paralysis, permanent disability and death.

RUBELLA (GERMAN OR THREE-DAY MEASLES): Respiratory disease characterized by a rash, fever, and swollen lymph nodes.

TETANUS (LOCKJAW): Disease caused by *Clostridium tetani.* Symptoms include a painful tightening of the muscles, which often includes "locking" of the jaw so that the patient cannot chew or open his mouth.

VACCINE: Weakened or inactivated version of a virus or bacteria, which, when injected, causes the body to produce antibodies against the disease.

with the disease, particularly today when so many diseases seem unlikely to occur. Mandatory vaccinations limit parents' freedom to take care of their children as they see fit.

All states provide for medical exemptions for children with diseases like leukemia or allergic reactions or weakened immune systems, since officials know that vaccination carries a greater risk for such children. Medical exemptions must be verified by a physician's report. Laws in 48 states allow for parents to request exemptions on the grounds of religious belief, and in 17 states exemption can be based on philosophical or personal reasons. Some studies indicate that a growing number of parents exercise that option. However, this option comes at a cost, both to the child and to society as a whole.

The nonvaccinated children are at a much higher risk for getting the disease. Some claim that should nonvaccinated children become infected, they also pose a greater risk to others in the population. Accordingly, those pressing for compulsory vaccination argue that exemptions undermine the effort to curb disease and threaten the public health.

Some studies do reflect a trend of increased childhood diseases as parents opt for no vaccination. But the studies are limited in scope and the trends are hard to interpret. One anecdote illustrates what can occur when parents forego vaccination for their children. In Philadelphia in the 1990s, a resurgence of measles infected 1,600 children and killed nine. Officials tied the outbreak—perhaps unfairly—to the decision made by a single religious group to forego vaccinations for the children in their midst. As children inside the religious community became sick, they inevitably infected children in the city as a whole.

The group exercised its rights to refuse vaccination under Pennsylvania law. However, the reaction of public health officials in assigning blame to the group illustrates the pressure such advocacy puts on parents who have reservations about vaccines based on the spotty safety record and manufacturing lapses presented in congressional hearings and documented in scientific literature.

No doubt the medical and public health professions advocate vaccination for good reasons. It is their professional and informed judgment that leads them to that position. However, their zeal only serves to blind them to the well established rights of minorities in our society. There are very sound arguments for vaccination of children. With proper information and education, parents themselves can weigh the benefits to their children and the need to take precautions so that other members of the community are not placed at risk of contracting the diseases. But even deeply held religious beliefs may not be enough to excuse parents from responsibility for the deaths of innocent children. Many health professionals turn a deaf ear to legitimate rights and concerns of the public. The record, as established in congressional hearings, documents a persistent disregard of the concerns raised by citizens and their elected representatives.

Many of the concerns raised by anxious parents and parents of those whose children have been injured betray the overreaction to alleged but unsubstantiated claims. Adverse events go underreported, but seldom are they serious. The

Immunization Safety Review Panel of the Institute of Medicine has reviewed and independently assessed most of the claims of major harms caused by vaccinations. Consistently, the Panel has not been able to find conclusive evidence that the harms are extensive or even linked to the vaccination. Of course, this is of little consolation to the parents of children with autism, epilepsy, learning disabilities, and multiple sclerosis.

Safety Problems and Quality Control Lapses What makes those claims worrisome to parents is the clear evidence of safety and quality control lapses. Batches of vaccines are found to be "bad," though that does not always result in actual harms to the children. For decades a preservative in vaccines was a substance, thimerosal, containing mercury, which is a known toxin, particularly dangerous to pregnant and nursing mothers and very young children because it can impair cognitive development.

Polio vaccine has led to an almost complete eradication of the disease. In fact, from 1979 to 1999 the only cases of polio in the United States were caused by the oral polio vaccine, which contains the live virus. Over the 20 years, there were eight to nine cases of polio each year. Meanwhile, a safer vaccine, the killed-virus Salk vaccine, was developed. Public health officials, however, were slow to recommend that communities begin using the killed-virus vaccine rather than the live-virus one.

Until authorities demonstrate steps to restore public trust in the promise of vaccines, it is reasonable to suppose that parents may have serious misgivings about vaccinating their children. Fortunately, there are signs of improvement in the oversight of vaccine administration, in quality control, and in efforts to involve the citizenry more directly in decisions about vaccine policies. The misunderstanding and misinformation that have contributed to the lost of trust in vaccination programs would not have gained such a strong foothold had authorities been more responsive and forthcoming in dealing with the public. In the meantime, is it fair to expect parents to subject their children to vaccinations?

There is no major, imminent threat, such as smallpox, that as a practical matter leaves little room for choice. The history of vaccination laws illustrates the reasoning that was behind the need to protect the public from the havoc of epidemics. Vaccines may prevent illness, but medicine has improved sufficiently that many of the highly contagious vaccine-preventable diseases do not threaten the whole population. It is perhaps time to reexamine the policy of compulsory vaccination.

The Model of Great Britain In Great Britain, for example, under a voluntary vaccination program, there are higher rates of immunization than in the U.S. under the compulsory program. In a recent report on vaccination, the core principle of voluntary childhood vaccination was reaffirmed. The public was informed that at the current rates of 90 to 95% voluntary vaccination, there was adequate protection for all.

It appears that public health authorities in the United States may have become complacent in their efforts to promote vaccination. As long as vaccination is compulsory, there is little need to inform and educate the public. A voluntary program in which public officials work hard to persuade private citizens to comply may yield a higher vaccination rate than a "mandatory" program in which there is little public advocacy.

Still, it will be argued, the need to protect other members of society from highly contagious diseases is undermined by a policy that allows for parental choice. However, this view substantially undervalues the individual's sense of civic and public responsibility. Decades of compulsory vaccination policy have deadened the awareness of individuals of their responsibilities and even of their own enlightened self-interest.

The underlying problem has been, as stated at the outset, a pattern of top-down decision-making based on circumstances that are no longer true. As recently as 1999, 15 state legislatures considered bills to reduce or eliminate requirements of vaccination for school entry. The public does not perceive immediate danger. Consequently, it uses another yardstick to assess the risks. Authorities have not been effective in articulating the value of vaccination because they have relied on the compulsory policy. A vaccine is developed, tested, and introduced. Its benefits are trumpeted, but its potential harms are not. The federal government has slowly come around to underwriting some or all of the costs for childhood vaccines for those at the poverty level, but the program is underfunded and too limited to be effective even among this group. Others see the expense of vaccination as a mandated way of increasing the profits of drug companies. Both groups see only the financial aspect and ask themselves why they should be forced to pay for something over which they have no say.

Other Considerations There is a broader set of considerations that need to be factored into the overall cost to society. First, there is the consideration of the effects of the disease vs. the effects of immunization. A parent will be willing to take steps to spare the child sickness or death. All of us have a clear interest in avoiding epidemics and in the huge costs that are entailed in loss of work and in medical expenses, not to mention personal harm. When we understand that immunization cannot work magic, but can reduce or eliminate the unwanted effects of diseases without on average any great harm, we can accept the choice. For those unfortunate enough to incur harm, society has established a fund to compensate. Thus, there is a sense of fairness to the system, although it will never be entirely fair to the injured parties.

A second area of consideration is the cost of taking away the freedom of choice of parents in the matter of deciding about what is best for their children. It may be more efficient to bundle vaccinations in an assembly-line manner in a mass inoculation program, as occurred when the polio vaccine first became licensed. But the majority of childhood vaccines are administered now in the course of well-baby care by private physicians. Parents are presented with information about potential side effects, but they have no choice about whether to participate. This does little to help parents understand the public benefits of the program or their private responsibility for taking part in it.

Compulsory vaccination has served us well in the past. But it now thwarts many of the values we cherish. As the catastrophic epidemics of decades ago become a distant memory, Americans would be well served by the public health system transferring vaccination responsibility back to parents and working vigorously to educate parents about the need for routine vaccinations. —CHARLES R. MACKAY

Viewpoint:

No, mandatory vaccinations have greatly reduced the incidence of many diseases and should be maintained.

As the twentieth century dawned, America's children were in grave danger. Each year, hundreds of thousands were killed, deformed, or crippled by a villain that could be neither seen nor heard. Tiny germs and bacteria had the ability to invade cities, one person at a time, spreading like wildfire with a simple cough or sneeze. During the first half of the century, the numbers of children killed or injured at the hands of serious diseases were staggering. More than 15,000 children succumbed to diphtheria every year; 20,000 were born with severe mental or physical defects caused by rubella; and as many as 57,000 were crippled by polio. Smallpox alone decimated more than 300 million people—more than the total casualties from all wars combined—before it was eradicated in 1979.

Now, in the early years of the twenty-first century, American children face far less of a threat from disease. Smallpox has been eradicated, polio has been wiped out in much of the world, and the incidence of diphtheria, pertussis, and measles is very low. Millions of children who would not have lived to see their first birthday or who would have been wheelchair-bound for life are now thriving, thanks to what has been called the most beneficial health program in the world: immunization.

It seems inconceivable that anyone would choose not to take advantage of such a lifesaving technology, especially when vaccines have been proven safe and effective by the leading medical experts in the world. But many parents, because of religious or philosophical beliefs, misconceptions about vaccine safety, or concerns over personal freedom, are choosing not to vaccinate their children.

Americans are among the most privileged citizens in the world because we have been afforded freedom of choice. But that freedom is not absolute. Imagine if people were allowed to disobey traffic signals, or shoot guns whenever they felt like it? The result would be great harm inflicted on many innocent people. Similar to traffic and gun control laws, the federal government has upheld compulsory vaccination laws as reasonable to protect the general public. As the Supreme Court stated in 1905 in *Jacobson v. Massachusetts,* "in every well-ordered society . . . the rights of the individuals in respect of his liberty may . . . be subjected to such restraint, to be enforced by reasonable regulations as the safety of the general public may demand."

Dispelling the Myths about Vaccines The single greatest concern parents have with regard to vaccines is over their safety. Some point to research linking vaccines with everything from sudden infant death syndrome (SIDS) to autism. Others say multiple vaccines can overwhelm a child's developing immune system or cause the illness they were designed to protect against. Experts counter that no vaccine is 100% effective, and just as with any medical treatment, there are risks, but the risks are minor when compared to the potential for death and disability from vaccine-preventable diseases.

Consider the statistics: For every 20 children infected with diphtheria, one will die. One in 200 children will die from pertussis, and one in 1,000 will die from measles. As many as 30% of people infected with tetanus will die. Compare that to vaccinations for these diseases, which have not been conclusively linked to a single death. When vaccines do cause side effects, they are usually minor. Some children will have

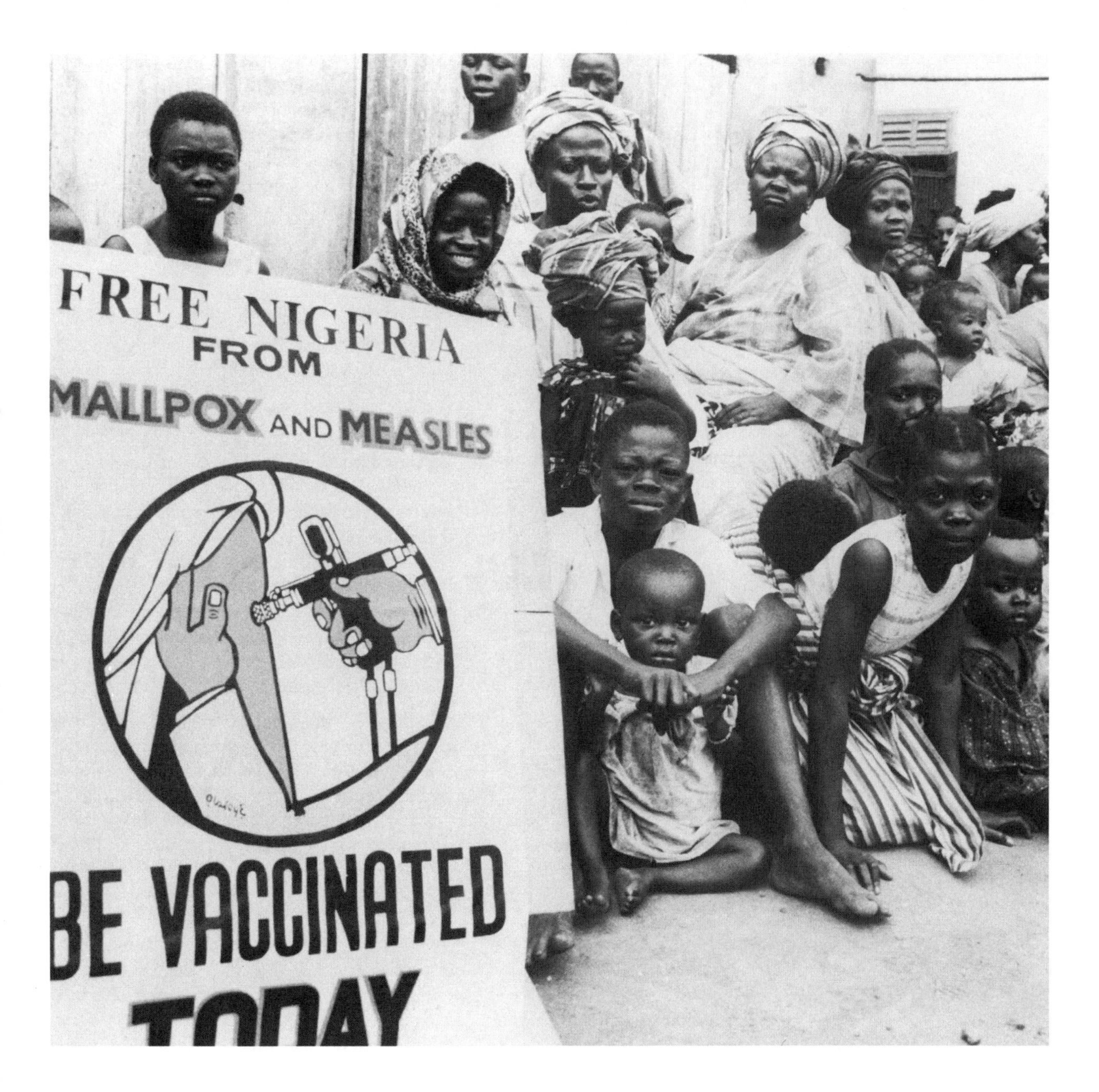

Nigerians await vaccinations, 1969. The use of vaccinations has helped eradicate many deadly diseases.

(© Bettmann/Corbis. Reproduced by permission.)

redness or swelling around the injection site, or run a slight fever.

Vaccine-induced Disease It's true that vaccines contain bacteria or viruses, but they have been either killed or weakened to confer immunity to the child without causing infection. As Louis Pasteur first discovered in the late 1800s, introducing these microbes is necessary to teach the body to regard diseases as foreign invaders. Normally, when the body senses invasion by a foreign body, for example a virus, it produces protein molecules called antibodies that move out into the bloodstream to attack and destroy the invaders. Once a person has been infected with an illness, the immune system learns to instantly recognize that particular virus as foreign, and will attack on sight, preventing future infection. Vaccines cause immunity by stimulating that same type of reaction without actually infecting the vaccinated person with the disease.

Vaccines and SIDS, Autism, and Diabetes Studies linking the DTP (diptheria-tetanus-pertussis) vaccine with sudden infant death syndrome (SIDS), the MMR (measles-mumps-rubella) vaccine with autism, and the Hib (*Haemophilus influenzae* Type b—the bacteria responsible for one type of meningitis) vaccine with diabetes, have fueled incendiary media reports and ignited anger and fear among parents. With all three vaccines, links were determined because the conditions emerged shortly after children were vaccinated. Using this same logic, then, it would be safe to assume from a study of early morning car accidents that people who drank orange juice were more likely to be involved in an accident than those who drank lemonade. Of course, we know that's not the case. More people drink orange juice in the morning than lemonade, but the connection ends there.

In the case of the DTP vaccine, shots are administered when a child is between 2 and 6 months old, the age at which a child is at prime risk for developing SIDS. When the Institute of Medicine at the National Academy of Sciences reviewed a number of studies comparing immunized versus non-immunized children, they found that the number of SIDS cases in the two groups was nearly identical, and they concluded that "the evidence does not indicate a causal relation between [DTP] vaccine and SIDS."

Concerns arose about a link between the MMR vaccine and autism after a 1998 British study described 12 children who displayed behavioral problems shortly after receiving the vaccine. Experts say the study was faulty for several reasons, among them the small number of cases investigated and the lack of control children for comparison. Investigators studying the link subsequently have found no increase in autism cases since the MMR vaccine was introduced in 1971. Most scientists believe that autism originates in the developing fetus or shortly after a child is born. They say the MMR vaccine may actually confer a benefit by protecting the mother against rubella, one of the few proven causes of autism.

A handful of studies have also suggested that Hib, among other vaccines, increases a child's risk of developing type 1 diabetes. In patients with diabetes, the body either does not produce enough insulin (which is needed to convert sugars from foods into energy), or cannot use it effectively. One study, conducted in Finland, compared multiple doses with single doses of the Hib vaccine. It found that 205 children in the multiple vaccine group developed diabetes as compared to 185 in the single dose group. Scientists and public health officials have examined all related studies and have thus far been unable to find clear evidence that any vaccine increases a person's risk of developing diabetes.

Over the years, teams of researchers around the world have tried to assert a link between various vaccines and febrile seizures, neurological disorders, and inflammatory bowel disease. In each case, the leading doctors and medical experts in the United States thoroughly researched and disproved the connection.

As for the contention that multiple vaccines overwhelm a child's immune system, experts have found no evidence to support this theory. Children are not only able to tolerate multiple vaccines at once, but vaccines actually protect a child's immune system from attack by serious bacterial and viral infections. What's more, children are receiving far fewer antigens (substances that stimulate the body's immune response) in today's vaccines than they did 40 years ago. By age 2, each American child will have received 7 vaccines in a series of 20 shots, as compared to just 3 shots in the 1960s (diptheria-tetanus-pertussis, polio, and smallpox). The smallpox vaccine (which is no longer given) alone contained about 200 antigens; the combined antigens in the total number of vaccines recommended today add up to fewer than 130.

How Safe Are Vaccines? Vaccines are among the safest forms of medicine ever developed, and the current vaccine supply in the United States is the safest on record, say health officials. Before any vaccine can be licensed by the Food and Drug Administration (FDA), it must go through extensive safety evaluations: first in the laboratory, then in animal trials, and finally in several phases of human trials, which may involve thousands of people. During the trials, vaccine manufacturers set the most effective dosage and look for any signs of adverse reactions. Even after licensing, manufacturers continue to monitor their vaccines for safety, submitting samples of each lot to the FDA for testing before the vaccine is released to the public.

In 1990 the Centers for Disease Control and Prevention (CDC) and the FDA established a database called the Vaccine Adverse Event Reporting System (VAERS). Health providers are required to report any adverse vaccine reactions to this system, and that information is combined with reports from vaccine manufacturers and the general public. Whenever an adverse reaction is reported into VAERS, the CDC and FDA act immediately, distributing safety alerts to doctors and the public, changing the wording on vaccine labels, and proposing changes to the vaccines themselves. For example, when concerns were raised in 1999 over thimerosal (a mercury-containing preservative used in vaccines since the 1930s), the Public Health Service and American Academy of Pediatrics called on health practitioners to avoid the use of these vaccines whenever possible, and urged manufacturers to remove the additive from all vaccines, which they did.

The Consequences of Ceasing to Vaccinate
Over the past four decades, several countries tried to decrease vaccinations, always with disastrous results. Great Britain, Japan, and Sweden cut back on pertussis vaccinations in the 1970s and 1980s out of concerns over the vaccine's safety. Soon after, a pertussis epidemic in Great Britain infected more than 100,000 people and killed 36. In Japan cases rose from 393 in 1974 (when the vaccine rate was about 80%) to 13,000 cases and 41 deaths in 1979 (when the vaccine rate fell to about 20%). In Sweden the annual incidence of pertussis cases among children ages 0 to 6 rose from 700 cases in 1981 to 3,000 in 1985.

Even the United States has suffered the results of reduced immunizations. Between 1989 and 1991, low coverage of MMR vaccinations among preschool children sparked a measles epidemic that infected 55,000 children and killed 120. Since then, coverage levels have risen to about 90%, and annual measles cases in the entire Western Hemisphere number fewer than 500 a year.

"If enough people refuse vaccination—and it can take a decline of only a few percentage points in the immunization rate—all children in the community are placed at greater risk," noted Dr. Bruce Gellin, executive director of the National Network for Immunization Information, in a statement released in 2000.

Modern vaccines are about 90 to 95% effective, which means that for every 20 children immunized, one or two may not become immune. Those one or two children are therefore vulnerable if they come into contact with the disease in school or at play. But if enough people in a community are vaccinated, it confers what is known as "herd immunity," protecting those at risk from disease.

If parents decide to stop vaccinating, health officials say we will see a resurgence of disease outbreaks, the likes of which we have not witnessed in several decades. Thousands of children will be at risk from the crippling effects of polio, the debilitating scourge of rubella, and the deadly toll of tetanus and diphtheria. —STEPHANIE WATSON

Further Reading

Centers for Disease Control and Prevention, National Immunization Program [cited July 29, 2002]. <http://www.cdc.gov/nip/>.

Dudley, William, ed. *Epidemics: Opposing Viewpoints.* San Diego, CA: Greenhaven Press, 1999.

Hyde, Margaret O. and Elizabeth H. Forsyth. *Vaccinations: From Smallpox to Cancer.* New York: F. Watts, 2000.

Immunizations: What You Should Know. Produced by Brian Peterson and Brian Wear. Cambridge Educational, 1994.

Immunization Safety Review Committee. Institute of Medicine [cited July 29, 2002]. <http:// www.iom.edu/ImSafety>.

Karlen, Arno. *Man and Microbes.* New York: G.P. Putnam, 1955.

National Network for Immunization Information [cited July 29, 2002]. <http://www.immunizationinfo.org/>.

National Academy of Sciences. *Overcoming Barriers to Immunization: A Workshop Summary.* Washington, DC: National Academy Press, 1994.

National Academy of Sciences. *Preliminary Considerations Regarding Federal Investments in Vaccine Purchase and Immunization Services: Interim Report on Immunization Finance Policies and Practices.* Washington, DC: National Academy Press, 1999.

National Academy of Sciences. *Immunization Safety Review: Thimerosal-containing Vaccines and Neurodevelopmental Disorders.* Washington, DC: National Academy Press, 2001.

National Academy of Sciences. *Calling the Shots: Immunization Finance Policies and Practices.* Washington, DC: National Academy Press, 2000.

National Academy of Sciences. *Immunization Safety Review: Multiple Immunizations and Immune Dysfunction.* Washington, DC: National Academy Press, 2002.

Salmon, Daniel A. and Andrew W. Siegel. "Religious and Philosophical Exemptions from Vaccination Requirements and Lessons Learned from Conscientious Objectors from Conscription." *Public Health Reports* 116 (July-August 2001): 289–95.

Historic Dispute: Is space filled with a medium known as the *aether*?

Viewpoint: Yes, prior to the twentieth century, many scientists believed in the existence of the aether.

Viewpoint: No, only with Albert Einstein's work in the early twentieth century did most scientists accept that the aether did not exist.

The nature of light and the possible existence of a vacuum have been subjects of philosophical speculation since the time of the ancient Greeks. Aristotle argued for the impossibility of a true vacuum, and the first efficient air pump was not developed until the time of Robert Boyle (1627–1691) While the ancients had acquired some practical knowledge of optics, the modern theory of light and color begin with the work of the Dutch scientist Christiaan Huygens (1629–1693) and the English scientist Sir Isaac Newton (1642–1727).

Huygens demonstrated how the reflection of light could be understood geometrically if light were treated as a wave traveling at finite speed. Newton claimed that light was the flow of particles. Boyle had demonstrated that sound could not travel through a vacuum, though light passed without difficulty. That observation, together with Newton's immense scientific prestige, worked against the wave hypothesis.

The wave character of light became more credible after Thomas Young (1733–1829) established that light waves could interfere with each other, and after the discovery of the polarization of light—which proved that light was a transverse wave—by the young French experimenter Etienne-Louis Malus (1775–1812) in 1808. Since sound waves, water waves, and waves traveling down a taught string all clearly require the vibration of a medium, it was only natural to assume that light waves were of the same character. Since light travels through a vacuum and through interplanetary space, it was thought that the light-carrying medium or luminiferous aether (sometimes spelled "ether") had to permeate all space. Attempts to assign mechanical properties to the aether were beset with problems. Because light traveled with immense speed and was a transverse wave, the aether had to be extremely rigid and solid rather than fluid. On the other hand, it apparently offered no resistance to the motions of the moon and planets.

The connection between light and electromagnetic phenomena was established by the Scottish mathematical physicist James Clerk Maxwell (1831–1879) in his *Treatise on Electricity and Magnetism,* published in 1873. Maxwell showed that the experimental results observed in the laboratory for systems of stationary charges and electrical currents could be summarized in a set of differential equations satisfied by the components of the electric and magnetic field. Maxwell's equations had solutions that described mutually transverse electric and magnetic waves traveling together through space with a speed that could be calculated from the force constants for the electrostatic and magnetic force, which in turn could be determined in the laboratory. Maxwell's theory received wide acceptance and underlined for physicists the importance of better understanding the aether, which was now apparently involved in electrical and magnetic phenomena as well as optical ones.

If an aether existed, it would also provide an absolute frame of reference for motion. As Earth completes its nearly circular orbit of the Sun, it should experience an "ether wind," the velocity of which would vary from one season to another. Light traveling in the direction of this wind and light traveling at right angles to it would then move at different speeds. Detecting the difference in speed was the basis of the Michelson experiment of 1881, later repeated more carefully as the Michelson-Morley experiment. The latter showed at most a small effect.

The inconclusive result of the Michelson-Morley experiment did not in itself spell the demise of the aether theory. Instead of the aether being perfectly stationary, perhaps Earth dragged a certain amount with it as it moved. Possibly the act of moving through the aether affected the measuring instruments.

It is generally considered that the special theory of relativity, as published by Albert Einstein in 1905, demonstrated the aether did not exist. What Einstein showed, however, was that an entirely consistent physics could be developed in which all observers in uniform motion with respect to each other would measure the same value for the speed of light, regardless of the speed with which they observed the light source to be moving. Einstein's theory met with a certain amount of opposition, in part because it did away with any need for a space-filling aether, but it gained general acceptance after observations made during a solar eclipse confirmed the effect the Sun's motion had on gravitational fields, as predicted by the expanded form of the theory published in 1912.

Twentieth-century physics was also forced to revisit the question as to the wave or particle nature of light. In 1901 Max Planck (1858–1947) proposed that the energy of a light wave could only increase or decrease by finite amounts, which came to be called *quanta.* In 1905 Einstein explained the characteristics of the photoelectric effect, the emission of electrons from a metal surface when exposed to light, as the transfer of quanta of energy from the electromagnetic field to single electrons. The quanta of light energy came to be thought of as particle-like "photons." Further insight into the particle properties of light was gained in the discovery of the Compton effect, in which electrons and x-ray photons collide, exchanging energy and momentum; and in the process of pair production, in which a high-energy photon is converted into an electron-positron pair.

While relativity theory made it possible to consider the existence of a totally empty region of space, the quantum field theory—developed to explain the interaction of high-energy particles with the electromagnetic field—requires that the vacuum be thought of as anything but empty on the shortest time scales. In fact, the vacuum is treated as a place where electron-positron and other particle-antiparticle pairs come momentarily into existence and disappear. The propagation of electromagnetic fields through empty space is aided by the so-called vacuum polarization that results from the behavior of these "virtual" pairs, which might in a sense constitute the latest incarnation of the aether. —DONALD R. FRANCESCHETTI

Viewpoint:

Yes, prior to the twentieth century, many scientists believed in the existence of the aether.

The belief in the aether (or ether) that was prominent in the nineteenth century is often described in scornful and derisive tones. The search for the aether, and the disbelief when it was not found, is often ridiculed, and portrayed as one of the follies of modern science. Yet there were very good scientific reasons for supposing the existence of a substance through which light could travel. The famous 1887 Michelson-Morley experiment was not as cut and dried as is often portrayed in the literature, and the controversy over the existence of the aether raged for decades after it was performed.

Early Studies of the Aether The aether is a hypothesized substance that allows the transmission of light throughout the universe. Its existence was postulated as far back as the ancient Greek philosophers, who offered a number of speculative theories on the nature of light and sight. The existence of the aether was seemingly set in stone when both René Descartes (1596–1650) and Isaac Newton (1642–1727) used it in their competing scientific philosophies. Descartes proposed a mechanical universe filled with matter, which he called the plenum, which was for all purposes the aether by another name. Newton was more cautious about the existence of the aether, but speculated that it filled the whole universe, and acted as "God's Sensorium," the means by which the universe was controlled. Newton's followers considered the aether essential, as it provided a stationary frame of reference for the laws of motion.

While both Descartes and Newton had used the aether mainly for philosophical reasons, there was some experimental support for the substance. In the 1660s Robert Boyle (1627–1691) succeeded in showing that the ringing of a bell could not be heard inside a vacuum. However, while sound was silenced in a vacuum, you could still see through it. Obviously, while sound required air to propagate, light did not. This suggested a more subtle substance than air was inside the supposed vacuum, and it was naturally assumed that it was the aether. This theory, by removing the notion of a complete vacuum, something that was unthinkable to many, found strong support.

Newton claimed that light was corpuscular in nature, consisting of tiny particles moving at infinite speed. However, at about the same time Christian Huygens (1629–1695) proposed that light was composed of waves, and that it had a finite speed. Waves travel through a medium, for example, a ripple in a pond moves through the water, and sound results from the vibrations of a body disturbing the air. It seemed only natural to assume that if light was a wave it must also travel through some sort of medium. Newton's corpuscular theory was dominant at first, but in the nineteenth century there were many experimental results that could only be explained if light was indeed a wave. The existence of the aether seemed more logical than ever.

Many properties of a wave are determined by the medium through which it travels. The mathematical relationship between pitch and tension in a piece of string has been known since the time of Pythagoras (sixth century B.C.). Galileo (1564–1642) refined the concept to relate to specific frequencies of vibration. The speed of the vibrations passing through a substance move more rapidly if the substance is stiffer. The speed a wave travels down a rope depends on the rope's tension and mass. The more tension or lighter the rope, the faster the wave travels. As details about the nature of light began to be found experimentally it became possible to work backwards and deduce the nature of the aether.

As early as 1746 Leonhard Euler (1707–1783) calculated that the density of the aether must be at least a hundred million times less than that of air, but its elasticity had to be a thousand times greater, making it more rigid than steel. In the 1840s it was proposed that the aether was an elastic-fluid, a theory that helped explain the phenomenon of double refraction, as well as circular polarization, and some thermodynamics problems. However, the work of Thomas Young (1773–1829), Augustin Fresnel (1788–1827), and François Arago (1786–1853) showed that light waves were transverse, not longitudinal as previously supposed. Transverse waves were unknown in any fluid medium, so the aether was therefore logically limited to being a solid rigid enough to allow the high speed of light waves to be transmitted, yet somehow offering no resistance to the motion of larger objects such as the planets, as the Newtonian laws of mechanics satisfactorily explained the motion of the planets. It was suggested that the aether was a stiff jelly-like substance that would act as solid for vibrations, but allow the easy passage of large objects.

René Descartes
(The Library of Congress.)

Despite the growing uneasiness in the strangeness of the aether, there was further support for its existence in 1850 when Armand Fizeau (1819–1896) and Jean Leon Foucault (1819–1868) revealed that light travels more slowly in water than in air, confirming that it travels as a wave, and that its wavelength is decreased in the medium of water. Fizeau also showed that the velocity of light can be sped up or slowed down by making the water flow with or against the light. The work of Michael Faraday (1791–1867) and James Clerk Maxwell (1831–1879) united electricity, magnetism, and light, and the aether became the proposed medium for magnetic force as well, yet appeared to have no other properties.

Michelson-Morley Experiment Towards the end of the nineteenth century a number of experiments were performed to measure the relative motion of Earth to the aether. In 1881 Albert Michelson (1852–1931) attempted to measure Earth's speed through the aether using a sensitive measuring device developed by Michelson called an interferometer. It is often erroneously stated that the purpose of the

KEY TERMS

AETHER DRIFT: The relative movement of Earth and the aether. The velocity of Earth through the stagnant aether creates an aether drift.

ELECTROMAGNETIC WAVE: A system of electric and magnetic fields traveling together in space.

LONGITUDINAL WAVE: The medium is displaced in line or parallel to the direction of travel, causing compressions and rarefactions. Sound waves in air are longitudinal.

PHOTOELECTRIC EFFECT: Release of electrons from an object exposed to electromagnetic radiation.

PHOTON: Particle of electromagnetic radiation.

RELATIVITY: The theory that the distances and time intervals measured by an observer depend on the state of motion of the observer in such a way that all observers measure the speed of light in vacuum to have the same value, regardless of the motion of the light source.

TRANSVERSE WAVE: Where the displacement of the medium occurs at right angles to the direction of travel of the wave motion. For example, a vibrating string, or ripples in a pond.

VACUUM: Space in which there is very few atoms or molecules.

WAVE-PARTICLE DUALITY: Light can be described as both consisting of electromagnetic waves, and particles of energy. Which description is more appropriate depends upon what aspect of light is being described.

experiment was to determine the existence of the aether. Michelson assumed the aether existed, and was only attempting to measure the expected effects it would produce. However, the experiment failed to produce the expected results.

It is also erroneously reported in many histories that this experiment, and the later refinements, showed no changes in speed in any direction. However, this first "null" result did actually find some change in the interference pattern, as did all the later tests, but the effect was so small as to fall within experimental error. Various reasons were proposed for the lack of a result, from errors in the experimental setup, to the possibility that the jelly-like aether was dragged along with Earth as it moved. No one seriously suggested that the aether might not exist.

While Michelson was satisfied that his interferometer was extremely precise, he nevertheless attempted to refine his technique. Teaming up with Edward Morley (1838–1923), who had a background in precise measurements, they repeated the interferometer experiment, and also confirmed Fizeau's experiment regarding the change in the speed of light in a flow of water. In 1887 they published another "null" result, stating that the change in the interference pattern was far too small to account for a stationary aether.

While many histories of science hail the 1887 result as a turning point in physics, and see it as the death of the aether, neither Michelson, Morley, or the wider scientific community of the time perceived the results in that way. Again, many plausible reasons were given for the result, all of which assumed the existence of the aether. There was simply too much previous evidence for the aether to allow one experiment to overthrow it. That same year Heinrich Hertz (1857–1894) demonstrated the first wireless transmission of electronic waves, helping to begin the radio revolution and providing strong support to the existence of the aether as a combined electromagnetic medium for light, magnetic, and electric waves.

The most promising explanation for the Michelson-Morley result was that Earth dragged a portion of the aether along with it as it traveled, hereby reducing the effective relative motion. However, Sir Oliver Lodge (1851–1940) performed an experiment that showed that this was not the case. Others, such as Sir George Stokes, insisted the aether-drag hypothesis was correct, as it explained a number of other optical phenomena.

Another explanation was provided by George FitzGerald (1851–1901) and Hendrick Lorentz (1853–1928) who both independently suggested that the experimental setup might shrink in the direction of Earth's movement. Maxwell had shown earlier that the force between two electric charges depends on how they are moving. FitzGerald and Lorentz suggested that the electromagnetic forces inside a solid may therefore contract by one part in a hundred million if moving in the direction of Earth's orbit. In the 1890s Lorentz calculated the transformations, which describe the length of a moving object, and its contraction when viewed by an observer at various speeds.

In 1905 Morley collaborated with Dayton Miller (1866–1941) to refine the aether experiment. While it returned a "very definite positive effect," it was still far too small to match the expected values of Newtonian physics. Then, that same year, Albert Einstein (1879–1955) published his special theory of relativity, which implied that there was no need for the mysterious and undetectable aether. All of the current experimental contradictions could be explained away by the new theory, and the aether could be forgotten.

Aether and the Theory of Relativity Relativity was not widely accepted at first, and one of the reasons for the strong resistance to it was the cherished notion of the aether. There were many eminent scientists of the time who refused to believe that waves could travel without a medium to propagate through. Such an idea went against centuries of scientific authority and experimental evidence. It was not until after World War I that a direct observational experiment of relativity could be made, when the light of stars behind the Sun was shown to be bent by the Sun's gravity during a solar eclipse. Relativity was pronounced triumphant, and support for the undetectable aether dwindled.

Einstein had been greatly influenced by the Lorentz transformations, but he admitted that he did not seriously consider the Michelson-Morley experiment, as it was only one of a whole series of puzzling aether experiments that had been done in the nineteenth century. Yet the Michelson-Morley experiment became one of the most celebrated means of explaining the ideas of relativity theory to the wider scientific community. Over time it became accepted that the experiments had been about proving or disproving the existence of the aether, and had formed a crucial logical step towards the theory of relativity.

Heinrich Hertz

(© Bettmann/Corbis. Reproduced by permission.)

In 1920 Einstein gave a speech titled "Aether and Relativity Theory" to an audience in Germany. Surprisingly the speech assumed the existence of the aether, even going so far as to say "space without ether is unthinkable." However, Einstein's aether was a far different substance than the strange solid-fluid substance that had proved to be impossible to find. Yet that old concept of the aether survived in many circles, and in 1920 Miller began a further series of aether experiments. In 1925 Miller announced that he had found a consistent, but small, positive result over many years, and he concluded that "there is a relative motion of the earth through the ether," contrary to relativity. Michelson and many others immediately began experiments of their own, none of which found positive results, and by 1929 even Miller was conceding that his results may have been in error.

The aether was consigned to the same fate of such other hypothesized substances as phlogiston and caloric and magnetic effluvia. Over time those who had championed the aether have become something of a laughing stock, and the story of the search for the aether is told as though it was folly. However, there were many good logical and scientific reasons for thinking the aether existed. Throughout the nineteenth century a number of key experimental results were obtained that seemed most easily explained by a electromagnetic medium. The controversial nature of many of the so-called "null" results in the aether-drift experiments did not immediately suggest the non-existence of the aether, and a number of innovative, but logical, explanations for the Michelson-Morley results were given, some of which helped inspire relativity theory. The aether was shown to be undetectable and unnecessary, however, that does not imply that it may not exist. There may indeed be a substance that acts as a medium for the transmission of light that has no other properties. John Bell (1928–1990), a mathematical physicist, even proposed that the theory of the aether be revived as a solution to a dilemma in quantum physics—although perhaps only half-seriously. The aether is still occasionally invoked by inventors of impossible perpetual motion machines and free-energy devices, but it should be remembered that it once had the support of solid evidence and logical arguments, and it only disappeared from scientific thought after many years of debate and experiment showed that it was simpler to abandon the medium than explain its weird properties and lack of detectability. —DAVID TULLOCH

Viewpoint:

No, only with Albert Einstein's work in the early twentieth century did

most scientists accept that the aether did not exist.

In the opening years of the twentieth century, the study of light was the central concern of most leading physicists. Mainstream physics agreed that light consisted of electro-magnetic waves, moving through a substance called the *aether*. However, the precise nature of the aether had not been identified and the resolution of this problem was regarded as the major task for physics in the new century. Then, due to his musings on the significance of the speed of light, Albert Einstein (1879–1955) developed a revolutionary framework for thinking about the relationship between space and time. In a few years, the special theory of relativity, as this framework was called, had changed the focus of physics. This transformed the understanding of light. Einstein showed that the aether, and the notion of absolute space and time that it represented, did not exist. The aether gradually ceased to be of any interest to the mainstream of inquiry in the physical sciences. Einstein's work on the photoelectric effect also transformed the understanding of light waves. This led to the idea of wave-particle duality as a way of characterizing the nature of light in different situations.

Waves of Light Prior to the nineteenth century, the mainstream of scientific thought regarded light as streams of tiny particles. This was the position of Sir Isaac Newton (1642–1727), whose authority had an enormous influence on all fields of scientific endeavor. It was not until the nineteenth century that serious consideration began to be given to the idea that light might consist of waves rather than particles. Successive experiments by physicists found that the wave theory best explained the behavior of light. The work of the British physicist James Maxwell (1831–1879) on electro-magnetic phenomena further strengthened this approach. Maxwell developed equations to describe the behavior of electro-magnetic phenomena. Thus, he was able to calculate the speed at which electro-magnetic waves moved in a vacuum, which coincided with the velocity of light. From this, Maxwell concluded that light was waves of oscillating electro-magnetic charges.

The Luminiferous Aether The triumph of the wave theory of light made the aether the center of attention. The aether had a long history in philosophical thought about the nature of the universe. It was believed to be the substance that filled up the realms of celestial space and was often identified as the mysterious fifth element, along with air, water, earth, and fire, the basis of all matter. The aether acquired additional significance in the nineteenth century, with its emphasis upon a mechanical approach in which all events could be attributed to local causes. Nineteenth-century physicists were opposed to the idea of action at a distance, the idea that an object at one point could affect an object at another point with no medium for the effect. For a lighted candle to influence another object, such as the human eye, there had to be something that could transmit the influence from one point to another. If light was a wave, it required a medium to move through. As it had been observed that light traveled through a vacuum, it was concluded that whatever the medium of light was, it was even thinner than air. The aether, the mysterious substance that filled space, was identified as the *luminiferous medium*.

Therefore, to understand light, it was necessary to understand the special characteristics of the light medium. Many physicists developed complex models in order to try and describe the aether, but none of these were able to capture its special and apparently contradictory qualities. The aether had to be very thin and elastic, so that solid objects, such as the earth, could pass through it without resistance. At the same time, it had to be very dense to allow for the transmission of the vibration of light from one part of the aether to the next at such high speeds. For the last two decades of the nineteenth century, the attempt to accurately describe the aether occupied some of the best minds in physics.

The Aether and Absolute Space One of the most crucial issues regarding the aether was the question of its immobility. Most physicists regarded the aether as fixed and stagnant, completely stationary. Objects might move through it, but no part of the aether was moving, relative to any other part. It stretched throughout space, acting as a framework against which the movement of light could be studied and measured. In this sense, the aether can be identified with the concept of absolute space. Most of the statements made about the movement of objects in the universe were relative; Earth was moving at a certain speed relative to the Moon, or another moving planetary object. Yet nineteenth-century science had at its foundation the idea of absolute space and time, a fixed reference system in nature from which Earth's absolute velocity could be measured. All movement in the universe was therefore movement relative to this stationary system. However, given that we participate in Earth's movement through the universe, some people were skeptical about the possibility of being able to step outside this and gain knowledge of our absolute position in the universe. As James Maxwell so eloquently wrote: "There are no landmarks in space; one portion of space is exactly like every other portion so that we cannot tell where we are. We are, as it were, on an unruffled sea, without stars, compass, soundings, wind or tide, and we cannot tell in what directions we are going. We have no log which we can cast out to take a reckoning by; we may compare our rate of motion with respect to the neighboring bodies, but we do not know how these bodies may be moving in space."

The Michelson-Morley Experiment However, not everyone shared this vision of humanity hurtling blindly through space. Many believed that the aether provided a fixed reference system against which all motion could be measured. Two American scientists, Albert Michelson (1852–1931) and Edward Morley (1838–1923), constructed a machine that could theoretically discern the speed of Earth through the aether. The Michelson-Morley experiment was set up to measure the impact of the aether drift on the velocity of light. The aether drift was the wind that was created by the velocity of Earth through the motionless aether, just as a person traveling in a car on a still day feels a wind if he or she put a hand out the window. Michelson reasoned that this wind should have an effect on the speed of light. He set up a machine that constructed a race between two rays of light, one moving in the same direction that Earth was moving, and the other in a perpendicular direction. The effect of the aether drift meant that one ray of light should reach the finish line before the other. The difference in time would be incredibly small, but Michelson devised a machine, based on earlier experiments, called an interferometer, that was sensitive enough to be able to measure the difference.

However, the experiment produced a null result. The rays of light were unaffected by the aether drift. This suggested that, relative to the aether, Earth was not moving at all. It was as if the car you were in was traveling at 40 mph (70 kph), but you could not feel the wind that should have been created by the velocity of the vehicle. This result was puzzling, because so much of what scientists knew about the movement of light depended upon the existence of the aether drift. Once again, the mysterious aether had appeared to evade detection. A variety of different explanations were developed that tried to account for Michelson and Morley's results. However, until Einstein, none directly questioned the existence of the aether itself. Conceptually, it was too fundamental to the fabric of physics as a medium for the movement of electro-magnetic phenomena.

One physicist who was particularly interested in the significance of the result of the Michelson-Morley experiment was H. A. Lorentz. (1853–1928) Through his calculations on the behavior of small electrically charged particles, Lorentz observed that their mass changed as they moved through the aether. Therefore, Earth's velocity through the aether caused a contraction in matter that was positioned parallel to Earth's movement. In the case of the Michelson-Morley experiment, everything that was traveling into the aether wind was contracted. This contraction canceled out the disadvantage of racing into the wind, producing the null result. Lorentz admitted that this theory was "very startling." However, he managed to incorporate these conclusions within the traditional framework of physics, retaining the importance of the aether as the fixed reference system against which to measure the movement of electro-magnetic phenomena.

Isaac Newton
(© Corbis. Reproduced by permission.)

Einstein and the Aether In 1905, Albert Einstein considered the problem from a different angle. The issue of the movement through the aether was not his prime concern. Instead, Einstein's theories were produced out of his consideration on the contradictions between two apparently valid principles. The first was the Galilean principle of relativity, which says that the laws regarding motion are the same regardless of the frame of reference. In other words, whether you are in a car traveling at a uniform speed of 100 kph, or in a train traveling at 10 kph, the laws of motion are the same in both cases. The second principle stems from Maxwell's equations, which established the law that light in a vacuum always moves with a velocity c = 300,000km/sec. According to the

ALBERT EINSTEIN AND THE NOBEL PRIZE

Albert Einstein was awarded the Nobel Prize for physics in 1921. That he received the prize was not surprising. By this time, the general theory of relativity, which applied his ideas about the relativity of space and time to the problem of gravity, had been confirmed by observational data. As a result, Einstein had achieved worldwide fame and was recognized as one of the greatest minds in the history of science. Indeed, the fact that he was going to receive a Nobel Prize was so widely anticipated that in 1919 the expected prize money for the award was included as part of his divorce settlement with his wife.

However, what is surprising is that Einstein was not awarded the Nobel Prize for his most famous achievement. His citation states that he received the prize "for services to theoretical physics, and especially for his discovery of the photoelectric effect." There was no direct mention of the special and general theory of relativity, which had gained him celebrity status and for which he is still best known today. The reason for this indicates something about the nature of the Nobel Prize and about the wider reception of Einstein's theory. In 1895, the Swedish industrialist Alfred Nobel had left provision for the Nobel Prize in his will, which specified that part of his fortune should be "distributed in the form of prizes to those who, during the preceding years, shall have conferred the greatest benefit on mankind." Theoretical physics was not strong in Sweden at this time, and to those judging the prize, the significance of the theory of relativity to "the greatest benefit of mankind" was not clear. It seemed more in keeping with the spirit of Nobel's instructions to cite a part of Einstein's work, such as his discovery of the photoelectric effect, that appeared to have obvious practical applications. Therefore, while it is the special and general theory of relativity that has granted Einstein a degree of immortality, at the time it was not regarded as appropriate for a Nobel Prize.

—Katrina Ford

principle of relativity then, this must be the speed of light in all circumstances, whether it is measured in a car or on a train, regardless of the speed at which these are moving. Within the existing framework of physics, this was impossible, because the speed of anything always depended upon the framework from which it was being measured. For instance, the speed of the car might be 100 kph as measured from the roadside, but it will only be 90 kph measured from a train traveling parallel in the same direction at 10 kph. The same should apply to light. Therefore, either the principle that states the speed of light, or the principle of relativity, had to be incorrect.

However, Einstein found that the problem lay with the concept of absolute space and time. The old approach assumed that space and time were the same for all observers. Between any two events, say the firing of a pistol and the bullet striking its target, there would be a spatial separation, perhaps 80 ft (25 m), and a time interval, perhaps 0.04 seconds, which all observers would agree on. Einstein was able to show that this was not the case; the observed spatial separation of and time interval between the events depend on the reference frame of the observer. By eliminating the assumptions of absolute space and time, Einstein was able to develop a consistent physics in which the speed of light was always c, regardless of the frame of reference. This resolved the apparent contradiction that had so troubled him. These calculations exactly mirrored those that Lorentz had earlier devised in his observations, but Einstein was the first to recognize their true significance for ideas about space and time. These conclusions form the basis of what is known as the special theory of relativity.

The relativity of space and time had enormous consequences for the role of the aether in physics. Einstein's statement at the beginning of his 1905 paper "On the Electrodynamics of Moving Bodies" indicates this. "The introduction of a Light aether will prove to be superfluous, for according to the conceptions which will be developed, we shall introduce neither a space absolutely at rest, and endowed with special properties. . . ." There was no fixed point, no stationary reference system within this new vision of the universe; everything was in movement relative to each other. As Einstein points out, the result of the Michelson-Morley experiment will always be null. "According to this theory there is no such thing as a 'specially favored' (unique) co-ordinate system to occasion the introduction of the aether-idea, and hence there can be no aether-drift, nor any experiment with which to demonstrate it." In other words, the Michelson-Morley experiment kept getting a null result because they were measuring the velocity of Earth against something that did not exist. Because of this shift away from notions of absolute space and time, it was gradually recognized that the aether, or any kind of medium for light, was unnecessary.

Einstein and Photons Einstein's work in 1905 was also significant for light in other respects. The wave theory of light appeared to have triumphed over the particle theory in the nineteenth century. But Einstein's work on the photoelectric effect showed that the wave theory was by itself insufficient to explain the behavior of light. He reintroduced the notion of particles of light into the vocabulary of physics, but this time in a much more sophisticated way. In terms of its interaction with matter, Einstein suggested light was composed of particles of energy. These particles interacted with matter, and could eject electrons from their original positions. The development of the idea of light photons, as they came to be known, challenged the earlier dichotomy between theories of light based on particles or waves. Whether the concept of light as a wave or as a particle is more appropriate depends on the context of what is being observed. The two models complement each other, and this is referred to as wave-particle duality.

Conclusion Einstein's work on light went beyond classical physics in two respects. He eliminated the need for the aether as the light medium, and he demonstrated that the description of light as a wave was inadequate to explain all the characteristics of light. As in most cases, the publication of Einstein's work in 1905 did not cause a sudden rupture with earlier ideas. The full ramifications were not immediately apparent to all, and many physicists ignored his work and carried on with their own research into the aether. But gradually, the impact began to filter through the profession of physics, and the cutting edge of physics was inquiry based on these new ideas. The aether was firmly tied to the most basic assumptions of nineteenth-century physics. With the death of absolute space, it ceased to be of any interest to most mainstream physicists. —KATRINA FORD

Further Reading

Baierlein, Ralph. *Newton to Einstein: The Trail of Light: An Excursion to the Wave Particle Duality and the Special Theory of Relativity.* Cambridge and New York: Cambridge University Press, 1992.

Einstein, Albert. *Relativity: The Special and the General Theory.* 15th ed. New York: Bonanza Books, 1952.

Gribbin, John. *Schrödinger's Kittens and the Search for Reality.* London: Weidenfield & Nicholson, 1995.

Purrington, Robert D. *Physics in the Nineteenth Century.* New Brunswick, NJ: Rutgers University Press, 1997.

Swenson, Lloyd S. *The Ethereal Aether: A History of the Michelson-Morley-Miller Experiment, 1880–1930.* Austin: University of Texas Press, 1972.

Zajonc, Arthur. *Catching the Light.* London: Bantam, 1993.

Is the "many-worlds" interpretation of quantum mechanics viable?

Viewpoint: Yes, the "many-worlds" interpretation of quantum mechanics is viable, as it provides the simplest solution to the measurement dilemma inherent in the standard model.

Viewpoint: No, the "many-worlds" interpretation of quantum mechanics is not viable for numerous reasons, including measurement problems and the inability to test it scientifically.

Quantum mechanics is one of the most successful theories of modern physics. Chemists, physicists, materials scientists, molecular biologists, and electrical engineers use the predictions and terminology of the theory to explain and understand numerous phenomena. Looked at closely, however, the theory has certain unsettling features. Particles such as the electron have wavelike properties, while electromagnetic waves behave, at times, like particles. One must abandon the notion that each particle has a definite location at all times. These counter-intuitive aspects of the theory led many physicists, most notably Albert Einstein (1879–1955), to reject it or at least regard it as temporary and inherently incomplete.

Among the strangest aspects of quantum theory is the sudden change in state that is alleged to occur when a measurement is made. If a beam of light is sent through a polarizing filter, it is found that the emerging beam is completely polarized, and thus it will pass through any subsequent polarizing filters with the same orientation as the first without any loss of intensity. If the intensity of the light source is reduced so that one can be sure that only one photon or quantum of light energy is passing through each filter at a time, the result is unchanged. If one of the filters is rotated by 45 degrees, however, only half of the photons pass through. For any particular photon, there is no way of determining whether or not it will pass through the rotated filter. This appears to be a truly random process—an example of God playing dice, as Einstein would have said.

A further problem in quantum measurement theory concerns the existence of so-called entangled states. An electron and an anti-electron can annihilate each other, creating two photons traveling in opposite directions. The polarizations of the photons will be highly correlated. If one of them goes through a polarizing filter, one knows with certainty that the other will not pass through a filter with the same direction of polarization, and this conclusion is unaffected by rotating both filters by an equal amount. It is as if the second photon knew what the first had decided to do, even though it could be light years away by the time the first measurement was made.

According to the standard version of quantum mechanical theory, the state of any system evolves continuously in time according to the so-called time-dependent Schrödinger equation until a measurement is made. The state is described by a wave function—sometimes called a *state vector,* referring to an abstract mathematical space, called *Hilbert space,* with an infinite number of dimensions. According to the theory, the wave function contains all the information that one can have about the system and gives the probability of each possible outcome of any measurement. Once the measurement is made, the wave function is claimed to change to a function that predicts with 100% certainty that the outcome of the same measurement performed immediately after the first will be the same. Before the measurement, the wave function is said to represent a superposition of states, each corresponding to a possible outcome of the measurement. When the measurement is made, the wave function is said to collapse, in that only the part of the original wave function corresponding to the observed result remains.

To highlight some of the paradoxical aspects of the standard quantum theory of measurement, Erwin Schrödinger (1887–1961) proposed a thought experiment now generally referred to as *Schrödinger's cat.* In this case we are asked to consider a cat enclosed in a cage along with a vial of poison gas, a Geiger counter, and a small amount of a radioactive material. At the start of the experiment, the amount of radioactive material is adjusted so that there is a 50-50 chance that the Geiger counter will record at least one decay in the next hour. The Geiger counter is connected to an apparatus that will break open the vial of poison gas if a decay is registered, killing the cat. The whole apparatus—cage, cat, vial, and counter—is covered for an hour. At the end of the hour, according to the quantum theory of measurement, the system will be in a superposition of states, half of which involve a live cat and the other half a dead one. Uncovering the cage constitutes a measurement that would collapse the wave function into one involving a dead or a live cat. The notion of a creature, half-dead, half alive, is inconsistent with human experience and perhaps even less acceptable than the blurring of particle and wave behavior that quantum mechanics describes.

While a graduate student in physics at Princeton University, Hugh Everett III proposed what he termed a "relative state formulation" of quantum mechanics, which might provide an alternative to some of the more paradoxical aspects of quantum measurement theory. Everett advocated thinking about a single wave function for the entire universe,which always evolved in time in a continuous fashion. The wave functions would describe a superposition of states involving observers and experimental apparatus. In the Schrödinger cat case the superposition would involve states in which the cat was dead and known to observers as dead, and states in which the cat was alive and known to observers as alive. Everett's proposal solves the entangled states paradox as well, since the wave function provides a superposition of states with each possible outcome of the two polarization measurements. Everett's original paper said nothing about multiple worlds.

The multiple-worlds idea emerged in subsequent work by Everett, his advisor, John Wheeler, and physicist Neill Graham. In this work it became apparent that the wave function for the entire universe envisioned by Everett was actually a superposition of wave functions for alternative universes. In each, the observers had observed definite outcomes of the Schrödinger cat experiment, but every time a measurement was made, the universe split into multiple universes, each with a separate outcome.

Opinions on the many-worlds interpretation vary greatly among physicists. Since there are no testable predictions, some feel that it falls outside the scope of scientific investigation and is merely an interesting speculation. Scientists generally reject theories that cannot be proven false by an experiment, but some exception is made for cosmological theories, those about how the whole universe behaves, since we cannot set up a test universe. Others feel that since the many worlds theory eliminates some counterintuitive aspects of the standard theory, it has some merit. The notion of alternate realities has long been a favorite of science fiction writers. The idea that all conceivable—that is, not self-contradictory—universes ought to exist also has appeal, in that it eliminates the need to explain why our universe has the particular properties (electron mass, gravitational constant, and so on) that it has—though this goes far beyond the original theory. The pro and con articles illustrate two of the many possible positions. —DONALD R. FRANCESCHETTI

Viewpoint:

Yes, the "many-worlds" interpretation of quantum mechanics is viable, as it provides the simplest solution to the measurement dilemma inherent in the standard model.

The Copenhagen, or standard, interpretation of quantum physics, as defined by Niels Bohr (1885–1962) and others during the 1920s, has proved to be one of the most successful scientific theories ever created. Its predications have been confirmed time and time again, and not a single experiment has contradicted the theory. Yet many scientists are profoundly unhappy with the philosophical consequences of standard quantum theory.

The Copenhagen interpretation of quantum theory contains a number of worrying paradoxes and problems. One of the strangest, and most worrying, is the role that measurement

KEY TERMS

AXIOM: Something that is accepted as true for the basis of an argument or theory.

OCKHAM'S RAZOR: A principle proposed by William of Ockham (c.1300–1349), who argued that the best answer was almost always the one with the fewest assumptions. It has become a guiding principle of scientific theory.

SCHRÖDINGER EQUATION: The quantum mechanical equation that describes the evolution of a physical system in time.

STERN-GERLACH DEVICE: An apparatus used by O. Stern and W. Gerlach in their experiment to measure the intrinsic spin angular momentum of silver atoms. They found that it takes only two discrete values, commonly called "spin up" and "spin down."

THOUGHT EXPERIMENT: An experiment that cannot, or should not, be performed (at least at the time proposed) in the real world, but can be imagined, and the consequences calculated. This technique was made famous by Albert Einstein, who used a number of thought experiments in his theories of relativity, and in his debates over the nature of reality and physics with other scientists.

WAVE FUNCTION: In quantum mechanics the mathematical function that describes the state of a physical system, sometimes also called the wave vector, as it can be represented as a vector in an abstract, many dimensional space.

plays in quantum physics. The act of measuring is outside quantum theory, and the standard explanation involves ad hoc assumptions and generates uncomfortable paradoxes. The simplest solution to this dilemma is the many-worlds (or many-universes) interpretation of quantum theory, first proposed by Hugh Everett III in 1957.

The Problem of Measurement Quantum mechanics is essentially a statistical theory that can be used to calculate the probability of a measurement. It is a epistemological theory, not an ontological theory, in that it focuses on the question of how we obtain knowledge, and does not concern itself with what is. It describes and predicts experimental observations, but as it stands it does not explain any of these processes. Yet despite the focus of the theory on the probabilities of measurement, the act of measurement is beyond the scope of the Copenhagen interpretation.

Imagine a single photon moving toward a strip of photographic film. Before measurement takes place the Copenhagen interpretation describes the photon as a wave function, as defined by Erwin Schrödinger's (1887–1961) wave equation. The wave function describes a broad front of probabilities where the photon could be encountered, and does not give any specific values to the position until measurement. When the wave function hits the film only a single grain of the film is exposed by the single photon. The wave function appears to have collapsed from a broad front of probability to a single point of actual impact. This collapse, or "reduction of the state," takes place instantaneously, even though the front of probability could be very large.

The collapse of the wave function, however, is not described by the Schrödinger wave equation. It is an addition to quantum mechanics to make sense of the fact that observations do appear to occur. Many-worlds theory solves the measurement problem of quantum physics, by allowing for all outcomes of the wave function to be correct, so the wave function does not collapse. Instead all outcomes exist, but in separate realities, unable to interact with each other. Each measurement results in the branching (or grouping depending on which many-worlds theory you use) of universes corresponding to the possible outcomes. No additions to quantum theory are required, and the Schrödinger wave equation as a whole is taken as an accurate description of reality. Many-worlds theory is therefore an ontological theory that explains what the quantum world is, and how the act of measurement takes place.

Measurement is even more of a problem in the standard quantum interpretation when you consider the notion of entanglement. A measuring device interacting with a quantum system to be measured, according to the Schrödinger wave equation, inherits the quantum measurement problem, ad infinitum. In effect the measuring device and the object to be measured just become a larger quantum system. According to the mathematics that quantum physics uses so effectively, there can be no end to this entanglement, even if more and more measuring devices are added to measure the previous entangled systems. Again, the standard interpretation of quantum physics seems to suggest that the act of measurement is impossible.

Cats, Friends, and Minds Several of the originators of quantum theory expressed their concerns over the direction in which quantum mechanics developed, including Albert Einstein (1879–1955) and Schrödinger. While Schrödinger may have provided quantum theory with some of its most important mathematical tools, namely his wave equation, he could not agree philosophically with Bohr and other proponents of the Copenhagen interpretation.

Schrödinger proposed a famous thought experiment involving a cat in a box that will be killed if a radioactive particle decays, in order to show how the assumptions of the quantum world fall down when applied to larger objects.

The experiment focuses on a typical quantum probability outcome, a radioactive particle that has a certain chance of decaying in a certain time. The Copenhagen interpretation suggests that the particle exists in a superposition of states until observed, and then the wave function collapses. However, while it seems reasonable to describe a quantum system as in a superposition of states, the idea that the cat is somehow both dead and not-dead until observed seems a little strange. If we replace the cat with a human volunteer, often referred to as Wigner's friend after Eugene Wigner (1902–1995), who first suggested it, then things become even more unreal. Until observed, quantum theory tells us that Wigner's friend is both alive and dead, in a superposition of states, but surely Wigner's friend himself knows if he is dead or alive? Wigner invoked the idea that the conscious mind is the key to observations. It is the observer's mind that collapses the wave function, not just its interaction. While this interpretation avoids the problem of entanglement, it makes the mind something that exists outside of quantum mechanics, and it further adds to the measurement problems. If we consider the exposed grain of film in the first example, Wigner would claim that until a conscious mind views the developed film the wave function has not collapsed. It remains as a broad wave front until observed, perhaps years later, and then instantaneously collapses. The power of the conscious mind seems to create reality. But what counts as a conscious mind? Can a cat determine reality? An insect or an amoeba? And what happened before the evolution of conscious minds in the universe? Was nothing real in the universes until the first conscious lifeform looked up at the sky and observed it?

Many-worlds theory is the simplest and most economical quantum theory that removes such problems. The cat and the human volunteer are both alive and dead, but in different universes. There is no need to invoke a special power of the mind, or to add in the collapse of the wave function. Many-worlds model proposes that the same laws of physics apply to animate observers as they do to inanimate objects. Rather than an ad hoc explanation for measurement, or the elevation of the mind to cosmic status, many-worlds theory restores reality to quantum theory.

Erwin Schrödinger

(© Bettmann/Corbis. Reproduced by permission.)

Some Criticisms The many-worlds interpretation had been challenged on a number of grounds, and like any other quantum interpretation it does have its problems. However, most objections are baseless and ill-informed. Some critics have charged that it violates the law of conservation of energy principle, but within each world there is perfect agreement with the law. Another common objection is that it removes free will, as all outcomes of a choice will exist in some universe. Yet, many-worlds theory actually has less problem with free will than some other interpretations, and its expression can be thought of as a weighting of the infinite universes.

Related to this is the criticism that many-worlds theory removes probabilities from quantum theory. When the odds of Schrödinger's imaginary cat dying are 50:50, the splitting (or grouping) of worlds seemed straight forward. But when the odds are tilted to 99:1, how many worlds are created? If just two worlds are formed, each relating to one of the possible outcomes, then there is no difference from the 50:50 situation. However, if you regard the worlds created as a grouping or differentiating of the universes, with 99% having one outcome and 1% the other, then it is obvious that probability is retained in the theory.

It has been argued that many-worlds theory violates Ockham's Razor, the basic principle of the simplest explanation usually being the correct one. Opponents claim that by introducing an infinite number of universes many-worlds model introduces far too much "excess metaphysical baggage" into quantum theory. How-

ever, the theory actually requires less additions to quantum mechanics than any other interpretation. Depending on the particular version of many-worlds theory, there are either no extra assumptions (it simply uses Schrödinger's wave equation), or there are a very small number of assumptions relating to how the many worlds are created or grouped.

One valid criticism of the many-worlds interpretation is that there is no agreed upon version of the theory. Indeed, it seems that every major proponent of the theory has formed their own, slightly different, take on the basic concept, such as whether the worlds physically split, split without a physical process, or if there is just a grouping or reordering of existing universes. There are spin-off theories, such as the many-minds interpretation, in which it is the mind of the observer that splits, rather than the universe. There is also the modal logic theory of possible worlds, which is derived from philosophical precepts rather than quantum theory. It should be possible, one day, to differentiate between all these theories experimentally, as they produce slightly different predictions. It is often claimed that many-worlds theory is untestable. However, there have been experiments conceived that would help determine between many-worlds and other interpretations of quantum theory, but currently the means are not available.

Local, Deterministic, and Universal The many-worlds interpretation has a number of advantages over other quantum theories. It is a local theory, in that it does not rely at any point on faster-than-light signaling or cooperation within quantum systems. It therefore retains the theory of relativity, unlike many other interpretations, which introduce non-local effects to explain measurement and cooperation between particles. Many-worlds theory enables the Schrödinger wave equation to retain its smooth, continuous, and deterministic nature by removing the collapse of the wave function. This is in marked contrast to the standard interpretation, in which the world is a very undetermined entity in which particles have no fixed positions until measured (and we have seen the problems associated with measurement).

While the many-worlds theory is unpopular with many physicists—after all, the standard interpretation works incredibly well as long as you do not consider the philosophical consequences—it does have strong support from cosmologists. Quantum theory implies that the entire universe can be described by a wave equation, and therefore treated as a single quantum system. However, if the observer must be outside the system to be viewed, then it would seem that the universe can never be real. The wave function that describes the universe as a whole could never collapse to an actual value. Many-worlds theory also explains why the universe has developed the way it has to allow life to evolve. The seemingly fortunate distribution of matter in the universe, and the fortuitous values of various constants, and the position of the Earth relative to the Sun all appear to conspire to allow life. However, if there are an infinite number of universes our fortunate circumstances are just a subgroup of the many possible worlds, most of which would not have developed life.

Multiple-worlds theory offers a number of subjective advantages over the standard, Copenhagen interpretation of quantum theory. It is at least as viable as any other interpretation of quantum theory, but has the advantage of requiring few, if any, additional assumptions to the mathematics of quantum physics. It also solves the problems of measurement, entanglement, and the suggestion that the mind is somehow outside physics. It restores the determinism of classical physics, while retaining the utility of the Copenhagen interpretation. Although it does suffer from a current inability to be tested, and a bewildering array of competing versions, the advantages are enough to make it a valuable theory until such time as experiments can be performed to test any differences in its prediction. —DAVID TULLOCH

Viewpoint:
No, the "many-worlds" interpretation of quantum mechanics is not viable for numerous reasons, including measurement problems and the inability to test it scientifically.

A standard plot device in science fiction is the existence of alternate realities in which another "version" of the self exists, usually in counterpoint to the "real" self. For example, Joe A is a nice guy who likes kids and animals and goes to church every Sunday. However, in the "other" world, Joe B may look and talk exactly like Joe A and even have the same genetic makeup; but he hates kids, kicks dogs, and only goes to church when he thinks he can pry open the contribution box and steal money. The "many-worlds" interpretation (MWI) takes this idea a step further by having countless Joes, A through Z and beyond, with a new Joe being created every time a measurement (or "decision") is made. As fodder for science fiction stories, it is a great idea. As an explanation of reality, even in the most theoretical realms of quantum physics, it is problematic at best.

At first look, the many-worlds theory seems to answer one of quantum physics' most nagging paradoxes in that it does not require the collapse of the wave function as set forth by the Copenhagen interpretation. Developed by Niels Bohr (1885–1962) and others, it has long been the generally accepted interpretation of quantum mechanics. The Copenhagen interpretation states that an "unobserved" system evolves in a deterministic way according to a wave equation. However, when "observed" the system's wave function "collapses" to a specific observed state (for example, from a diffuse wave not restricted to one location in space and time to a particle that is restricted in space and time and thus observable). The problem arises because this interpretation gives the observer a special status unlike any other object in quantum theory. Furthermore, it cannot define or explain the "observer" through any type of measurement description. As a result, detractors say that the Copenhagen interpretation describes what goes on in the observer's mind rather than a physical event. Some physicists have espoused that the wave function is not real but merely a representation of our knowledge of an object.

The many-worlds interpretation is an unorthodox alternative that promises to model the complete system (both the macro and the micro world) while relegating the observer to a simple measuring device. By leaving out the wave function collapse, the MWI in quantum physics says that a quantum system that has two or more possible outcomes does not collapse into one outcome when observed but branches out or realizes all the outcomes in other universes or worlds. So, whether an atom might decay or not decay, the world "branches" so that both possibilities come to pass. Furthermore, each world and the observers in it are unaware of the existence of the other worlds. Unlike the Copenhagen interpretation, MWI does not depend on our observation of it to create a specific reality but chooses one of two or more possible worlds, all of which are real. Although enticing and interesting, the viability of this model fades upon closer examination. It does not really solve the problems facing many of the theories in quantum physics, which seems to describe microsystems accurately but encounters numerous problems when applied to larger, classical systems like the material world that we taste, touch, see, and smell everyday.

One of the most fundamental problems with MWI is that it violates the law of conservation of energy. This law states that the universe always has the same amount of energy, which can neither be increased nor decreased. Mainstream physicists generally agree that energy cannot be created from nothing or annihilated into nothing. As a result, conservation of energy places significant constraints on what states the world can pass through. But the propagation of ever-expanding multiple worlds violates this law. For example, when a world diverges, where does the extra energy come from to create this divergent world? As energy (in the form of fundamental particles like electrons, protons, neutrons, and photons) is exchanged between these worlds, the amount of energy in them would fluctuate, resulting in one of the worlds having more or less energy. This result contradicts the conservation of energy. In fact, the ever-expanding universe in the many-worlds interpretation would not conserve but dissipate energy. To represent a valid or unified cosmology, the theory would require an unlimited amount of energy or the creation of energy out of nothing.

The Measurement Problem One of the goals of the MWI was to create a formal mathematical theory that corresponded to reality in a well defined way and solved the measurement problem of quantum mechanics. To account for the wave-like behavior of matter, quantum mechanics has incorporated two primary laws. The first uses a linear wave equation to account for time trajectory of all unobserved systems and their wave-like properties. The second is the standard collapse theory that accounts for the deterministic qualities we see in systems when we observe them. The two laws, however, are counterintuitive and, to a certain extent, mutually exclusive in that one describes a system when we are not looking and the other when we are. Although dropping the standard collapse theory solves the ambiguity and, some might say, logical inconsistencies in these two laws, it is the collapse theory that ensures we end up with determinate measurements. Without the theory, there is no plausible explanation for these measurements or our perception of reality.

Since the theory claims to represent our reality as well as other classical approaches in physics and mathematics, then it should be rigorously defined in the same manner. In an article in the *International Journal of Modern Physics,* A. Kent puts it this way, "For any MWI worth the attention of physicists must surely be a physical theory reducible to a few definite laws, not a philosophical position irreducibly described by several pages of prose." Such a theory, says Kent, must include "mathematical axioms defining the formalism and physical axioms explaining what elements of the formalism correspond to aspects of reality." These criteria are concretely defined by other theories, including the theories of general relativity and electrodynamics. According to Kent and many others, the many-worlds model fails to produce such axioms for various reasons, primarily the fact that the axioms have an extremely complicated notion of a measuring device.

To work, the MWI says that we must choose a preferred basis to explain the determinate measurement. Rudimentarily, the preferred basis refers to the fact there are always many ways one might write the quantum-mechanical state of a system as the sum of vectors in the Hilbert space (for purpose of simplicity, defined as a complete metric space). Choosing a preferred basis means choosing a single set of vectors that can be used to represent a state. However, what basis would make the observers' experiences and beliefs determinate in every world? The correct preferred basis in the MWI would depend on numerous factors, including our physiological and psychological makeup. Taking this a step further, a detailed theory of consciousness would have to be developed to validate the MWI. To put it another way, it would be necessary to classify all known measuring devices to make MWI axioms consistent with the known outcomes of experimental results. Since most theories can describe physical phenomena in precise and relatively simple mathematics, the need for such an extensive classification precludes MWI axioms from being accepted as fundamental physical laws.

MWI also raises questions about any meaning for statistical significance or "weight" since statistical predictions are subjective experiences made by an observer. In fact, the theory has no validity in making any type of statistical predictions for the standard collapse theory. For example, say the probability of an electron spinning up or down is one-half after passing through the Stern-Gerlach device. But what is the significance of "one-half" if the world splits into another world, resulting in the electron spinning both up and down? In the MWI the same branching would occur whatever the odds—50:50, 60:40, or 70:30; all would result in the electron spinning both up and down. Furthermore, the standard predictions of quantum mechanics concerning probabilities are made irrelevant because it is unknown which of two future observers is making the measurement. As a result, it means nothing to say that an experience will be "this" rather than "that" when "that" is happening in another world or perhaps in our world. Who can tell? If the MWI is valid, nothing occurs in our world according to probabilistic laws.

To Be or Not to Be? In many ways, the MWI directly invalidates the foundation of our conscious interpretations and interactions with the world. It does this by insisting on a deterministic reality in a microphysical world that leads to branching of various "macro" realities that all exist but are not accessible. As a result, our consciousness and experiences are meaningless in terms of having a functional role. There is no free will because ultimately there are no important questions to ask since everything that can happen will happen. If the theory itself is designed so that our minds have no effect on the outcome of an event, why then does such a theory grapple with the question of mind? It also does not address the notion that peoples' personal experiences are unified in that we perceive our experiences as a whole following only one path and, in turn, never "experience" the branching worlds. As a result the MWI is more of a metaphysical philosophy than a science. It presents us with a view of nature that is vague, complex, and schizophrenic.

Another fundamental argument against the many-worlds model is that it violates Ockham's Razor, which is science's way of saying "keep it simple stupid." Developed in the fourteenth century by English philosopher William of Ockham (c. 1300–1349), the maxim states that hypotheses or entities should not be multiplied beyond necessity. It proposes that scientists should favor the simplest possible explanation as opposed to more complex ones. It is hard to imagine any theory more complex than one that relies on the existence of numerous worlds that are unseen and unknowable to explain what we do see in this world. In other words, if we want to explain our world and our experiences, then we should do so via the world or worlds we know. So far, there is only one world we know of, and that is the one we live in. If it is to be considered viable, quantum physics must rely on a quantum conception of reality based on the empirical data available to us and that fits a formula of physical theory. This is the only way that we would have a true "place" in a reality that we could both know and act upon in a way that meant anything.

Inherent in the phrase "scientific theory" is the notion that such a theory can ultimately be proven correct or incorrect through experiments. In the case of the many-worlds model, since the inhabitants of these many worlds are not aware of each other, to verify the theory experimentally is impossible. Does that mean we should accept the model on faith? This sounds more like religion than science. Even if it were a logical deduction, which it is not, it would still need to be proven.

Proponents of the many-worlds model say that this is the only solution that "fits" so it must be true because no one has come up with a better alternative. For science, this is as close to blasphemy as you can get. It is not that no alternatives exist—it is just that we have not found them yet in terms of theories, axioms, and mathematics. It is like the story about the nineteenth-century U.S. patent office director who suggested that the patent office be closed because everything that could be discovered had already been discovered. There are other alternatives to the many-worlds model, and time will surely reveal them. —DAVID PETECHUK

Further Reading

Bell, J. S. *Speakable and Unspeakable in Quantum Theory.* Cambridge, MA: Cambridge University Press, 1987.

Bohm, D., and B. J. Hiley. *The Undivided Universe: An Ontological Interpretation of Quantum Theory.* London: Routledge, 1993.

Davies, Paul. *Other Worlds.* Harmondsworth, England: Penguin, 1988.

Deutsch, David. *The Fabric of Reality: The Science of Parallel Universes.* New York: Allen Lane, 1997.

DeWitt, Bryce S., and Neill Graham, eds. *The Many-Worlds Interpretation of Quantum Mechanics.* Princeton, NJ: Princeton University Press, 1973.

"Everett's Relative-State Formulation of Quantum Mechanics" *Stanford Encyclopedia of Philosophy* [cited July 22, 2002]. <http://plato.stanford.edu/entries/qm-everett/>.

Greenstein, George, and Arthur G. Zajonc. *The Quantum Challenge.* Sudbury, MA: Jones and Bartlett Publishers, 1997.

Kent, A. "Against Many-Worlds Interpretations." *International Journal of Modern Physics* 5, no. 9 (1990): 1745–62.

Leslie, John. "A Difficulty for Everett's Many-Worlds Theory." *International Studies in the Philosophy of Science* 10, no. 3 (October 1996): 239.

Wheeler, J. A. and W. H. Zurek. *Quantum Theory and Measurement.* Princeton, NJ: Princeton University Press, 1983.

Can computational chemistry provide reliable information about novel compounds?

Viewpoint: Yes, properly used, computational chemistry can be a reliable guide to the properties of novel compounds.

Viewpoint: No, computational chemistry is not a reliable guide to the properties of novel compounds.

In 1897, British physicist J. J. Thomson announced the discovery of the electron. Fourteen years later, New Zealander Ernest Rutherford, working at the University of Manchester, astounded the physics world by demonstrating that all the positive charge in the atom is concentrated in a tiny nucleus. Rutherford's discovery was unsettling because the laws of Newton's mechanics and the more recently discovered laws of electricity and magnetism were utterly inconsistent with the existence of stable arrays of electrons either standing still or moving around a nucleus. Over the next 20 years, physicists, notably Bohr, Heisenberg, Schrödinger, Dirac, and Pauli, developed quantum mechanics. Quantum mechanics is a comprehensive theory of motion that agrees with Newtonian mechanics for macroscopic objects but describes the motion of subatomic particles. The theory was confirmed by detailed agreement with experimental results and allows for stable electron motion around a point-like nucleus.

Although modern chemistry texts almost invariably begin with a discussion of atomic structure and explain chemical phenomena in terms of the sharing or exchange of electrons, it would be a serious mistake to believe that modern chemistry began with the discovery of the electron. A list of chemical elements can be found in Lavoisier's treatise of 1789, and Mendeleyev in 1870 systematized the elements by their chemical properties in his periodic table. To late-nineteenth-century chemists, molecules were composed of atoms, thought of as small, hard balls held together by somewhat springy chemical bonds that had a definite orientation in space.

Although most chemists now accept that quantum mechanics provides an accurate description of the motions of the electrons and nuclei that underlie chemical phenomena, many chemists differ markedly on the extent to which quantum mechanics actually provides insight that can guide the progress of chemical research and invention. Although the quantum mechanical equations that govern a system of electrons and nuclei are not difficult to write, they are extraordinarily difficult to solve.

It is an unfortunate fact of mathematics that the equations governing the motion of more than two interacting particles, whether those of Newtonian or of quantum mechanics, cannot be solved exactly. In quantum mechanics, the system is described by a wave function that according to the standard, or Copenhagen (after Bohr), interpretation, gives the probability of finding electrons and nuclei at any set of points in space. The possible wave functions for the system are the solutions of the Schrödinger equation for the system, which includes terms describing the kinetic energy of the particles and the electromagnetic attraction between every pair of charged particles. For molecules with more than one electron, the wave function has to satisfy the rather esoteric criterion of changing algebraic sign (plus to minus or vice versa) whenever the coordinates of any two electrons are interchanged. This strange mathematical requirement is the basis for the Pauli exclusion principle, which limits the electrons in atoms to only two in each atomic orbital.

Because the Schrödinger equation cannot be solved exactly, except for systems of only two particles, such as the hydrogen atom, techniques have been developed for finding approximate solutions. Approximate methods can be characterized as either ab initio or semiempirical depending on whether they make use of physical data other than those contained in the Schrödinger equation. Both types of calculations, along with certain other modeling techniques that generally require the use of high-speed computers, now constitute the field of computational chemistry.

Ab initio calculations are generally done only for molecules composed of a relatively small number of electrons. When most ab initio techniques are used, it is assumed that the overall electron wave function can be expressed as a sum of terms, each describing an assignment of electrons to molecular orbitals that are combinations of the orbitals of the atoms that make up the molecule. The interaction between electrons consists of an averaged electrical repulsion plus so-called exchange terms that are a consequence of the Pauli principle and the mathematical requirements it places on the overall wave function. The calculation of the exchange terms is generally the most time-consuming part of an ab initio calculation.

For molecules such as insulin, the hormone that regulates the metabolism of sugar in the body, which contain several hundred atoms and several thousand electrons, accurate ab initio calculation might take centuries or longer, even on the fastest supercomputers. Fortunately for biology and medicine, the molecules in living organisms involve a relatively few bond types, and the forces between the atoms can be modeled with so-called molecular mechanics programs. Such an approach represents, in spirit, a return to the nineteenth century view of the molecule as a collection of atoms held together by localized bonds. The results are generally satisfactory for biological molecules and candidate drugs, partly because only a few types of bonds have to be considered and partly because molecules in living cells generally recognize other molecules on the basis of their shape and the distribution of electrical charge within them.

For chemical purposes, the wave complete function actually contains more information than is needed. The strengths of the chemical bonds, their vibrational characteristics, the charge distribution within the molecule, and the shape of molecules overall are completely determined if the accurate time-averaged distribution of electrons, the so-called electron density, is known. Unfortunately, quantum theory does not provide a method by which electron density can be directly computed. The so-called density functional methods, honored with the awarding of the 1998 Nobel Prize to Walter Kohn, actually involve use of a simple formula to approximate the exchange terms in the calculation of the molecular orbitals according to the calculated electron density. In practical density function calculations, the formula chosen is that which best describes a related series of compounds. This blurs the boundary between ab initio and semiempirical methods, causing some chemists to question whether it is quantum theory or chemical insight that is responsible for the success of the methods.

Computational chemistry as currently practiced seems to be a reliable guide to the behavior of biological molecules and some families of catalysts. It has reduced greatly the cost and time associated with developing new drugs. On the other hand, chemists continue to debate the value of computational methods in explaining the character of compounds, such as aluminum monoxide (AlO) and carbon monophosphide (CP), that stretch traditional notions of bonding. It is these chemical novelties that can be counted on to keep the debate alive for some time. —DONALD R. FRANCESCHETTI

Viewpoint:

Yes, properly used, computational chemistry can be a reliable guide to the properties of novel compounds.

The cartoon image of a chemist standing amid boiling, fuming beakers and test tubes in a cluttered chemistry laboratory needs to be redrawn. Many twenty-first century research chemists are likely to be seen in front of high-resolution computer displays constructing colorful three-dimensional images of a novel compound. These scientists may be using computer-assisted drug design (CADD) software that incorporates computational methods to develop pharmacophores, sets of generalized molecular features that are responsible for particular biological activities. There is also a good possibility the scientist has access to sophisticated molecular modeling programs to make virtual designer molecules, as for drug candidates, or possibly catalysts that could accelerate production of advanced chemical compounds.

KEY TERMS

AB INITIO: Latin for "from the beginning." A term that describes calculations that require no added parameters, only the most basic starting information. The properties are worked out from the beginning. In practice such methods can be overly time consuming, and some parameters often are introduced to increase the speed of calculation.

CATALYST: A chemical added to a reaction to accelerate a process but that is not consumed by the process.

COMBINATORIAL CHEMISTRY: Technology designed to greatly accelerate the rate at which chemical discoveries can be made. With robots and other advanced techniques, hundreds or thousands of different chemical compounds can be produced simultaneously in rows of minute wells called *microarrays.*

ELECTRON CLOUD: The physical space in which electrons are most likely to be found.

INFORMATICS: The application of computational and statistical techniques to the management of information.

MOLECULAR MODELING: Any representation of a molecule, from a two-dimensional pencil drawing to a three-dimensional wire-frame construction. The term has become linked to the computational methods of modeling molecules and is sometimes uses synonymously with *computational chemistry.*

QUANTUM MECHANICS: A physical theory that describes the motion of subatomic particles according to the principles of quantum theory. Quantum theory states that energy is absorbed or radiated not continuously but discontinuously in multiples of discrete units.

SCHRÖDINGER'S WAVE EQUATION: A mathematical formula that describes the behavior of atomic particles in terms of their wave properties for calculation of the allowed energy levels of a quantum system. The solution is the probability of finding a particle at a certain position.

In the past two decades, computational power and software have advanced to the point at which mathematical models can produce three-dimensional structures of molecules as complex as novel proteins and can relate receptor structures to drug candidates. Chemists can work "in silico" using computational techniques to perform virtual experiments to narrow options for the ideal drug or catalyst before moving into a "wet" laboratory to test the designs.

Overview of Computational Chemistry When theoretical chemistry, the mathematical description of chemistry, is implemented on a computer, it is called *computational chemistry.* Some very complex quantum mechanical mathematical equations that require approximate computations are used in theoretical chemistry. Approximate computations produce results that are useful but should not be considered "exact" solutions. When computations are derived directly from theoretical principles, with no experimental data used, the term *ab initio* is used to describe the results. Ab initio is Latin for "from the beginning," as from first principles. Because the energy of the electrons associated with atoms accounts for the chemistry of the atoms, calculations of wave function and electron density are used in computational chemistry calculations.

Wave function is a mathematical expression used in quantum mechanics to describe properties of a moving particle, such as an electron in an atom. Among the properties are energy level and location in space. Electron density is a function defined over a given space, such as an electron cloud. It relates to the number of electrons present in the space. Ab initio schemes include the quantum Monte Carlo method and the density functional theory method. As the name implies, the density functional theory method entails electron density. The quantum Monte Carlo method uses a sophisticated form of guessing to evaluate required quantities that are difficult to compute.

Modifications of ab initio schemes include tactics such as semiempirical calculations that incorporate some real-world data from the laboratory and simplify calculations. Semiempirical calculations work best when only a few elements are used and the molecules are of moderate size. The method works well for both organic and inorganic molecules. For larger molecules it is possible to avoid quantum mechanics and use methods referred to as *molecular mechanics.* These methods set up a simple algebraic expression for the total energy of a compound and bypass the need to compute a wave function or total electron density. For these calculations all the information used must come from experimental data. The molecular mechanics method can be used for the modeling of enormous molecules, such as proteins and segments of DNA. This is a favorite tool for computational biochemists. Powerful software packages are based on the molecular mechanics method. Although there are some shortcomings in this method on the side of theory, the software is easy to use.

1998 Nobel Prize in Chemistry The 1998 Nobel Prize in chemistry was jointly awarded to Professor Walter Kohn of the University of California at Santa Barbara and Professor John A. Pople of Northwestern University. The laureates were honored for their contributions in developing methods that can be used for theoretical studies of the properties of molecules and the chemical processes in which they are involved. The citation for Walter Kohn notes his development of the density functional theory, and the citation for John Pople notes his development of computational methods in quantum chemistry.

A press release by the Royal Swedish Academy of Sciences pointed out that although quantum mechanics had been used in physics since very early in the 1900s, applications within chemistry were long in coming. It was not possible to handle the complicated mathematical relations of quantum mechanics for complex systems such as molecules until computers came into use at the start of the 1960s. By the 1990s theoretical and computational developments had revolutionized all of chemistry. Walter Kohn and John Pople were recognized for being the two most prominent figures in this movement. Computer-based calculations are now widely used to supplement experimental techniques.

The Swedish Academy press release notes that Walter Kohn showed it was not necessary to consider the motion of each electron. It is sufficient to know the average number of electrons located at any one point in space. This led Kohn to a computationally simpler method, the density functional theory. John Pople was a leader in new methods of computation. He designed a computer program that at a number of points was superior to any others being developed at the time. Pople continued to refine the method and build up a well-documented model of chemistry. He included Kohn's density functional theory to open up the analysis of even complex molecules. Applications of Pople's methods include predicting bond formation of small molecules to receptor sites in proteins, which is a useful concept in drug research.

Computer-Assisted Drug Design CADD has gained in importance in the pharmaceutical industry because drug design is a slow and expensive process. According to the Pharmaceutical Research and Manufacturers of America (PhRMA), it takes 12 to 15 years to bring a new drug to market, the first six and one-half years being discovery and preclinical testing. The Tufts Center for the Study of Drug Development estimates the average cost of developing one drug is approximately $802 million. That figure includes the cost of failures, and far more candidate drugs do not make it to market than do.

It is important to eliminate bad candidates early in the drug development process. A particular compound may be effective against a disease-causing agent outside the body but be ineffective when the agent is in the body. That is, the drug candidate may not meet ADMET criteria. ADMET is an acronym for absorption, distribution, metabolism, excretion, and toxicity. It refers to how a living creature's biology interacts with a drug. Animal testing has not been eliminated in drug design, but it has been greatly reduced with the new in silico methods.

John Pople
(The Gamma Liaison Network. © Liaison Agency. Reproduced by permission.)

Computational chemistry is used to explore potential drug candidates for lead compounds that have some activity against a given disease. Techniques such as modeling quantitative structure-activity relationships (QSAR), which help to identify molecules that might bind tightly to diseased molecules, are widely used. At the beginning of drug design, computational chemistry is used to establish what are called virtual libraries of potential drug candidates. Large collections of these potentially useful chemical structures are called *libraries.* Collections of computer-designed potential compounds are called *virtual libraries.* Computational chemistry is used in conjunction with combinatorial chemistry, a grid-based synthesis technique that can simultaneously generate tens, hundreds, even thousands of related chemical compound collections, also called libraries.

The chemical libraries have to be screened for real usefulness for a particular drug application. All of the first-screened most likely candidates are called *hits.* When the hits are optimized, they are called *leads. Optimizing* means refining the molecules to maximize the desired properties. Once a number of lead compounds have been found, computational chemistry is

WHAT ARE CHEMICAL LIBRARIES?

Chemicals are not what one usually expects to find in libraries. But in the sense that libraries are organized collections of useful information, the term *chemical library* is an appropriate name for the collections of systematically produced chemical compounds that are the result of combinatorial chemistry.

Combinatorial chemistry is a relatively new way to "do" chemistry. Whereas traditional research chemists follow the conventional scientific method of mixing reactants in test tubes, beakers, or flasks to produce one compound at a time, combinatorial research chemists use miniaturized equipment and robotics to prepare large numbers of compounds in grid-based plates (called *microtiter sheets*) that have very small wells arranged in rows in which many experiments can be run simultaneously.

Based on the work of Robert Bruce Merrifield, winner of the 1984 Nobel Prize in chemistry, combinatorial chemistry was the Eureka! discovery of Richard Houghten in 1985. Houghten wrote about it later, "I woke up in the middle of the night with the 'flash'. . . if I could make mixtures in a systematic manner . . . " (Lebl, 1999).

Most combinatorial chemistry reactions are performed on tiny beads of plastic called *microspheres*. One reactant is attached to each bead, and the beads are placed systematically one to a well. One row of wells has the same reactant on each bead, but the reactants are slightly different on each successive row. The other chemical needed for the reaction is added in solution to the plate column by column. The wells in a column have the same reactant, but each column is systematically slightly different from the previous column. That is why combinatorial chemistry is described as grid-based. The method is also called *parallel synthesis*. There is another way to do combinatorial chemistry in which the beads are contained in small mesh containers. That was the way Houghten first did it. He called it "tea bag" synthesis.

Making a large number of related compounds quickly is only the beginning. Sorting them and identifying the products that are useful is called *high throughput screening and assaying*. This process also is automatic. Running combinatorial chemistry laboratories requires sophisticated software, called *data management and mining.*

There are two major applications for combinatorial chemistry. Drug design is the first use, and right behind it is research to find new and better catalysts for cleaner, greener chemical manufacturing and fossil fuel burning.

—M. C. Nagel

used to check the ADMET criteria. One of the challenges in using CADD and combinatorial chemistry for drug design is the data overload produced that has to be mined (sorted) and managed. These processes would be impossible without a computer. An example of the size of libraries can be seen in the libraries produced by a biotechnology company called 3-Dimensional Pharmaceuticals (3DP). 3DP has a probe library of more than 300,000 individually synthesized compounds that have been selected for maximum diversity and compound screening utility. The company also has a synthetically accessible library of more than 4 billion compounds that have been developed through computational chemistry and can be synthesized on demand.

Computer Chemistry for Catalyst Development The second largest application of combinatorial chemistry, which generally includes computational chemistry, is the development of catalysts for industrial applications that include the vast chemical and petrochemical industries. It also includes emission controls to meet the growing world pressures for cleaner, greener processing of fuel and chemicals. BASF, one of the world leaders among chemical industries, uses molecular modeling as the starting point in its catalyst research. However, the challenges in the development of catalysts are different from the challenges in drug design, except for data management and mining. Complex interrelations between composition and processing variables make the development of catalysts demanding. Intense research and development are being conducted internationally.

In the United States, Symyx, a high-tech company in California, is a leader in combinatorial chemistry. The company's closest competition is a company in The Netherlands, Avantium, that was formed in February 2000. Avantium has produced software under the trademark of VirtualLab with informatics and simulation capability. By starting with computational chemistry, catalyst research saves laboratory time and chemicals. Symyx representatives estimate their approach is up to 100 times faster than traditional research methods and reduces the cost per experiment to as little as 1% of the cost of traditional research methods. Eliminating chemical waste in research and development is a valuable added benefit.

Software Solutions For computational chemistry to be widely used by chemists it has to be available in familiar formats that create an interface between sophisticated computational chemistry and state-of-the-art molecular modeling and analysis tools and the desktop. A number of companies specialize in software for the scientific community. Tripos, headquartered in St. Louis, has been in the business since 1979 and so has "grown up" with the advances in computational chemistry. Since 1998, Tripos has been developing a flexible drug discovery software platform that has attracted the attention of Pfizer, the world's largest drug maker. In January 2002, Pfizer joined Tripos to jointly design, develop, and test a range of methods for analyzing and interpreting drug design data.

The drug design company Pharmacopeia in June 2000 spun off a wholly owned subsidiary, Accelrys, to produce software for computation, simulation, management, and mining of data for biologists, chemists, and materials scientists. Pharmacopeia has had a software segment for some time to meet its computational chemistry needs, but the needs grew and so did the software segment with the acquisition of a number of small specialty companies. Accelrys offers a suite of approximately 130 programs that include data for the analysis of DNA. The company also has software to annotate protein sequences, a capability that aids in the use of genomic data for drug discovery. Those data are becoming available from the Human Genome Project.

Is computational chemistry a reliable guide to the properties of novel compounds? The research and development results of every major chemistry and biochemistry laboratory indicate that it is. —M. C. NAGEL

Viewpoint:

No, computational chemistry is not a reliable guide to the properties of novel compounds.

In recent years computational chemistry and the modeling of molecules have revolutionized chemistry and biochemistry. Both the 1998 and the 1999 Nobel Prizes in chemistry were awarded in the fields of molecular modeling. Some chemists have even foreseen a time when the lab-coated chemistry researcher will be a thing of the past and will be more computer scientist than laboratory experimenter. Despite the rhetoric and much heralded successes of molecular modeling techniques, computational chemistry has many limitations and is not a reliable method when applied to novel compounds.

Even the Best Model Is Only a Model The most accurate and complete theoretical model available for computational chemistry is quantum mechanics and the Schrödinger wave equation. Unfortunately, Schrödinger's equation is too complex to be solved completely for all but the simplest situations, such as a single particle in a box or the hydrogen atom. To reduce the problem of calculating the properties of atoms and molecules to something possible, approximate methods must be used. When this is done, errors and strange glitches often are introduced in the models calculated.

A number of models approximate the Schrödinger wave equation, some of which are more exhaustive than others in their calculations and so can suffer from extremely long computation times. The so-called ab initio methods of molecular orbital calculation can lead to accurate approximations of a molecule. Such methods are used for can calculation of the properties of a molecule with knowledge of only the constituent atoms, although in practice information often is added or approximated to reduce calculation times. For example, the Hartree-Fock method works on the principle that if the wave functions for all but one electron are known, it is possible to calculate the wave function of the remaining electron. With an average of the effects of all but one electron, the movement of one electron is estimated. The process is repeated for another electron in the previously averaged group until a self-consistent solution is obtained. However, a number of approximations are introduced in the practical calculations. For example, the method ignores the Pauli principle of electron exclusion—that no more than one electron can exist in the same space—because it effectively omits calculations of particle spin. Relativistic effects resulting from heavy nuclei or high-speed particles are ignored in ab initio methods. More general approximations are used to increase the speed of calculation, but at the expense of accuracy. However, the method can still result in computations that would be far too long for practical purposes, even when run on a supercomputer.

The Born-Oppenheimer approximation ignores correlation between the motion of elec-

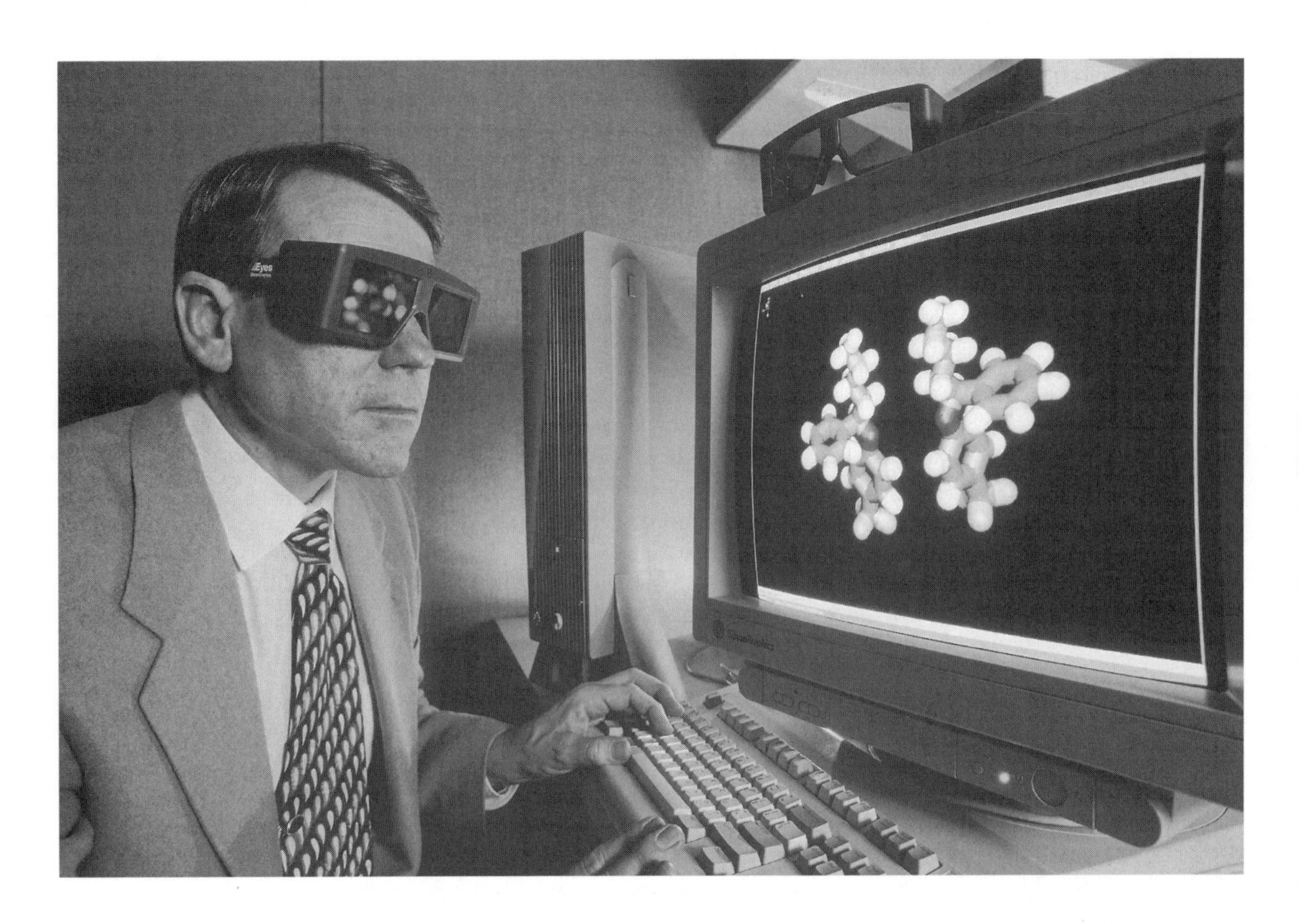

A scientist views a computer model of Fluoxetine, or Prozac, with the help of 3-D glasses.
(© Roger Ressmeyer/Corbis. Reproduced by permission.)

trons and that of nuclei because the electrons move much faster. It is a good approximation but still requires exhaustive computations of the electron orbitals. The Huckel theory, developed in 1931 uses a more extreme approximation by ignoring electrons not directly involved in chemical bonds and by assuming that some electron orbitals can be treated as identical. Application of this theory greatly reduces the number of calculations. There is always a trade-off between the accuracy of the model and the speed at which it can be generated. In general, the larger the molecule to be modeled, the less accurate the model can be, given reasonable time constraints.

Even with the great advances in computational speed, parallel processing, and other innovations in computer technology that have occurred, such approximations can take many days, months, or years to run for a single molecule. Yet even if speed issues are overcome with new generations of super-fast computers, ab initio models are only models, and they still suffer from errors caused by the approximations used in calculation. A number of common, simple molecules resist ab initio computation, and results are produced that do not agree with results of experiments. For example, ab initio calculations of AlO, CP, and nitric oxide (NO) all give vibrational frequencies that wildly disagree with results of experiments. Methods exist to correct the errors for well-studied molecules such as these. However, for novel compounds that have no experimental confirmation, there can be no knowledge of what the errors may be, or of how to compensate for them.

Nonquantum Models Quantum mechanical models give much more information than is needed for chemical analysis, especially if molecular geometry is the main focus. Many of the computer programs used to model molecules ignore quantum effects and reduce the problem to classic calculations, which can be performed much faster. There is a limit to the information that can be gained from such methods, and faster turnaround times come with greater risk of inaccuracies and errors in the model. Force-field methods, which evolved out of the study of vibrational spectroscopy, are related to the old method of building three-dimensional wire-frame real-world models. The models generated can be thought of as a collection of balls (the atoms) connected by springs (the chemical bonds). With calculation of the energy and bond angles in the molecule, a computer-generated model of the likely geometry of the molecule is produced. The calculations are fast in comparison with quantum methods, and the results are sufficiently accurate to be useful, but they cannot be relied on completely and give poor results in some circumstances, such as non-bonded interactions and metal-carbon bonds. As with quantum methods, the final results usually are adjusted with additional parameters so the generated results agree with those of experiments. In the case of novel compounds, such a process can only be a best guess.

There are less accurate modeling methods that are faster still. Previously generated data often are used as a starting point. In the case of novel compounds, this method can only be a general guide. Minimization algorithms are used to attempt to find a best fit to an approximate molecular geometry. Small adjustments in the likely geometry of a molecule are computed to find the lowest energy configuration. For rigid molecules, this is likely to be a good fit to the real geometry, but for flexible molecules, other computations should be added to give validity to the computer models. One textbook compares the process of minimization to "walking around a hillside in a thick fog, trying to find the way down hill." Whereas going down to a local minimum is not much of a problem, if there are isolated valleys that do not connect, how can you be sure you have reached the lowest point? Conformation searches are a more systematic method of finding the lowest point; they are akin to having the fog-bound walker go down many paths rather than only one. However, such an algorithm takes many times longer, and so again many computer programs use approximate methods to increase the speed of calculations.

More than a Question of Speed The problem of speed in chemical modeling is not just a matter of waiting until computers become faster. In many cases the theoretical estimate for performing a rigorous modeling computation on a large molecule or a complex molecular reaction on an ideal computer would still take longer than the time the universe has been estimated to exist. Unfortunately, the types of molecules that fall into this category are precisely those that chemists, biochemists, and medical researchers are most often interested in modeling. As a result, only the most simple and quick computational methods can be used. Experimental data often are plugged into the model to increase the speed of calculation. Once again, the problem with novel compounds is that such additions are unavailable or unreliable, so the entire method can only be considered approximate.

Because full confirmation modeling is impossible for complex molecules, further shortcuts and approximations often are made to provide useful, but rough, guides to the potential properties of a new molecule. One method of molecular analysis developed for use with novel compounds in the drug industry is to compare a new compound with a similar well-studied compound. In many drug development studies, small modifications are made to a molecule in an effort to improve the overall effect of the drug and remove unwanted side effects. A computer can quickly check (in less than one minute) whether a newly constructed chemical acts by a previously recognized mechanism. This method focuses on key areas of the molecule, such as receptor sites, but can only recognize constructions that fit a predetermined pattern. Although such algorithms have had some success, they cannot identify possible drugs that function with a new mechanism. The only way to be sure a potential molecule will work is with experimental methods.

Experimental Confirmation Needed Many computer models agree closely with experimental data because the programs have been tweaked. Parameters and corrections have been added to compensate for errors in the approximations used to generate the model. These parameters can be guessed for novel compounds when such new structures closely resemble previously studied molecules. However, such guesses must be checked against experimental results, and the process falls down for novel compounds that differ too greatly from known compounds.

Computational chemistry is a rapidly changing field. As computers increase in speed and the modeling programs used to analyze molecules are adjusted and improved, the range of molecules and interaction between molecules that can be represented increases. There are, however, limits to the accuracy of a computed model, even with an ideal computer. As a result, approximate methods must be used to generate chemical models in a reasonable time. The errors often found in such models are corrected by the addition of values derived from experimental data. With novel compounds, such data do not exist, and further approximations and guesses must be made, limiting the accuracy of any computed model. The accurate modeling of large and novel molecules represents a great challenge to computational chemistry. Although great strides are being made in the field, theoretical limits continue to make such models unreliable. The final analysis must always be experiments in the real world. —DAVID TULLOCH

Further Reading

Cramer, Christopher J. *Essentials of Computational Chemistry: Theories and Models.* New York: Wiley, 2002.

Goodman, Jonathan M. *Chemical Applications of Molecular Modelling.* Cambridge, England: Royal Society of Chemistry, 1998.

Grant, Guy H., and W. Graham Richards. *Computational Chemistry.* Oxford, England: Oxford University Press, 1995.

Leach, Andrew R. *Molecular Modeling: Principles and Applications.* 2nd ed. London, England: Pearson Education Corporate Communications, 2001.

Lebl, M. Parallel Personal Comments on "Classical" Papers in Combinatorial Chemistry. *Journal of Combinatorial Chemistry* 1 (1999): 3–24.

"Press Release: The 1998 Nobel Prize in Chemistry." Royal Swedish Academy of Sciences. October 13, 1998 [cited July 26, 2002]. <http://www.nobel.se/chemistry/laureates/1998/press.html>.

Schlecht, Matthew F. *Molecular Modeling on the PC.* New York: Wiley, 1998.

Wilson, Elizabeth K. "Picking the Winners." *Chemical and Engineering News,* April 29, 2002 [cited July 26, 2002]. <http://pubs.acs.org/cen/coverstory/8017/8017 computers. html>.

———. *Computational Chemistry: Applying Computational Techniques to Real-World Problems.* New York: Wiley, 2001.

Young, David. *Introduction to Computational Chemistry.* Dallas: Cytoclonal Pharmaceutics [cited July 26, 2002]. <http://server.ccl.net/cca/documents/dyoung/topics-orig/compchem.html>.

INDEX

F

G

H

I

J

K

L

M

N

O

P

T

W

X

Y

Z

ISBN 0-7876-5767-0